MODERN ANALYTICAL TECHNIQUES

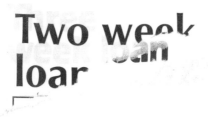

Modern Analytical Techniques

Frank Owen
Ronald H. Jones

POLYTECH PUBLISHERS LTD STOCKPORT

First Published 1973

© Copyright Polytech Publishers Limited,
  36 Hayburn Road, Stockport, SK2 5DB.

ISBN 0 85505 010 1

Reproduced and printed by photolithography and bound in Great Britain at The Pitman Press, Bath

Preface

It is becoming increasingly obvious that, if business in this country is to be controlled effectively, there is a need for both established businessmen, and those who are studying to become business men, to obtain an understanding of the mathematical and statistical techniques used to provide a basis for decision taking. We are rapidly entering an era when it is necessary for the manager to be numerate.

Unfortunately, the modern techniques are often based on concepts, such as matrix arithmetic or inequalities, which have not been commonly taught in the general school curriculum until very recently. In this book we explain such mathematical concepts step by step as a basis for practical problems taken from the business world.

Our approach has developed from courses of lectures given at Liverpool Polytechnic to students who have proved to be willing, (though often unwitting), guinea pigs. Our experience has shown that there are few serious students who cannot obtain an understanding and appreciation of mathematical techniques provided that they are presented in a logical, relevant and interesting fashion. For this reason we have tried always to use as examples problems drawn from those that the businessman faces daily, rather than using the usual abstract examples or those drawn from the natural sciences.

The necessary mathematical concepts have been, as far as possible, introduced at the point where they become relevant to the problem under consideration. There are, however, certain concepts which are fundamental, and these have been gathered together and treated in isolation in the Introductory chapter. The remainder of the book is entirely objectively orientated.

In our teaching we have found the need for frequent tutorial work, and each chapter contains a number of tutorials. Whilst the text can be used as a basis for lectures, the teacher will find the tutorials useful, not merely as a set of exercises, but also a bridging point between concepts and application. The student who works conscientiously through them will find his knowledge of mathematics deepened and enriched.

Analytical techniques is increasingly becoming a compulsory subject in many examinations. We have borne in mind the needs of students reading for the new Foundation course, and the new Final Part II of the Institute of Chartered Accountants; the Analytical Techniques paper of the I.M.T.A.; stages I and IV of the I.C.M.A., as well as the mathematical and statistical content of Higher National Diplomas in Business Studies, C.N.A.A. degrees in Business Studies, and the Diploma in Municipal Administration. At the same time we believe that the practical businessman who wants a clearer understanding of the way his 'specialists' operate, will also find much that is useful.

The mathematician will readily recognise the debt we owe to many writers, both general and specialist. Almost every book we have read has helped to formulate and extend our ideas. Our major debt, however, remains to our students. Such merits as this book amy have owes much to their criticism of and comment on our lecture material. Its weaknesses, of course, remain our personal responsibility.

To those students of Liverpool Polytechnic who have found mathematics to be exciting, this book is dedicated.

Frank Owen
Ronald H. Jones

Liverpool Polytechnic
March 1973

CONTENTS

Introduction

The straight line

Walk into the office of any numerate manager and you will probably be impressed by the number of graphs and diagrams you will see on his walls or littering his desk. Some will show curves, others blocks or rectangles, while still others will be 'merely' straight lines. Do not be misled; the simple straight line can be of tremendous help in solving problems very simply and efficiently. Furthermore straight lines play a significant role in many of the techniques of analysis used in this book. Let us then, examine the nature of the straight line.

EXAMPLE 1
Let us suppose that a transport manager is instructed to hire a car for the use of the company directors. He approaches a car hire firm who quote three different weekly tariffs:

Tariff 1.   A fixed charge of £20 per week.
Tariff 2.   A charge of 7½ pence per mile travelled.
Tariff 3.   A fixed charge of £5 per week with an additional charge of 5 pence per mile travelled.

The first thing you should notice is that if the second or third tariff is chosen the weekly charge will vary with the mileage travelled. In the language of the mathematician we could say that the weekly charge is a *function* of mileage. If the transport manager decides to select the first tariff then he pays £20 per week irrespective of mileage.

Let us now graph the three tariffs:

The line $y = c$

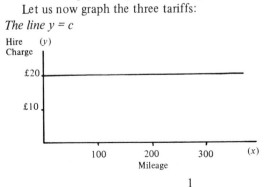

1

The graph shows the weekly charge if the first tariff is chosen. Notice that the vertical axis (called the *y* axis) measures the weekly charge, while the horizontal axis (called the *x* axis) measures the mileage per week. The relationship between the weekly charge and mileage per week is represented by a horizontal straight line, indicating that mileage does not affect the charge which is a constant £20 per week.

As we are using the symbol *y* to stand for the weekly charge, we can state that for any mileage

$y = 20$

This is the *equation* of the line. Now it is not necessary of course that the hire charge be £20 per week — it could be any amount £*c*. However, the relationship will always be shown by a horizontal straight line with the equation $y = c$.

Also the line will always cut the *y* axis at a point *c* units above zero.

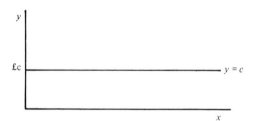

The line y = mx
Now let us graph the situation if the second tariff is chosen, i.e. a mileage charge of 7½ pence per mile. We could draw up a table showing how the weekly charge varies according to mileage like this:

Mileage (x)	0	50	100	150	200	250	300
Weekly charge £(y)	0	3.75	7.50	11.25	15.00	18.75	22.50

We deduce that if 200 miles a week are covered then the weekly charge is £15. We could locate 200 on the *x* axis and imagine a line drawn vertically upwards from this point. Again, we could locate 15 on the *y* axis and imagine a line drawn horizontally from this point.

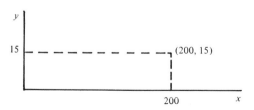

The lines intersect at the point with the *coordinates* (200,15). If we do this for the remaining points (0, 0), (50, 3.75) . . . etc. we find that it is possible to join the points with a straight line and we can read off the weekly charge (*y*) for any mileage (*x*) up to 300 miles.

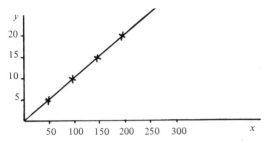

You should notice that we can calculate the weekly mileage charge (*y*) if we multiply the mileage (*x*) by 7½ pence, i.e.

$y = 7.5x$ (in pence)
or $y = 0.075x$ (in pounds)

As we can also read off the weekly charge from the graph it follows that the equation of the line is

$y = 0.075\,x.$

Now, the mileage charge may not be 0.075 pounds per mile. Suppose it is £*m* per mile. The equation of the line then becomes

$y = mx$

Any straight line which passes through the origin of the graph, i.e. the point (0, 0), has the equation $y = mx$.

In the diagram below, three such straight lines have been drawn: $y = \frac{1}{2}x$, $y = x$, and $y = 2x$. Notice that the greater is the value of *m* the steeper is the slope of the line. In fact, *m* measures the *gradient* of the line.

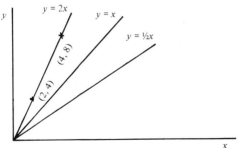

Let us now select two points on the line $y = 2x$, e.g. (2, 4) and (4, 8) and use them as a basis for calculating the gradient of the line.

Now what do we mean by a gradient? Suppose we saw a sign warning us that we were approaching a hill with a 'one in ten' gradient — we could represent the hill like this:

A mathematician, however, would not call this a 'one in ten' gradient, rather would he say that the line had a gradient of $\frac{1}{10}$. In other words, the mathematician calculates the gradient of a line by the formula:

$$\frac{\text{vertical distance}}{\text{horizontal distance}}$$

Now if we consider a line drawn on a graph, then clearly the vertical distance is the change in the value of y and the horizontal distance is the change in the value of x. Hence we can state that the gradient of a straight line joining any two points on a graph is

$$\frac{\text{Change in } y}{\text{Change in } x} \text{ or (as we shall express it later) } \frac{dy}{dx}$$

Let us now return to the points (2, 4) and (4, 8). The change in y is $8 - 4 = 4$ and the change in x is $4 - 2 = 2$. So the gradient of the line joining two points is $\frac{4}{2} = 2$. But we already know that both points lie on the line $y = 2x$ and so we have verified that $y = 2x$ has a gradient of 2 and that m does indeed measure the gradient.

The line $y = mx + c$

Finally let us consider the third tariff — involving a hire charge of £5 per week and a mileage charge of 5 pence per mile. Again, we could draw up a table showing how the weekly charge varies with mileage:

Mileage (x)	0	50	100	150	200	250	300
Weekly Charge £(y)	5.00	7.50	10	12.50	15	17.50	20

Graphing this we have:

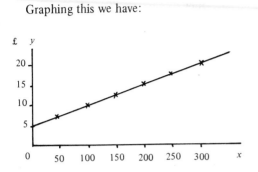

Again, we can read off the weekly charge (y) for any mileage (x) up to 300 miles. For any mileage (x) the weekly charge in pounds is

$0.05x + 5$

so the equation of this line is

$y = 0.05x + 5$

We can see that the gradient of this line must be 0.05. Select any two points on this line and verify it for yourself.

Suppose that the hire charge was £c and the mileage charge £m per mile. The equation connecting weekly charge and mileage would then be

$y = mx + c$

Can you spot the difference between the line $y = m$ and the line $y = mx + c$? The first line passes through the origin whereas the second line does not. What then can you conclude about the significance of c when drawing $y = mx + c$? Surely it is obvious that c determines the point at which the line cuts the y axis.

TUTORIAL ONE
1. List relationships drawn from the real world that could be described by the relationships $y = x$; $y = mx$; and $y = mx + c$.
2. On the same piece of graph paper draw straight lines to show how weekly charges for hiring a car would vary according to the three tariffs discussed in the text (extending the graph to cover the possibility of travelling 400 miles a week). Prepare a set of operational rules for deciding which tariff should be chosen. How many points need you calculate to draw a straight line graph?
3. You decide to have a Spanish holiday and convert your holiday savings into pesetas. The rate of exchange obtained from your bank

is 148 pesetas to the pound. Draw a graph which would enable you to convert pesetas into pounds for amounts up to £5.00. Read off the sterling value of 115 pesetas; 326 pesetas; and 428 pesetas (answers to the nearest ½ penny). What is the equation of the line?

4.   Draw the graph of the line $y = -2x + 4$. What do you notice? What is the significance of the sign of m in the equation $y = mx + c$?

5.   What is the gradient of the line joining the points (3, 45) and (7, 77)? Can you find the equation of this line?

   Hint: The equation of this line can be represented by $y = mx + c$. We know that when $x = 3, y = 45$ so $45 = 3m + c$. As you have already calculated the value of m, you should now have no trouble in evaluating c.

6.   So far we have been concerned with straight line graphs (called Linear graphs). Now we can tell by examining the equation whether or not a function is linear. If y is a function of x^n, then a linear relation would occur *only when $n = 1$*. Verify this for yourself by graphing

$$y = x^2 \text{ and } y = \frac{1}{x}$$

for values of x between -4 and $+4$.

Indices

Suppose you stopped the first man you met in the street and asked him to list the greatest scientific advances made over the last three decades. It is likely that one of his answers would be the exploration of space . The Space Centre at Houston has taught us a whole new vocabulary, and encourages us to contemplate the vast distances involved. Even these distances become insignificant when we enter the science of cosmology. The distance of the rim of our galaxy is so vast that it is beyond our comprehension. Our *number system* cannot efficiently cope with such quantities. A star that is 5,865,000,000,000 miles away is very close indeed, but this distance means little to us written in this form.

   We can overcome this problem in two ways. Firstly we could abandon the mile as our unit of astronomical distance, and this is precisely what the astronomers have done. They measure distances in *light years* – a light year being the distance travelled in one year at the speed of light (186,000 miles per second). The star we called our close neighbour would be one light years away. We will examine the second option which is to retain the mile but introduce a new number system based on powers of ten.

Instead of writing a million like this: 1,000;000
we could write it like this: 10 X 10 X 10 X 10 X 10 X 10,
or better still like this: 10^6 – called ten to the power six

You will probably remember this from your school mathematics. We can write two million like this: 2×10^6. Written in this way, a number is said to be in the *standard form*. Now let us write one light years in standard form.

$5,865,000,000,000 = 5.86 \times 10^{12}$

Obviously, standard form is a compact and convenient way of writing very large numbers. You should note that the standard form is always written as

$a \times 10^n$

where a is greater than one but less than 10, and where a is given correct to three decimal places. The power of ten (n) tells us to move the decimal point n places to the right, or multiply a by ten n times.

EXAMPLE 2

$4.52 \times 10^6 = 4,520,000$

You will meet the standard form on a number of occasions as you work your way through this book. Also you must be able to manipulate indices. (An index is merely the power of a number.) In all probability, you are well acquainted with the rules of indices, but in case you have forgotten them they will be restated now.

Suppose we are required to multiply 100 by 10,000

$100 \times 10,000 = 1,000,000$

Alternatively we can write it like this:

$10^2 \times 10^4 = 10^6$

In other words if you wish to multiply two numbers together, you can express both numbers as powers of some base number and add the indices. This is precisely what you do when you use logarithms. For example log 5 is 0.6990 or we can say that $5 = 10^{0.6990}$.
More generally, we can state that –

$\underline{x^n \times x^m = x^{n+m}}$

Likewise we can deduce that

$\underline{x^n \div x^m = x^{n-m}}.$

EXAMPLE 3

Suppose we wished to evaluate

$$\frac{32 \times 64 \times 4}{128}$$

We can express these numbers as powers of two

$$\frac{2^4 \times 2^5 \times 2^2}{2^6}$$

$$= 2^{4+5+2-6}$$

$$= 2^5$$

$$= 64$$

We have solved this by taking logarithms to the base of two.

TUTORIAL TWO

1.   Mean distance of the planets from the sun

Mercury	36,000,000	Jupiter	483,300,000
Venus	67,200,000	Saturn	886,200,000
Earth	92,820,000	Uranus	1,783,000,000
Mars	141,600,000	Neptune	2,793,500,000

Write these numbers in standard form.

2.   We know that x^3 is $x \times x \times x$ but what is the meaning of x^{-3}?
     Firstly consider this; $x^4 \div x^7 = x^{4-7} = x^{-3}$.

     Also, $x^4 \div x^7 = \dfrac{x^4}{x^7} = \dfrac{1}{x^3}$

     So it follows that $x^{-3} = \dfrac{1}{x^3}$

     What is the value of (a) 2^{-2}, (b) 5^{-3}, (c) 10^{-2}, (d) 10^{-5}.

3.   The last question enables us to use standard forms for numbers less
     than one. Consider, for example, the diameter of a molecule of
     water — 0.0000000339. We can write the diameter like this:

     $$3.39 \times 100,000,000^{-1}$$

     or $3.39 \times \dfrac{1}{100,000,000}$

     i.e. 3.39×10^{-8}

     You can now write the following in the standard form. (a) 0.001
     (b) 0.000123 (c) 0.456 (d) 0.0000000123 (e) 0.000000678

4. Can you deduce a value for x^0? (Hint: use an analysis similar to question 2.)

5. If

$$x^{\frac{1}{2}} \times x^{\frac{1}{2}} = x^1 = x,$$

what can you conclude about $x^{\frac{1}{2}}$?

The Sigma (Σ) Operator

When you learn French you readily accept the need to learn a completely new vocabulary; if you are training to be a reporter you will accept the need to learn shorthand. Anyone who is beginning to study a subject at anything but the most elementary level is faced by the fact that a new language has to be learned and that much of this language is expressed in a shorthand form. The economist talks of Marginal Cost and Marginal Revenue and customarily uses M.C. and M.R. as a convenient way of writing these items; the lawyer refers to a statute by a complex system of reference such as 33 & 34 Vict. c97 — this being the 97th act passed in the 33rd and 34th year of the reign of Queen Victoria; and the accountant discusses the implementation of V.A.T. rather than Value Added Tax.

So it is with mathematics. Can you imagine how cumbersome it would be if you were constantly faced by the following instructions or something similar: 'Subtract the figures in the column headed $\bar{x}$ from the corresponding figures in the column headed x. Square the result in each case and insert the figures in a column headed $(x - \bar{x})^2$. Now add together all the figures you have in this last column and insert the total'.

Is it not easier — much easier — to give the instruction $\Sigma (x - \bar{x})^2$? This new sign Σ (Sigma) is one of the most useful shorthand forms in mathematics. It means simply, 'Take the sum of'. It is our summation sign, and instructs us to add together certain quantities which appear immediately after the sign.

Thus Σx (Sigma x) means, 'add together all the figures in the column headed x', Σx^2 means square each figure in the column headed x and add together the result, and Σfx means, 'multiply the figures in the column headed f by the corresponding figures in the column headed x and add the resultant products'.

You must, of course exercise care and use your knowledge of the rules of mathematics in determining precisely what it is you are going to add. Before we go any further make absolutely certain that you can master the following tutorial.

TUTORIAL THREE

1. Distinguish carefully between:

    a. Σx^2 and $(\Sigma x)^2$

    b. Σxy and $\Sigma x \, \Sigma y$

    c. $\dfrac{\Sigma fx}{\Sigma y}$ and $\Sigma \left(\dfrac{fx}{y} \right)$

    d. $\Sigma \, 2d^2$ and $2 \, \Sigma \, d^2$

2. Imagine a class containing 40 students aged between 18 and 26. List in one column headed x the appropriate ages, 18, 19, 20 .... 26. Now insert in a second column headed f the (imaginary) number of students of each age. What is the value of Σx; Σf; Σfx? Calculate the value of $\dfrac{\Sigma fx}{\Sigma f}$. Look carefully at what you have done.

    Can you appreciate that you have calculated the average age of the class? We will always use $\bar{x}$ (bar x) to mean the arithmetic average of the values in the x column.

3. Using this value of $\bar{x}$, find the values of $(x - \bar{x})$; $(x - \bar{x})^2$; and of $\Sigma (x - \bar{x})^2$. The figure $x - \bar{x}$ is the *deviation* of the item x from the arithmetic average $\bar{x}$. Now calculate the value of

    $$\sqrt{\frac{\Sigma (x - \bar{x})^2}{\Sigma f}}$$

    You have now calculated the *standard deviation* of the distribution. We will have much more to say about this later.

General Properties of Summations

Although Σx is usually used alone in books of this type you will occasionally come across expressions such as $\sum\limits_{x=1}^{n} x$. Σx you already understand; the n and the $x = 1$ are called the limits of summation, and mean that we should replace x first by 1, then by 2, then by 3 and so on up to n and then sum the results.

Thus

$$\sum_{x=1}^{5} 2x = 2 + 4 + 6 + 8 + 10 = 30$$

and

$$\sum_{a=2}^{6} (3 + 4a) = 11 + 15 + 19 + 23 + 27 = 95$$

and generalising

$$\sum_{i=1}^{n} ax_i = ax_1 + ax_2 + ax_3 + \ldots + ax_n$$

This last expression $\equiv$

$$a(x_1 + x_2 + x_3 + \ldots x_n)$$

and since

$$(x_1 + x_2 + x_3 + \ldots + x_n) = \sum_{i=1}^{n} x_i$$

we may say

$$\sum_{i=1}^{n} ax_i = a \sum_{i=1}^{n} x_i$$

In other words, if the variable we are considering is multiplied by a constant we may place the constant in front of the summation sign.

More often than not the limits of summation need not be stated in the problem we are handling and, from this point, we will not state them. But remember that they exist. Thus we will write the equality above simply as:

$$\Sigma \, ax = a \, \Sigma \, x.$$

Sometimes we have to deal with expressions such as $\Sigma \, (ax + b)$, where b is a constant and x variable. (Distinguish carefully between this and $\Sigma \, ax + b$.) Now,

$$\Sigma \, (ax + b) = (ax_1 + b) + (ax_2 + b) + \ldots + (ax_n + b)$$

As there are n terms in this expression

$$\Sigma \, (ax + b) = (ax_1 + ax_2 + ax_3 + \ldots + ax_n) + nb$$

$$= a(x_1 + x_2 + x_3 + \ldots + x_n) + nb$$

$$= a \, \Sigma \, x + nb$$

Thus

$$\Sigma \, (ax + b) = a \, \Sigma \, x + nb$$

In many expressions it is not a constant b we are adding but a variable. Thus taking both x and y as variables:

$$\Sigma \, (x + y) = (x_1 + y_1) + (x_2 + y_2) + (x_3 + y_3) + \ldots + (x_n + y_n)$$

$$= (x_1 + x_2 + x_3 + \ldots + x_n) + (y_1 + y_2 + y_3 + \ldots + y_n)$$

$$= \Sigma \, x + \Sigma \, y$$

Thus:

$$\Sigma(x + y) = \Sigma x + \Sigma y$$

and we can extend this result to

$$\Sigma(x + y + z) = \Sigma x + \Sigma y + \Sigma z$$

or any other similar expression irrespective of the number of terms.

Finally, one of the most common of all summations is $\Sigma(x - y)^2$. Since

$$(x - y)^2 = (x^2 - 2xy + y^2)$$

$$\Sigma(x - y)^2 = (x_1^2 - 2x_1y_1 + y_1^2) + (x_2^2 - 2x_2y_2 + y_2^2) + \ldots$$
$$+ (x_n^2 - 2x_ny_n + y_n^2)$$
$$= (x_1^2 + x_2^2 + \ldots + x_n^2) - (2x_1y_1 + 2x_2y_2 + \ldots + 2x_ny_n)$$
$$+ (y_1^2 + y_2^2 + \ldots + y_n^2)$$
$$= \Sigma x^2 - \Sigma 2xy + \Sigma y^2 = \Sigma x^2 - 2\Sigma xy + \Sigma y^2$$

Thus

$$\Sigma(x - y)^2 = \Sigma x^2 - 2\Sigma xy + \Sigma y^2$$

You will meet all these expressions from time to time in this book, so you are advised to master them now.

TUTORIAL FOUR

1. Using the Sigma notation write in an alternative form:

   a. $\Sigma(x + y)^2$
   b. $\Sigma(x^2 - y^2)$
   c. $\Sigma(x + y)^3$
   d. $\Sigma(2x + 4y + b) - b$ being a constant
   e. $\Sigma(3x^2 + 2y + 6)$
   f. $\Sigma(x - y)^3$

Averages and Dispersion

Statistical Data

Imagine for a few moments that you have been asked to obtain some statistical information about, say, the level of wages in a medium sized factory employing 300 people. How would you go about it? Provided that you have the cooperation of management, the raw material is readily available in the form of wages records. What then is the problem?

Well, in the first place, what do you mean by wages? Think for a minute and you will see that it could mean gross or net pay, per week or per year, including or excluding overtime and bonus payments. This may be merely a matter of simple definition, but before we begin to look for information we must know what it is we are looking for.

This, however, is only the start of our problem. When we see the wages book we will see that individual wages differ and themselves vary week by week, so we cannot merely say that the wage paid in this factory is £x per week. Admittedly there are many ways we could get over this difficulty. We could state the highest and the lowest wage and consider the *range*; we could state the most common level of wages or the *mode* of the series (if it is readily discernible); we could list every wage received and argue that this is the only way in which we can present a fair picture of the level of wages. More commonly we would simplify the data by presenting it in the form of a *frequency distribution* and showing how many workers receive a wage within given limits; or simplify still further by using the data to obtain an *average* and a measure of the way in which individual items are scattered (or dispersed) around the average.

Thus the process of presenting statistical data is essentially a process of simplification. We begin with 300 figures of wages, reduce these to perhaps a dozen in our frequency distribution, and end up with just two figures — a measure of average and a measure of dispersion. If we understand what we are doing we can probably get as much information (if not more) out of the latter two than from the original 300.

The Frequency Distribution
It is probably true that most people tend to think numerically in terms of 'blocks'. We can easily visualise the magnitude of the difference between say £5 and £7, or £30 and £50. But can we so readily appreciate the difference between say £5.5723 and £5.5726 or between £1257320 and £1257420. In considering the latter examples you may be quite sure that most people would not differentiate between them because in the one case the difference is 'too small' and in the other because £100 in over a million pounds 'does not matter'. They would tend to lump them together in a class which they might describe loosely as 'about £5.50' or about £1¼ million'. There is of course a loss of arithmetic accuracy; but do you not think that there may be a great gain in understanding?

It is this process of grouping similar figures into broad categories that gives us the so-called Frequency Distribution. This is a table which tells us how many times (*the frequency*) the item we are examining is found to be of a particular order of magnitude.

We might decide, for example, that it would be sufficiently detailed if we analysed the wages in our factory in groups of £50 width. Our resultant frequency distribution might appear like this:

Annual Wage £	No. of workers
1200 and under 1250	13
1250 and under 1300	42
1300 and under 1350	51
1350 and under 1400	63
1400 and under 1450	50
1450 and under 1500	42
1500 and under 1550	25
1550 and under 1600	13
	300

This is the type of statistical table you will meet in most of the work you are going to do. But do not forget that although the frequency distribution is a simple way of presenting a large number of individual figures, in the process we have lost the only means we have of ensuring absolute arithmetical accuracy. The errors are not large, but a few years ago a colleague of ours was adamant in describing statistics as 'the science of approximation', and there is a large measure of truth in this.

The information contained in a frequency distribution can be presented visually very simply in a 'bar diagram' or HISTOGRAM.

This is simply a diagram in which the horizontal axis measures the magnitude of the variable — in this case wages — and bars are drawn vertically the height of which represents the frequency of occurrence of each magnitude — in this case the number of workers. Since you will use the histogram in later chapters it is worth examining it closely now.

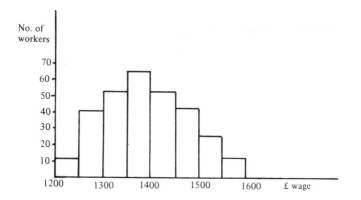

You will see that the x axis is scaled in £'s from £1200 to £1600 and that the width of each rectangle we have drawn represents a range of £50. Our frequency distribution presents no problems since each group of wage earners we consider receives a wage within a £50 range, and so the width of the rectangles remains the same throughout. But you will appreciate that had we varied the range of wages we were considering in our distribution throughout the table the width of the rectangle would have had to be altered accordingly. The y axis is scaled in terms of the number of workers and the height of the rectangle varies according to our frequencies. Now an important characteristic of such a histogram is that not only does the height of the bar indicate the frequency, but also the area of the bar is the same percentage of the total area of the histogram as the frequency of the group it represents is of the total frequency. Thus the group receiving £1400 and under £1450 comprises one sixth of the total number of workers, and if you examine it the bar representing these workers is one sixth of the total area of the histogram.

You will readily see that this can create a problem if we were to change the width of one particular group of wages. Suppose we had put together in our distribution the last two groups so that our table finished £1500 and under £1600 − 38 workers. We have doubled the width of the bar and if we maintain the same scaling on the y axis we will have doubled the area. If we had bars one inch wide, for example in our original histogram and one inch of height represented one worker our two bars would have had an area of 38 square inches. If now we double the width of the bar and still maintain a height of 38 inches this group now has an area of 76 square inches and the relationship between area and frequency is destroyed. In practice this involves cumbersome adjustments of the height of the bars, but so far as you are concerned we need not use frequency tables which have this characteristic and what we would ask you to concentrate on is that the area of any bar is the same proportion of the total area of the histogram as the frequency it represents is of the total frequency.

The Arithmetic Mean (or Average)

Every schoolboy is familiar with the idea of an average, and you will all know that to find the average of a group of items we add the items together and divide the result by the number there are. Thus the average of 7, 9, 14, 8 and 12 is $\dfrac{7 + 9 + 14 + 8 + 12}{5} = 10$.

Symbolically we may represent the items by $x_1, x_2, x_3, \ldots x_n$, without being specific as to their magnitude. The number of items we may call n, and our arithmetic average, which we always call $\bar{x}$ is:

$$\frac{x_1 + x_2 + x_3 + \ldots + x_n}{n}$$

or using the notation you have just learned

$$\bar{x} = \frac{\Sigma x}{n}$$

Now, all this is very simple, but the arithmetic mean above has exactly the same properties as all other arithmetic means, and one of these is so useful as an aid to computation that it is worth while spending a little time to grasp it fully.

Obviously it is not very likely that more than a small proportion of the items we examine will have exactly the same value as the arithmetic mean. The others will all differ (or *deviate*) from it by greater or lesser amounts. If we can list these deviations, taking note of the sign, we will find that their sum will always be zero. The rule is that the deviation is always the original data minus the arithmetic mean, which you may find easier to remember as

Deviation $= (x - \bar{x})$

Thus in our simple example we have:

x	7	9	14	8	12	$\bar{x} = 10$
$x - \bar{x}$	-3	-1	$+4$	-2	$+2$	$\Sigma (x - \bar{x}) = 0$

How will this help us? Suppose we do not know that the average is 10. We may proceed by assuming it to be some other figure, say 12. Listing our deviations as before, but this time from 12, we now have

x	7	9	14	8	12	$\bar{x}$ assumed to be 12
$x - \bar{x}$	-5	-3	$+2$	-4	0	$\Sigma (x - \bar{x}) = 0$

It is easy to see that the effect of raising $\bar{x}$ from 10 to 12 was to cause every deviation to differ by the same amount but in the opposite direction. Thus we raised our average by 2 and changed the deviations by -2 each, with the result that $\Sigma (x - \bar{x})$ became -10. Note however, that this is the result of 5 (or n) deviations, all changing by the same amount, and the amount by which they each change is $\dfrac{\Sigma (x - \bar{x})}{n}$, or

$\dfrac{-10}{5} = -2$. Note too that the change in the magnitude of the deviation is a result of an equal but opposite change in the magnitude of the average. Thus the average of the deviations from an assumed mean tells us by how much we have to adjust the latter, and the sign tells us in what direction; (just follow the sign — that is all). We have then:

$$\bar{x} = \text{assumed average} + \frac{\Sigma (x - \bar{x})}{n} = 12 + \frac{(-10)}{5} = 10,$$

which is, of course, the correct solution.

But, you ask, are we not complicating a simple problem that anyone can do? You will find later that, with complicated figures, you can save yourself a great deal of time and effort by using this principle.

EXAMPLE 4 The Arithmetic Mean of a Frequency Distribution

Let us go back to our original problem, and from the frequency distribution we compiled try to calculate the arithmetic average. An initial problem that presents itself is that, although the frequency distribution is easier to absorb than 300 numbers, we have lost the detail that would have enabled us to calculate Σx with absolute accuracy. How can we overcome this? When faced with this difficulty we have to accept some element of approximation and assume that all the items found in a particular group have a value equal to the mid point of that group. This is not as arbitrary as it sounds. Some items will have a value greater than this, others will have a lower value, and over a large number of items these differences will tend to offset each other. Try an experiment. Add up the ages of everybody in your class. Now take the average of the oldest and the youngest person. Again obtain the total age assuming that everyone is the average number of years old. You will be surprised how close the two totals are — especially if there is a reasonable number in the group you take. Thus we have a means of estimating fx for each group. You must be careful however when you choose the centre point to use the exact upper and lower limits of the group. The group £1200 and under £1250 includes all wages between £1200.00 and £1249.99 and the centre point is £1224.995 or £1225. But if we are considering the number of seats in a cinema the range 1200 and under 1250 means in fact limits of 1200 and 1249 and the mid point is 1224.5.

In our calculation then:

1 Annual Wage (£) x	2 No. of Men f	3 Mid Point c	4 $\dfrac{c-x}{50}$	5 $\dfrac{f(c-x)}{50}$
1200 and under 1250	13	1225	-3	-39
1250 and under 1300	42	1275	-2	-84
1300 and under 1350	51	1325	-1	-51
1350 and under 1400	63	1375	0	
1400 and under 1450	50	1425	$+1$	$+50$
1450 and under 1500	43	1475	$+2$	$+86$
1500 and under 1550	25	1525	$+3$	$+75$
1550 and under 1600	13	1575	$+4$	$+52$
	300			$+89$

Step 1 Determine the mid point of each group (Column 3).

We will use the principle outlined in the previous section and list our mid points as £1225, £1275 etc.

Step 2 Select an assumed mean.

It is immaterial what figure you choose since the method will work using any assumed mean, but you will find it easier to choose one of the centre points you have listed and preferably one somewhere near the middle of the distribution. In this example we will use £1375 as an assumed mean.

Step 3 List the deviations.

We will now list the deviations from this assumed mean in column 4 still using the centre points as representing the value of their group. You will note that these are $-150, -100, -50, +50$ etc. Each of these is divisible by 50 and we can simplify the arithmetic still further by dividing throughout by this common factor. Thus our deviations appear as $-3, -2, -1, 0$ etc.

Step 4 Multiply the deviations by the frequency (Column 5).

We must remember that in each group there are a number of items each of which has this deviation from the assumed mean. Since we are trying to find the total deviations we must multiply each deviation by the appropriate frequency. Remember that in this step you must take careful note of the sign of the deviation.

Step 5

We are now in a position to determine what adjustment we must make to our assumed average. Firstly do not forget that we divided our deviations by 50 so the true value of our total deviations is $+89 \times 50$ or $+4450$.

Since we are dealing with 300 items the deviation attributable to each is $\dfrac{+4450}{300} = +14.83$, and this is the amount by which we must adjust our assumed mean.

$$\bar{x} = £1375 + \frac{89 \times 50}{300} = £1375 + 14.83 = £1389.83$$

For those of you who prefer algebraic formulae:

$$\bar{x} = \text{Assumed Mean} + \frac{\Sigma \, [f(c - \bar{x})]}{\Sigma f}$$

It is better however, to understand the method rather than to learn a formula parrot fashion.

TUTORIAL FIVE

1. The expression $\dfrac{\Sigma x}{n}$ tells you how to calculate the arithmetic mean. It does not define it. Can you attempt such a definition?
   (Hint: Think of the essential properties of the arithmetic mean.)
2. Liverpool is 200 miles from London. You journey by car from Liverpool to London at an average speed of 40 miles per hour. On the return journey your average speed is 60 miles an hour. Why is your overall average NOT 50 miles an hour?
3. Suppose you are told on applying for a job that the average wage paid was £30 per week. Would this be sufficient information for you? What other information about wages would you wish to have?
4. Criticise the following classification:

Age	No. of Workers
15–20	17
20–25	34
25–30	31
30 and over	27

5. Suggest suitable group widths that might be used in preparing frequency distributions showing:

   a. Wages in a particular industry
   b. The number of seats in London cinemas
   c. The age of students

   Give an example in each case of the centre point of one of the groups you have chosen.
6. Take the last example in the text and calculate the arithmetic mean using frequencies of 1, 4, 5, 6, 5, 4, 2, 1 respectively (reading the groups downward). What can you say about the impact of frequencies on the arithmetic mean? Why do you think the above frequencies were chosen?

7. Over a period of three years employees of a major company
    obtained pay rises of 8%, 10%, and 9%. Why was the average pay
    rise not 9% a year?

Standard Deviation

If you have thought conscientously about the tutorial above you will
have realised that in calculating the arithmetic mean we have com-
pressed a great deal of data into a single figure — and in so doing have
lost a great deal of useful information. One particularly disturbing loss
is that we do not know how the items are scattered around the arithmetic mean. Or in statistical language, we do not have any knowledge
of the dispersion of the distribution.

This may be a matter of great importance to those who use our
figures. An average wage of £1000 a year in a particular factory may be
a result of 100 men each earning £1000 a year. But it may equally
result from 98 men each earning £600 a year and 2 men each earning
£20,600. If you are an applicant for a job at that factory the differences
will mean a great deal to you. What we need is some measure of the
way the items are dispersed. Are they all close to the average, or are
they scattered far and wide?

There are many ways in which we might measure this spread, but
statisticians are generally agreed that far and away the best measure is
the *Standard Deviation*. This is a measure which you will meet constantly in a wide range of techniques and it is vital that you grasp it
thoroughly at this early stage.

What is Standard Deviation? Expressed in words it sounds cumbersome. It is the square root of the average of the deviations squared.
Perhaps it is easier to grasp symbolically:

$$\text{Standard Deviation } (\sigma) = \sqrt{\frac{\Sigma (x - \bar{x})^2}{n}}$$

Why, you may ask, is it a question of squaring the deviations and
then taking the square root of the average deviation squared? This is
something that you will meet time and time again in statistics. Partly
the answer lies in the fact that deviations may be positive or negative
and no statistician likes to ignore signs. Yet, if we do not ignore them,
$\Sigma (x - \bar{x})$ must always be equal to zero. There is more to it than this,
however. We can use the standard deviation to give us a vast amount of
information which no other measure of dispersion can possibly give us,
and you will be surprised at how many times you will meet it in this
book.

EXAMPLE 5

Suppose we have a very simple series as below. We could calculate the Standard Deviation as follows:

X	5	7	10	11	12	$\bar{x} = 9$
$(X - \bar{X})$	−4	−2	+1	+2	+3	
$(X - \bar{X})^2$	16	4	1	4	9	$\Sigma (X - \bar{X})^2 = 34$

The sign we use for Standard Deviation is σ (Sigma — but this time the small letter)

$$\sigma = \sqrt{\frac{\Sigma (X - \bar{X})^2}{n}} = \sqrt{\frac{34}{5}} = 2.608$$

This calculation is, of course, simple so long as the arithmetic mean is a whole number. It becomes more difficult if the mean is a decimal, say, 9.24. If in addition the deviations are large and the frequencies equally so, it becomes both cumbersome and frustrating. Imagine the problem of calculating $326 \times (174.29)^2$, and then performing a dozen or so similar calculations. You would soon be willing to accept any simplification we had to offer.

Fortunately, we can extend the method we used to calculate the arithmetic mean using an assumed average. In fact, once you have calculated the mean in this way you will need only one more simple column to calculate the standard deviation.

What we do is to take the average of the deviations squared as if the assumed mean were the true mean, adjust in much the same way as we adjust the assumed mean and take the square root of the result. The expression is

$$\sigma = \sqrt{\frac{\Sigma (fd^2)}{\Sigma f} - \left(\frac{\Sigma fd}{\Sigma f}\right)^2} \times \text{Simplification factor}$$

where d is the deviation of the item x from the assumed mean and f is the frequency of that item.

Let us take each of these elements in turn:

a. $\dfrac{\Sigma (fd)^2}{\Sigma f}$ This is merely the calculation of the average of the deviations squared, taking deviations from the assumed mean not the true mean. You will remember too that we are taking the centre point of each group to represent the group as a whole.

b. $\left(\dfrac{\Sigma fd}{\Sigma f}\right)^2$ Can you see that $\dfrac{\Sigma fd}{\Sigma f}$ is the adjustment we make to the assumed mean to obtain the true mean? Since we have squared our deviations we must also square the adjustment. You will appreciate that the sign will always be negative.

c. Having taken the square root of this adjusted figure we may or may not have to multiply by a simplification factor. If you have divided the deviations by 25 or 50 or any other convenient number to simplify the arithmetic it is at this point that we must remultiply.

Let us go back to our last example and calculate the Standard Deviation:

Annual Wage £x	Employees f	Mid Point c	$\dfrac{c - \bar{x}}{50}$	$\dfrac{f(c - \bar{x})}{50}$	$f\left(\dfrac{c - \bar{x}}{50}\right)^2$
1200 and under 1250	13	1225	-3	-39	117
1250 and under 1300	42	1275	-2	-84	168
1300 and under 1350	51	1325	-1	-51	51
1350 and under 1400	63	1375	0	0	0
1400 and under 1450	50	1425	$+1$	$+50$	50
1450 and under 1500	43	1475	$+2$	$+86$	172
1500 and under 1550	25	1525	$+3$	$+75$	225
1550 and under 1600	13	1575	$+4$	$+52$	208
	300			$+89$	991

$$\bar{x} = 1375 + \frac{89 \times 50}{300} = £1389.83$$

$$\sigma = \sqrt{\frac{991}{300} - \left(\frac{89}{300}\right)^2} \times 50 = \sqrt{3.3 - .0877} \times 50$$

$$= \sqrt{3.2123} \times 50 = £89.6$$

What does this figure of £89.6 mean? We will show later that it has a precise meaning — namely that if the population we are examining is large enough and not abnormal:

Between $\bar{x} + 89.6$ and $\bar{x} - 89.6$ lie about 68.2% of the items examined.
Between $\bar{x} + (1.96 \times 89.6)$ and $\bar{x} - (1.96 \times 89.6)$ lie about 95% of the items examined.

We can calculate the percentage of items falling within any given range in this way and you will find this of the utmost importance in assessing the results of sample surveys later on. For the present it is enough to remember that

$\bar{x} \pm 1.96\sigma$ includes 95% of the items
$\bar{x} \pm 2.58\sigma$ includes 99% of the items
$\bar{x} \pm 3.29\sigma$ includes 99.9% of the items

TUTORIAL SIX

1. What is the difference between:

   a. $\Sigma f \times d^2$ b. $\Sigma (f \times d)^2$ c. $\Sigma (f \times d^2)$ d. $\Sigma f \times \Sigma d^2$

2. If we square the Standard Deviation we get a statistical measure called the VARIANCE. So

   $$\text{Var}(x) = \frac{\Sigma (x - \bar{x})^2}{n}$$

   Use the rules already formulated with respect to summations to show that

   $$\text{Var}(x) = \frac{1}{n} \left[\Sigma x^2 - \frac{(\Sigma x)^2}{n} \right]$$

3. Using the formula derived in the last example in the text we can see that

   $$\sigma = \sqrt{\frac{\Sigma f x^2}{\Sigma f} - \left(\frac{\Sigma f x}{\Sigma f} \right)^2}$$

   if the items x occur with a frequency f. Use this formula to recalculate the standard deviation of the distribution you have constructed in Tutorial 3, Question 2.

4. In Factory A the average hourly earnings are 75 pence per hour with a standard deviation of 7 pence. In Factory B the statistics are 80 pence and 8 pence respectively. Which factory shows the greatest variation in earnings. (Hint: express σ as a % of the mean — this is called the Coefficient of Variation).

EXERCISES FOR THE INTRODUCTION

1. Since the abolition of scale charges for conveyancing a solicitor bases his fees on the sale price of property plus a fixed charge to cover the cost of postages, telephone calls, forms etc. His charge comprises a fixed cost of £15 plus 2% of the sale price plus £2 to cover postages etc.

   a. Given that C is the sale price of the house and F the fees charged, express F in terms of C in the form $y = mx + c$.

   b. Calculate the fees charged on houses sold at a price of £4000 and at a price of £10,000 and hence draw the graph of the function you have stated above.

   c. From the graph determine the fees charged on houses sold for £3000, £7500, and £12,000.

2. The relation between the price (p) of a commodity and the quantity bought per week (q) in thousands is:

$$p = 90(q + 9)^{-1}$$

   a. How many units are bought when price is 10p, 5p, 2p?
   b. Draw the graph relating price and quantity.
   c. The equation above shows price as a function of quantity bought. What is the equation showing quantity bought as a function of price?
   d. Graph the second function. Will you now put p or q on the y axis?

3. Consider the functions

$$y = 35x + 150$$
$$y = 25x + 200$$

   Graph both these functions and hence determine the value of x which satisfies both equations.

4. The cost of production of a particular commodity is given by the equation $C = x^{\frac{2}{3}} + 1000$, where x is the output per week. What equation would you use to show the output that could be produced at a given cost C?

5. If £P is invested at the beginning of a year and a further £a at the end of each year for n years the total sum invested after n years is given by

$$S = \left(P + \frac{a}{r}\right)(1 + r)^n - \frac{a}{r} \text{ where } r \text{ is the } \frac{rate \ of \ interest}{100}$$

   How would you change this equation if £a per year were *withdrawn* at the end of each year? Given that after n years during which £a per year is withdrawn the value of $S = 0$, find an expression showing How much you need to invest at the beginning of the first year to withdraw £a per year for n years.

6. A small tailor produces two commodities only, Jackets and Overcoats. He knows that it takes 2 hours labour to make a jacket and hours labour to make an overcoat. Experience tells him that to make a jacket he needs 3 yards of cloth and that to make an overcoat he needs 5 yards of cloth. He has available 225 yards of cloth and he and his sons can work for 200 hours each week.

   a. Obtain an expression showing how many hours labour are needed to produce x overcoats and y jackets. Draw a graph

showing all possible combinations of overcoats and jackets that
can be produced with 200 hours labour per week (ignoring the
question of cloth).

b. Obtain an expression showing how much cloth is needed to
produce x overcoats and y jackets. Draw a graph showing all
possible combinations of overcoats and jackets that can be
produced with 225 yards of cloth per week. (Ignore the question
of labour.)

c. From your graphs deduce that combination of overcoats and
jackets which will maximise his use of resources.

7.   The following is a list of marks obtained in a recent mathematics
examination by 40 students. Construct a frequency distribution and
from it calculate the arithmetic mean and standard deviation. Compare the arithmetic mean so obtained with that obtained from the
original data.

51	41	71	49	48	54	63	52	63	52	68	57	38	
32	54	43	54	61	43	56	37	51	49	58	64	59	
51	42	56	69	41	59	47	68	76	45	35	57	42	46

8.   Distribution of Salaries of qualified Accountants
                              one year after qualifying

Salary Range £	No. of persons
1300 and under 1350	13
1350 and under 1400	44
1400 and under 1450	111
1450 and under 1500	157
1500 and under 1550	92
1550 and under 1600	45
1600 and under 1650	16
1650 and under 1750	2

Calculate the arithmetic mean and the standard deviation of the
above distribution.

Chapter One

Inequalities

Inequalities

If you come to think about it, one of the most important results of a system of numbers is the ability it gives us to range things in order of magnitude. We know that a man 6 ft. tall is taller than a man 5 ft. 8 ins. tall without ever having seen either of them. See how far you can get describing your girl friend (or boy friend) without the use of numbers, and you will soon find that, however important the qualitative aspects, a really accurate description soon involves the use of numbers.

Moreover, the number system enables us to measure differences in magnitude. When we say that the man six foot high is four inches taller than the man five foot eight inches high, we are in fact making a number of mathematical statements such as:

$5'\ 8'' + 4'' = 6'\ 0''$; $6'\ 0'' - 4'' = 5'\ 8''$;
$5'\ 8''$ is less than $6'\ 0''$; $6'\ 0''$ is greater than $5'\ 8''$;
The difference in height is 4 inches; and so on.

Now for centuries mathematicians have concentrated on statements such as the first two above, which express the equality of two or more magnitudes. Perhaps this is because it is easier to visualise precisely, things which are the same. But surely the important thing about measurement is the difference between one set of measurements and another. In the last thirty years or so a whole new branch of mathematics has been developed based on the fact that quantities are unequal, and it is this that we must first turn our attention to.

To do this you must first understand two basic signs:

(a) $>$, meaning greater than (e.g. $x > y$, x is greater than y) and its allied $\geqslant$ – is greater than or equal to (e.g. $x \geqslant y$, x is greater than or equal to y).

(b) $<$, is less than (e.g. $y < x$, y is less than x). Can you see the meaning of $y \leqslant x$?

Suppose we say that $y < 2 < x$. We interpret this as y is less than 2, which is less than x, and can illustrate it in this way.

range of y values	range of x values

2

$-\infty$ .... -3 -2 -1 0 1 3 4 5 6 7 ... ∞

Now in this example neither x nor y is equal to 2, and so it should be apparent that y must always be less than x.

Is it equally obvious to you that if $x > 2 > y$, then $x > y$. This is really saying the same thing in reverse but it expresses a very important feature of the 'greater than', 'less than' relationship: it is transitive.

Simple Rules for Inequalities

All other relationships are equally simple:

1. Adding a number to both sides of an inequality does not change the sense of the inequality. Thus if $a > b$, $a + 2 > b + 2$; and if $a < b$, $a + 2 < b + 2$.

   You will realise of course that this rule is also true if the number we add is a negative number. If $a > b$, $a + (-1) > b + (-1)$ i.e. $(a - 1) > (b - 1)$.

2. Multiplying both sides of an inequality by the same positive number does not change the sense of the inequality. Thus if $c > d$. $2c > 2d$. More generally if $c > d$ and $s > 0$, $cs > ds$.

3. Multiplying both sides of an inequality by the same negative number reverses the sign of the inequality. Thus if $a > b$, $-2a < -2b$.

   If you find this one confusing, try it with numbers rather than letters. $5 > 3$ but $-2 \times 5 < -2 \times 3$ i.e. $-10 < -5$.

This is all you need to know to go a long way in the mathematics of inequalities.

Simple Linear Inequalities or Linear Inequalities with a Single Variable

EXAMPLE 1
Suppose we have an equation in the following form.

$2x + 4 > 3x + 2$
Subtracting $2x$ from both sides we have
$4 > 3x - 2x + 2$ or $4 > x + 2$.

Subtracting 2 from both sides

$4 - 2 > x + 2 - 2$ or $2 > x$

More normally we would say $x < 2$.

Now there is no single value of x, but a range of numbers within which the value must fall; but nevertheless the inequality is solved and every value of x which is less than 2 satisfies it.

Graphing an Inequality

Just as it is possible to draw the graph of an equation, so it is possible to draw the graph of an inequality. In this section at least most of the work we do on inequalities will be graphical.

Ordered Pairs

EXAMPLE 2

Consider the inequality $2x + y > 1$. It describes all combinations of x and y that satisfy the condition $2x + y > 1$.

We know already how to graph $2x + y = 1$. It will be the straight line joining the points $(\frac{1}{2}, 0)$ $(0,1)$. Remember that every combination of values of x and y on this line will satisfy the equation $2x + y = 1$.

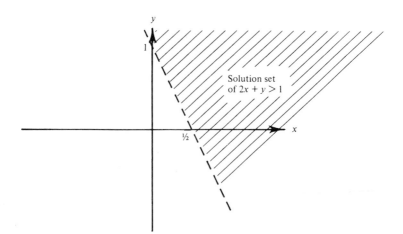

But if the inequality $2x + y > 1$ is to be satisfied, either the value of x must be greater than the value indicated by the graph, or the value of y must. Thus any point lying above the line $2x + y = 1$ gives a pair of values of $x + y$ which will satisfy $2x + y > 1$. We have many such pairs of values, and as a group we may call them the *solution set* of $2x + y > 1$.

Remember however that the solution set does not include any pair of points on the line $2x + y = 1$ – only above it.

EXAMPLE 3
Let us now depict geometrically the inequality $3x + 2y \leqslant 6$.

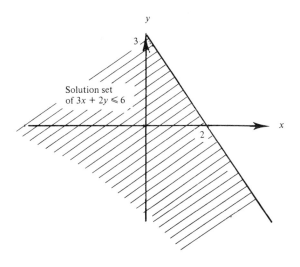

We proceed exactly as before, graphing the straight line $3x + 2y = 6$.

In this case however the value of $3x + 2y$ must be less than, or equal to 6 and this would be true if the value of x or y were to be less than, or equal to the value shown on the line. In other words, the solution set for $3x + 2y < 6$ consists of all pairs of values of $x + y$ falling on, or below the line $3x + 2y = 6$. Notice that when we graph a 'less than' or 'more than' relationship we use a broken line. If a broken line is drawn, then the combinations on that line do not occur in the solution set.

Simultaneous Linear Orderings

It is seldom we have to deal with a single inequality. Often there are two or more. Provided that they do not involve more than two variables $x + y$ they can easily be handled by simple graphical methods.

EXAMPLE 4

Suppose we have a pair of linear inequalities.

$$x + y > 1$$
$$2x - y > 1$$

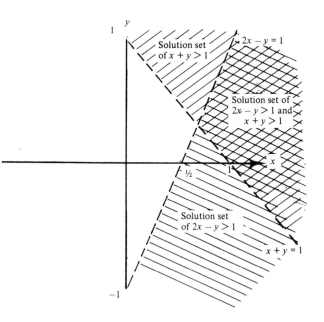

The diagram above represents the two inequalities. If both in-equalities are to be satisfied there must be an area where the solution sets coincide i.e. pairs of values of $x + y$ will satisfy both inequalities. The doubly shaded area above represents this coincidence of the solution sets.

TUTORIAL ONE

1. Give as many mathematical interpretations as you can of the statement that a given volume of water weights 10 pounds while half this volume weighs 5 lbs.
2. If $x > 3$ and $y > 4$ find the greatest integer less than $3x + 4y$.
3. If $x < 3$ and $y > -3$ find the least integer greater than $2x - 3y$.
4. Show that if $a > b$ then $\dfrac{1}{a} < \dfrac{1}{b}$.
5. Depict geometrically the solution set of $-x + y > 1$.
    Would the solution set be different if $-x + y \geqslant 1$?

6. Sketch the solution sets of the pair of inequalities

    $x + y < 1$.
    $3x + 2y < 6$.

    What would you conclude?

7. Sketch the solution set for the triple inequalities,

    $x + y \geqslant 2$.
    $x + 4y \leqslant 4$.
    $y > -1$.

8. Consider the non linear inequality $x^2 - 6x - 16 < 0$, then
   $(x + 2)(x - 8) < 0$. If $(x + 2) < 0$ what can you conclude
   about $(x - 8)$? Is your conclusion reasonable?
   If $(x + 2) > 0$ what you conclude about $(x - 8)$? Is your con-
   clusion reasonable?
   Deduce the solution to the inequality.

Linear Expressions and Linear Programming

Those of you trained in traditional mathematics where the normal
relationship is of the type $a = b$ or $2x + 4y + z = 26$ may be wondering why you have had to spend some time mastering expressions
which are basically inequalities. Once you begin to try to apply
mathematical techniques to problems of production you will not
wonder much longer. Most problems you will meet are of the type
which involve some *restrictions* as a result of having resources limited
in quantity or with some limitation on quality attainable. The rest of
this chapter is concerned with a simple application of the technique
you have just learned to the type of problem the businessman is
constantly facing. Possibly without realising it you have been studying a part of the technique known as 'linear programming'.

This expression is merely management jargon for the technique of
reducing a practical problem to a series of linear expressions and
then using those expressions to discover the optimal, or best, solution to achieve a given objective. That objective may be the maximisation of profit, the minimisation of cost, or possibly even the
best proportions in which to blend Scotch whisky. Whatever the
objective, the method tends to be the same.

Of course not all industrial problems are amenable to these
methods. There are times when it is quite impossible to express a
given situation in linear form. For these, other, and more complex
methods of analysis must be used. But you would be surprised how

many problems are capable of being solved by relatively simple techniques.

We will firstly concentrate on the simplest type of problem, those involving only two variables. The reason for this is that such problems are usually capable of simple graphical solution. You will readily appreciate that three variables involve three dimensional graphs which few people really understand, and with more than three variables we enter the realm of graphs which rapidly become unmanageable. For solutions to problems of this type you must wait until a later chapter.

EXAMPLE 5

Suppose a small tailor is producing two articles only, overcoats and jackets. He has a contract to supply a department store with 10 overcoats and 20 jackets per week. His labour force is limited to five men working a 40 hour week and due to a shortage of working capital he cannot purchase more than 225 yards of cloth per week.

To produce one overcoat takes 5 yards of cloth and 5 hours labour.
To produce one jacket takes 3 yards of cloth and 2 hours labour.

What combination of coats and jackets is it feasible for him to produce?

This is a typical problem that can be expressed mathematically by means of inequalities. We will first examine the *product-mix* that can be produced and then develop the argument to ask what product-mix should be produced.

Firstly we state the problem facing the tailor as a series of linear inequalities.

Let x = the number of overcoats produced.
and y = the number of jackets produced.

How does the contract affect his output? He must produce at least 10 overcoats and 20 jackets to meet this contract. Hence

$x \geqslant 10$ and $y \geqslant 20$ (These are graphed in diagram 1.)

There are however limits imposed on the volume of production by the fact that he has available only 200 hours labour and 225 yards of cloth. Let us consider labour first. If we produce x overcoats per week it will take $5x$ hours of labour. Similarly y jackets will take $2y$ hours of labour.

But we have only 200 hours of labour available so $5x + 2y$ cannot be greater than 200. Thus we may say

$5x + 2y \leqslant 200$. (The labour restriction is drawn in diagram 2.)

Considering now the cloth restriction, we can see that x overcoats take $5x$ yards of cloth, and y jackets take $3y$ yards of cloth. With only 225 yards available it is apparent that

$5x + 3y \leqslant 225$. (Which is drawn in diagram 3.)

We have now expressed mathematically the situation of the tailor. It is simply:

If he produces x overcoats and y jackets,
$x \geqslant 10$ $y \geqslant 20$.
$5x + 2y \leqslant 200$. $5x + 3y \leqslant 225$.

Before we can go further we must determine the combinations of x and y which are feasible, given the restrictions on labour and raw materials. By far the easiest way of doing this is to sketch the four inequalities, and by examining the solution sets determine what values of x and y satisfy all the inequalities — which we have done in diagram 4.

In this case we have shaded the combinations of output which are not feasible. It is immediately apparent that there is a polygon ABCD which is unshaded. Within this area production is feasible since neither the contract nor the limited labour force nor the restricted supply of cloth prevents it. This is why such an area is known as a feasibility polygon.

We have not yet of course solved any real problem for the tailor. All he has yet is a number of outputs of overcoats and jackets which he knows are possible. It could be 40 jackets and 20 overcoats, or 30 jackets and 25 overcoats. So does the feasibility polygon really help?

Profit Maximisation — Two Dimensional Model

Think once again of the problem we have already partially examined. We have expressed the restrictions on his output in the form of linear inequalities viz.

$x \geqslant 10$. $y \geqslant 20$.
$5x + 2y \leqslant 200$. $5x + 3y \leqslant 225$.

You have learned how to graph and interpret expressions of this type and should understand how we obtain the feasibility polygon

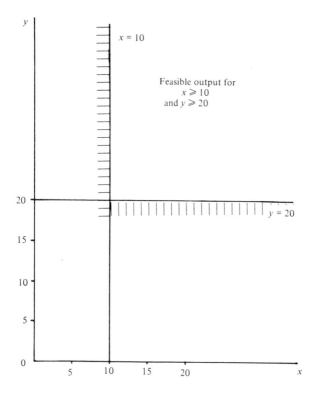

Diagram 1. Contract Restrictions

showing all possible combinations of jackets and overcoats the tailor
can produce within the limits of the restrictions imposed on him by
the availability of labour and working capital. (If you do not, turn
back now before reading further.) This feasibility polygon is the
area enclosed by heavy lines in diagram 5.

  To use what we have obtained we must now consider the objective desired by the tailor. It is highly probable that an important
objective will be to maximise his profit and we will concentrate on
this. The tailor knows that he makes a profit of £2 on each overcoat
he sells and a profit of £1 on each jacket he sells. If, as before, he
sells y jackets and x overcoats we may express his total profit as
Profit $= y + 2x$ and the problem we are posing is what combination
of jackets and overcoats will maximise this profit, and still satisfy the
restrictions. It is essentially the basic economic problem of how best

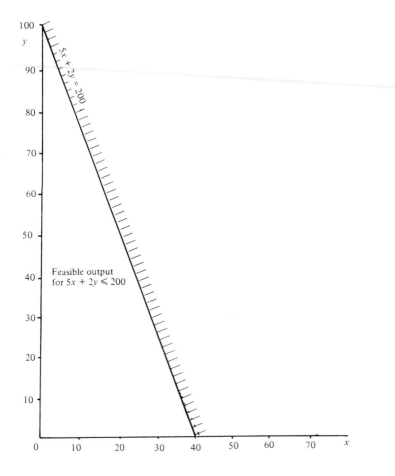

Diagram 2. Labour Restriction

to use limited resources which are capable of being used in more than one way.

If we state the problem this way:

Maximise $P = y + 2x$

  subject to $x \geqslant 10$ $2y + 5x \leqslant 200$.

  $y \geqslant 20$ $3y + 5x \leqslant 225$.

We have constructed a *mathematical model* of the problem we wish to solve. Constructing such models is an essential feature of all linear programming problems. The graphical solution is approached like this:

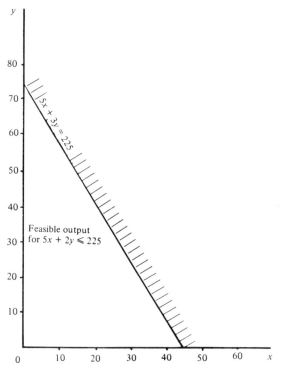

Diagram 3. Cloth Restriction

firstly we must graph the expressions derived from the restrictions and obtain the area of feasibility (which we have already done). (Diagram 5).

Now we superimpose on this graph a profit line representing Profit = $y + 2x$. To do this we must of course specify some value of P. Suppose we assume that the tailor produces merely to satisfy his contract and nothing else. In this case his output will be 10 overcoats and twenty jackets per week, giving a total profit of £40. (Always approach this problem by assuming minimum possible profit.) If he had no contract to deliver to a department store minimum possible profit would be zero, and this would result from no production i.e. we would start from the origin of the graph. However, in this case he could not make a smaller profit than £40. He could however make this same profit by producing either 20 overcoats or 40 jackets. If we now join the points representing 40 jackets on the vertical axis and 20 overcoats on the horizontal axis by a straight line, this line will give us all combinations

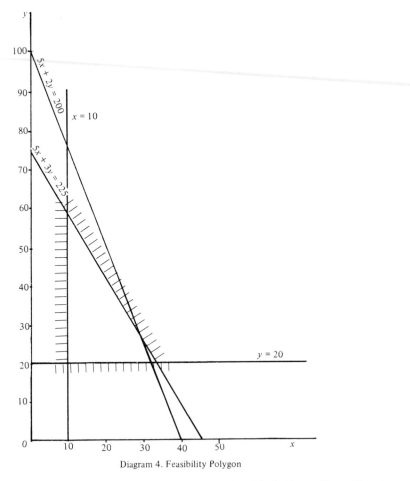

Diagram 4. Feasibility Polygon

of coats and jackets which will yield a profit of £40. You will readily see that in practice all but one of these combinations are barred to him by the existence of his contract. Graphically this is shown by the fact that our Profit = £40 line cuts the feasibility polygon at one point only, the point A, representing a production which just satisfies his contract demand and nothing else.

You should also be able to deduce that the feasibility polygon tells us that he can produce more than this and so increase his profit by producing more overcoats, by producing more jackets, or by producing more of both. In our approach to profit maximisation it seems obvious that he would first produce more overcoats since they yield a greater profit than the jackets. One might assume that he would wish to produce

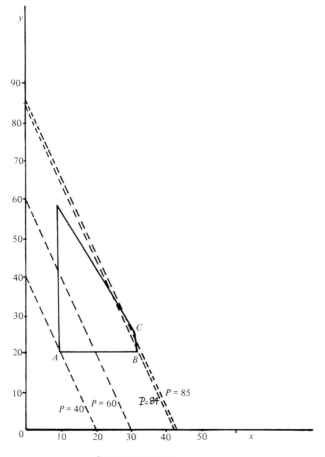

Diagram 5. Profit Maximisation

as many overcoats as possible, by moving to point *B* and producing 32
overcoats. This restricted output is determined by the restriction on
labour supplies since he is already using 40 hours labour to produce the
20 jackets needed for his contract. There is no doubt that this would
increase his profit. It would rise in fact to £84.

Before we consider this alternative however, let us look at one of the
intermediate stages in this move — the point at which the tailor can
make £60 profit by producing 20 overcoats and 20 jackets, or 10 overcoats and 40 jackets or 15 overcoats and 30 jackets. All these combinations on the profit line *P* = 60 are feasible.

If you think carefully about this line, several things are obvious. It does not give a unique solution to the problem we posed, since the profit line gives many different combinations of output all of which lie within the feasibility polygon. Secondly it is immediately apparent that this is not maximum profit. The area of the feasibility polygon lying to the right of the line $P = 60$ indicates that we could, with the resources available, produce more of both goods and so raise profit. It is apparent that to obtain a unique solution, the profit line must cut the feasibility polygon at an apex, and also all other combinations of output which give the same profit as this one must lie outside the area of feasibility.

We have then a very limited number of profit lines to consider and since a movement of the profit line to the right implies a rise in total profit the number is still further reduced.

Go back now to the point B and the line $P = y + 2x = 84$. This satisfies one of the criteria — it cuts the polygon at an apex — but it is not the solution we are looking for, since several other combinations of output all of which are feasible will give this profit.

We cannot raise profit by producing more overcoats since the output of 32 is restricted by the availability of labour supplies. But there is 5 yards of available material left unused. Is it possible to produce fewer overcoats and use the resources released in such a way as to produce sufficient jackets to raise profit. We will have to produce more than two jackets for every overcoat we sacrifice if we are to succeed.

If we give up one overcoat, we will have available 5 hrs. labour and 10 yards of material. Thus we could produce a theoretical 2½ jackets before our labour force is exhausted. But who will buy half a jacket? The maximum we can produce realistically is 2, and to replace one overcoat by two jackets does not raise profit. We must try producing two overcoats less. This will leave us with 10 hours of labour and 15 yards of cloth, which is enough for 5 jackets. So we can raise profit by £1. We have in fact moved to point C with a profit of £85. Why can we not continue this process, giving up two more overcoats and raising profits further? The possibility of this depended on our having had five yards of cloth unused. Now this has been used up, giving up an overcoat gives us sufficient cloth to produce 1.67 jakcets — which is non profitable.

By this time most of you will be thinking that point C was so 'obvious' in the first place that we could have moved to it immediately. The solution is seldom as obvious as this, and in the tutorial you will see that our solution will vary according to the profit ratio of the two articles. Master the general approach and you will never go wrong — well hardly ever.

TUTORIAL TWO

1. What output would the tailor produce if the profit was

   (a) Overcoats £4   Jackets £1.
   (b) Overcoats £1   Jackets £1.

2. What would be the effect of there being available an additional 5 hours labour?
3. What would be the effect of an additional 10 yards of material?
4. Deduce the effect of there being available an additional supply of both 10 yards of cloth and 5 hours of labour.
5. A firm is producing two brass ornaments, a standard and a de luxe model. The manufacturing process consists of two machines only

Machine 1	*Machine 2*
Standard model 1 hour	1 hour
De-luxe model 2 hours	5 hours

   There are 20 hours time available on machine 1 and 35 hours on machine 2. The firm makes 50 pence profit on the standard model and £1.50 on the de-luxe,

   (a) The first restriction is on machine 1 time and the second on machine 2 time. What is the third?
   (b) If the aim is to maximise profit, state the objective function.
   (c) Obtain by graphical methods the output which will achieve the objective.

Cost Minimisation

One of the earliest applications of linear programming was the determination of the most economical use of raw materials to achieve a given objective.

EXAMPLE 6

Let us suppose that a manufacturer of animal feeding stuffs is asked to supply 40 tons of feed per week to a large farmer. The farmer specifies that the food must contain a minimum of 25% fat and 20% protein, but that the total quantity may otherwise be made up by any bulk feed. Two convenient raw materials are available, A and B.

   A contains 50% fat and 25% protein.
   B contains 25% fat and 40% protein

The manufacturer has in stock 12 tons of A which he wishes to use, but further supplies are readily available. On the other hand he is able to obtain only 20 tons of B per week. The price of both A and B is £25 per ton. In what proportions should he mix A and B in order to satisfy the farmers requirements and minimise his cost of production?

The problem may be expressed mathematically as follows:

1. $A \geqslant 12$ because of the need to use existing supplies.
2. $B \leqslant 20$ because of the limitation on supplies. Also $B \geqslant 0$ as we cannot add a negative quantity.
3. The desired fat content of the mixture is 10 tons (25% of 40 tons) minimum, so 50% of A + 25% of $B \geqslant 25\%$ of 40. That is $2A + B \geqslant 40$.
4. The desired protein content is 8 tons (20% of 40 tons) minimum, so 25% of A + 40% of $B \geqslant 20\%$ of 40. That is $5A + 8B \geqslant 160$.
5. The restriction on demand is 40 tons per week so $A + B \leqslant 40$.
6. The cost of the mixture $C = 25A + 25B$.

Our full instructions read:

Minimise $C = 25A + 25B$ subject to
$A \geqslant 12 \quad 0 \leqslant B \leqslant 20 \quad A + B \leqslant 40$
$2A + B \geqslant 40 \quad 5A + 8B \geqslant 160$

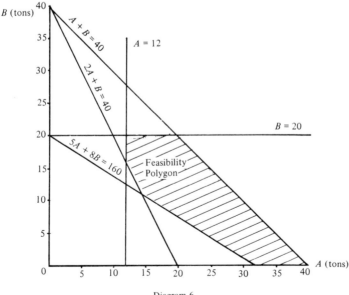

Diagram 6

The only one of these restrictions that may confuse you is that of the demand restriction. Why do we say that $A + B \leqslant 40$ rather than $A + B = 40$? The point is that we may be able to meet the specification of the feed by using quantities of A and B which total less than 40 tons and make good the difference by using a cheap edible bulk material (sawdust perhaps) rather than the more expensive raw materials.

The production feasibility polygon should now be easy to draw. Draw it and compare your result with diagram 6. Any combination of A and B within the shaded area will satisfy the fat and protein requirements and will not exceed the maximum 40 tons per week which is demanded.

We will approach the problem of minimisation in this way. Let us take an arbitrary cost figure of say £275. This sum is the cost of 11 tons of A, or 11 tons of B, or any combination of $A + B$ on the straight line joining the two points. This cost line is illustrated in Diagram 7.

As you can see this cost line does not touch the feasibility polygon at any point at all. What does this mean? Merely that it is not possible to produce the feed to the requested specification at a cost as low as this. So — let us take a different cost level, £750, and add this to our diagram. Certainly at this level of costs there are many combinations of $A + B$ which we can use which will satisfy our needs. We could combine 20 tons of A with 10 of B; or 15 tons of A with 15 of B. But this is not enough — we wish to minimise cost, and the existence of a part of the

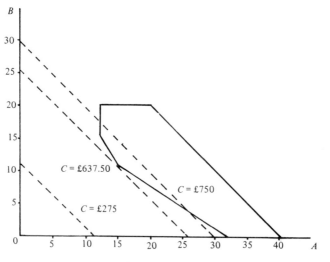

Diagram 7

feasibility polygon to the left of the line C = £750 indicates that we can achieve perfectly satisfactory results at a lower cost. We wish the product mix to be as far to the left as it can be. As in profit maximisation the need for a unique solution indicates that the cost line must cut the feasibility polygon at an apex. A study of the polygon and the slope of the cost curve shows that we minimise cost at point X, mixing 14½ tons of A with 11 tons of B at a cost of £637.50 and adding 14½ tons of cheap bulk to make up the required weight of feed.

It might be useful at this point to ask yourself why we rejected the point Y as a minimum cost point, and then ask in what circumstances would we be likely to move to point Y. (Consider relative prices.)

The problem we set ourselves is solved, but please remember you cannot solve such problems without the complete battery of curves. You need to know the slope of the cost curve to determine which apex of the feasibility polygon satisfies the objective.

Blending

EXAMPLE 7
For our last example in this chapter let us take a problem that those of you who are amateur winemakers may well have met — the problem of blending wines (which in some respect do not meet our requirements) in such proportions that the resultant blend is satisfactory in every respect. This is a problem constantly faced by, for example, the oil companies in attempting to produce a satisfactory commercial product.

As you may know, a good dry table wine should have certain characteristics. It should be about 30 degrees proof to give a satisfactory alcohol content; it should contain at least 0.25% acid, otherwise it is insipid; and its specific gravity should be greater than 1.06. There are probably other characteristics that the experts would consider to be of importance, but these will suffice. Let us suppose that a winemaker has a stock of three wines each of which is unsatisfactory in some respect. Their characteristics are as follows:

	Proof	Acid %	Specific Gravity	Stock (Gallons)
Wine A	27	0.32	1.07	20
B	33	0.20	1.08	34
C	32	0.30	1.04	32
Desired	30–31	0.25+	1.06+	Greatest possible

It should be evident to you that since we have three wines we ought to graph on a three dimensional basis, but since this is difficult we will take wine A of which we have the minimum stock and blend as much as possible of the other two wines with it so as to produce the required result.

It is necessary also to state that the resultant characteristics of the blend will follow linear laws. If, for example, we blend two gallons of wine A with three of wine B the acid content of the resultant blend would be:

$$\frac{(2 \times 0.32) + (3 \times 0.20)}{2 + 3} = \frac{0.64 + 0.60}{5} = 0.248\%$$

Firstly we will examine the restrictions, assuming we blend B gallons of B and C gallons of C with 20 gallons of A.

a. Restrictions on Proof

The resultant degrees Proof of such a blend would be

$$\frac{(20 \times 27) + (B \times 33) + (C \times 32)}{20 + B + C}$$

and this must be equal to or greater than 30, and equal to or less than 31.

Thus in respect of proof we have two restrictions to consider.

Firstly, $\dfrac{540 + 33B + 32C}{20 + B + C} \geqslant 30$; so $540 + 33B + 32C \geqslant 600 + 30B + 30C$;

Thus $3B + 2C \geqslant 60$ . . . . Restriction a.

But equally $\dfrac{540 + 33B + 32C}{20 + B + C} \leqslant 31$; so $540 + 33B + 32C \leqslant 620 + 31B + 31C$.

Thus $2B + C \leqslant 80$ . . . . Restriction b.

These two restrictions on Proof are graphed in Diagram 8(a).

b. Restrictions on Acid Content

We can deal similarly with the restrictions on acid content.

i.e. $\dfrac{(20 \times .32) + (.20B) + (.30C)}{20 + B + C} \geqslant 0.25$

$\therefore 6.4 + .20B + .30C \geqslant 5.0 + .25B + .25C$

$\therefore -.05B + .05C \geqslant -1.4$

i.e. $5B - 5C \leqslant 140$

$B - C \leqslant 28$

This is graphed in Diagram 8(b).

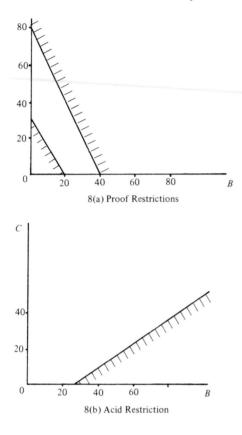

8(a) Proof Restrictions

8(b) Acid Restriction

c. *Restrictions on Specific Gravity*
   Likewise the specific gravity restriction is

$$\frac{(20 \times 1.07) + 1.08B + 1.04C}{20 + B + C} \geq 1.06$$

$$\therefore 21.4 + 1.08B + 1.04C \geq 21.2 + 1.06B + 1.06C$$

$$\therefore .02B - .02C \geq -0.2$$

or $C - B \leq 10$

This is graphed in Diagram 8(c).

d. *Quantity Restrictions*
   Finally we have the quantity restrictions which can be simply stated as
$B \leq 34$ and $C \leq 22$. Remember of course $B \geq 0$ and $C \geq 0$. This is
graphed in Diagram 8(d).

   The resultant feasibility polygon is a little more complex than those
you have met so far. Before you turn to diagram 9 where it is drawn,

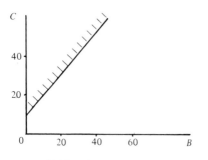

8(c) Specific Gravity Restriction

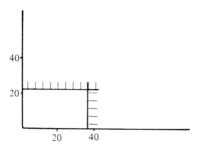

8(d) Quantity Restrictions

try to obtain for yourself from the restrictions we have obtained on previous page.

Now let us turn to the objective function. We want to obtain as much of the final blend as possible. Remember, however, that our blend is not merely $B + C$. It is (20 gallons of A) + $B + C$. The objective function then is

Maximise $Q = 20 + B + C$

and our instructions read:

maximise $Q = 20 + B + C$
         Subject to $3B + 2C \geqslant 60$ $2B + C \leqslant 80$
                       $B - C \leqslant 28$ $C - B \leqslant 10$
                       $0 \leqslant B \leqslant 34$ $0 \leqslant C \leqslant 22$

How do we obtain this objective? Let us take first a given level of our blend, say 60 gallons. This is a blend of 20 gallons of A plus 40 gallons of $(B + C)$. Since A is given we will bother only with $B + C$. The 40 gallons we require may be 40 gallons of B, or 40 gallons of C or any

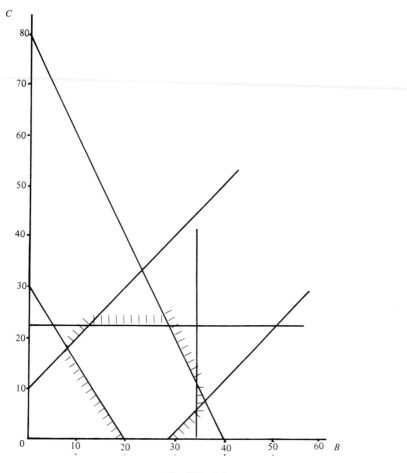

Diagram 9 Feasibility Polygon

combination of the two totalling 40 gallons. All such combinations are shown on the straight line joining 40*B* and 40*C*. In other words this line represents $Q = 60$ gallons. If you look at diagram 10 this line passes through the feasibility polygon and there are many combinations of $B + C$ which will blend with A to satisfy us. But equally we could blend more A or more B, or both and still remain within the restrictions. (How do we know?) Hence,

$Q = 20 + B + C$ is not maximised

You will realise that once again we need the line Q to be as far to the right as possible, touching an apex of the polygon but not passing through it. Inspection should show that this is satisfied when $Q = 71$ gallons. To obtain this gallonage we would take 20 gallons of wine A and blend with it 29 gallons of wine B and 22 gallons of wine C.

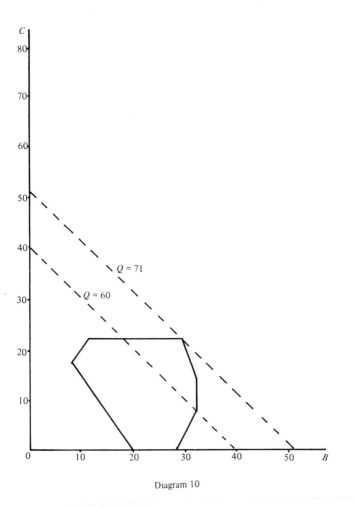

Diagram 10

These simple concepts of linear programming have carried us a long way. You can now obtain inequalities expressing restrictions, set up an objective function, and (so long as we deal in two variables) obtain a solution.

But how many firms produce only two commodities or have only two inputs? As soon as you try to introduce some realistic problem, simple graphical method breaks down. In the later chapters you will learn how to handle mathematically models involving several variables (although if there are say twenty or thirty you would be well advised to seek the aid of a computer).

TUTORIAL THREE

1. Distinguish carefully, between an objective and an objective function.

2. Interpret:

    (a) maximise profit   = $7x + 2y$
    (b) minimise cost   = $3a + 2b$
    (c) maximise Q = $40 + A + B$

3. Suppose that in Example 6 the cost of A was £20 per ton and the cost of B £10 per ton:
    (a) Write down the new objective function.
    (b) Obtain the optimum solution graphically.
    (c) Why is it not possible to obtain a unique solution?

4. In Example 7 we have blended wine $B + C$ with 20 gallons of wine A. Would we obtain the same solution if we blended A and C with 34 gallons of B. If not, why not? Is a solution possible? Would we obtain the same solution by blending $A + B$ with 22 gallons of C?

5. Compare Examples 5, 6, 7.
Suppose we change the objective function in each case to
    Example 5.   Maximise $P = 2y + 2x$
    Example 6.   Minimise $C = 20A + 25B$
    Example 7.   Maximise $Q = 30 + B + C$

Is it now possible to obtain optimal solutions?
What changes must you effect to obtain solutions?
In what ways do the changes affect the three types of problems?

EXERCISES TO CHAPTER ONE

1. Graph each of the following or shade the areas which do not satisfy the inequalities:

    (a) $y \geqslant 0$ (b) $x \leqslant 0$ (c) $y \geqslant x$
    (d) $y \geqslant 2$ (e) $x + y \geqslant 4$ (f) $3x - y \leqslant 9$
    (g) $5x - 4y \geqslant 6$ (h) $5x + 4y \leqslant 24$

2. A haulage contractor owns 6 three ton lorries and 16 thirty hundredweight vans which he is prepared to hire at a charge of £6 per day for a lorry and £4 per day for a van. A manufacturer wishes to transport a minimum of 900 brass castings a day but is unwilling to spend more than £48 a day on hiring transport. A lorry can carry 150 castings and a van can carry 60. Using L to represent lorries and V to represent vans:

    a. Express the above information in the form of a series of inequalities.

    b. Given that the objective is to minimise cost, what combination of vans and lorries should the manufacturer hire?

    c. List all combinations of vans and lorries which will satisfy the restrictions. What is the largest number of castings that can be carried in one day subject to these restrictions?

3. A sports manufacturer produces both tennis rackets and cricket bats. The inputs required for their production are:

	Machine time	Labour time
Bats	1 hour	4 hours
Rackets	2 hours	5 hours

He has available 26 hours of machine time and 92 hours of labour time per day. His sales manager informs him that his maximum daily sales of cricket bats is 20 and of tennis rackets is 11; and that the profit on sale is likely to be £2 on each bat and £3$\frac{1}{3}$ on each racket. What combination of bats and rackets will maximise profit and what will that profit be?

Should profit change to £1.30 on each bat and £2.60 on each racket, what conclusions could you come to regarding the best product mix?

4. A private airline operates two types of aircraft, the Bee and the Wasp, owning 8 of the former and 7 of the latter. The carrying capacities of the two aircraft are:

	Bee	Wasp
Passengers	50	80
Cargo	20 tons	10 tons

Traffic potential is unlimited but the firm has assurances that it will be called upon to carry at least 640 passengers and 190 tons of cargo per day. The estimated cost per journey is £800 for a Bee and £1600 for a Wasp. What is the most economical 'mix' of aircraft?

5. A manufacturer is making two products X and Y with the following input requirements.

Product	Raw Materials	Machine Time	Labour
X	4 tons	2 hrs.	1 hr.
Y	1 ton	1 hr.	1 hr.

Inputs are available in the following quantities per week:

90 tons of raw materials
50 hours of machine time
40 hours of labour

The profit is £4 per unit on X and £3 per unit on Y. What product mix will maximise profit?

6. Suppose that by law a certain health food must contain at least 3% of Vitamin A and at least 7% of Vitamin B. A manufacturer has stocks of two compounds with the following vitamin content.

	Vitamin A	Vitamin B
Vito	2%	10%
Slam	5%	6%

He wishes to blend Vito and Slam to produce a new product Wow, which meets the minimum vitamin requirements. The sales manager reckons that he can sell up to 60 lbs. of Wow per week. If Vito can be produced at £4 profit and Slam at a £5 profit, in what proportions should they be mixed to produce the 60 lbs. of Wow?

7. An oil company put additives to petrol to give improved performance and to reduce engine wear. Ideally each 1000 gallons should contain at least 40 mgs. of additive p, 14 mgs. of additive q, and 18 mgs. of additive r. The company can obtain stocks of ingredients which have the following weights of additive per litre:

Ingredient	mgs. p	mgs. q	mgs. r
y	4	2	3
x	5	1	1

Both ingredients cost £1 per litre. The company wishes to know the quantity of each ingredient which will satisfy the conditions and minimise cost.

Chapter Two

Matrix Arithmetic

In the last chapter we saw how linear programming problems could be solved using graphical methods. The system could cope with any number of restrictions, but only two unknown quantities could be handled. Of course, it would be possible to solve a 'three unknown' problem using a three dimensional graph, but this is very cumbersome. Suppose, however, the problem involved more than three unknown quantities — how could we solve it without the aid of a graph? Mathematicians have evolved the *Simplex Method* to deal with all linear programming problems. It is an extremely powerful tool of analysis, and yields a considerable amount of useful information in addition to satisfying an objective. The technique uses a system of arithmetic which is probably new to many of you.

Matrix Arithmetic

In the last chapter, we considered the case of a small tailor producing overcoats and jackets. We saw that it required 2 hours labour to make a jacket, and 5 hours labour to make an overcoat. If x overcoats and y jackets were produced, then the labour requirement would be given by the expression $2y + 5x$. An alternative way of writing this expression would be

$$[2 \quad 5] \begin{bmatrix} y \\ x \end{bmatrix}$$

The left hand side of this expression i.e. $[2 \quad 5]$ is called a *row vector* and gives the coefficients of y and x. The components of the vector are called *elements*. The right hand side is called a *column vector* and describes the variables to which the coefficients refer. The whole expression means multiply the row vector by the column vector. Thus

$$[2 \quad 5] \begin{bmatrix} y \\ x \end{bmatrix} = [2y + 5x]$$

Putting the vectors in this form enables us to derive rules for vector multiplication: multiply the first element in the row vector by the first element in the column vector, multiply the second element in the row vector by the second element in the column vector, then add the products. Using this form of arithmetic we can calculate the labour requirements if, say, 10 jackets and 5 overcoats were produced

$$[2 \quad 5] \begin{bmatrix} 10 \\ 5 \end{bmatrix} = [2 \times 10 + 5 \times 5] = [45]$$

Now suppose we have a series of combinations of outputs, like this

y	6	12	14	15
x	8	10	16	15

We could write the combinations as a series of column vectors

$$\begin{bmatrix} 6 \\ 8 \end{bmatrix} \quad \begin{bmatrix} 12 \\ 10 \end{bmatrix} \quad \begin{bmatrix} 14 \\ 16 \end{bmatrix} \quad \begin{bmatrix} 15 \\ 15 \end{bmatrix}$$

Or, better still, like this

$$\begin{bmatrix} 6 & 12 & 14 & 15 \\ 8 & 10 & 16 & 15 \end{bmatrix}$$

Written in this form the combinations form a *matrix*. The first row gives the quantities of y and the second row the quantities of x. The columns give the actual combinations. If we want to know the labour required to produce these combinations, then we *premultiply* the matrix by the row vector. We use the multiplication rule taking the columns of the matrix in turn.

$$[2 \quad 5] \begin{bmatrix} 6 & 12 & 14 & 15 \\ 8 & 10 & 16 & 15 \end{bmatrix} = [52 \quad 74 \quad 108 \quad 105]$$

Now let us consider the material requirements: 3 yards for a jacket and 5 yards for an overcoat. For x overcoats and y jackets the material requirements are given by the expression $3y + 5x$. We could combine the coefficients of the material expression with the row vector of labour coefficients to obtain a matrix like this

$$\begin{bmatrix} 2 & 5 \\ 3 & 5 \end{bmatrix}$$

The first row refers to labour, and the second to material requirements. The first column refers to jackets and the second to overcoats.

This matrix gives the labour and material requirements for the combination vector $\begin{bmatrix} y \\ x \end{bmatrix}$. Suppose for example, it was decided to produce 8 jackets and 5 overcoats, then the labour and material requirements would be given by

$$\begin{bmatrix} 2 \\ 3 \end{bmatrix}\begin{bmatrix} 5 \\ 5 \end{bmatrix}\begin{bmatrix} 8 \\ 5 \end{bmatrix} = \begin{bmatrix} 41 \\ 49 \end{bmatrix}$$

or 41 hours of labour, and 49 yards of material. Notice that the same rules of multiplication have been used, taking the rows of the coefficient matrix in turn. More generally.

$$\begin{bmatrix} a & b \\ c & d \end{bmatrix}\begin{bmatrix} y \\ x \end{bmatrix} = \begin{bmatrix} ay + bx \\ cy + dx \end{bmatrix}$$

The product is a vector, not a matrix. If we wished to calculate the requirements for two combinations, then the information would be given in a matrix. For example

$$\begin{bmatrix} 2 & 5 \\ 3 & 5 \end{bmatrix}\begin{bmatrix} 6 & 12 \\ 8 & 10 \end{bmatrix} = \begin{bmatrix} 52 & 74 \\ 58 & 86 \end{bmatrix} \quad \begin{matrix} \leftarrow \text{labour} \\ \leftarrow \text{material} \end{matrix}$$

jackets ↑ combinations ↑ for combination 2

overcoats   for combination 1

We can now state a general rule for the matrix multiplication in the form $AB = C$, where A and B are both 2×2 matrices.

$$\begin{bmatrix} a & b \\ c & d \end{bmatrix}\begin{bmatrix} e & f \\ g & h \end{bmatrix} = \begin{bmatrix} ae + bg & af + bh \\ ce + dg & cf + dh \end{bmatrix}$$

In fact, there is a very simple rule for matrix multiplication: multiply row by column. Suppose we had a 2×2 product matrix, we could identify its rows and columns like this

$$\begin{bmatrix} a & b \\ c & d \end{bmatrix} \begin{matrix} \leftarrow \\ \leftarrow \end{matrix} \begin{matrix} r_1 \\ r_2 \end{matrix} \quad \text{or} \quad \begin{bmatrix} r_1\,c_1 & r_1\,c_2 \\ r_2\,c_1 & r_2\,c_2 \end{bmatrix}$$

↑ ↑
c_1 c_2

The element b is the element in the first row and second column. It is obtained by multiplying the elements in the first row of the first matrix by the elements of the second column of the second matrix.

We can extend the size of the matrix to any dimension and use the same multiplication rules, but beware! Not all matrix multiplications are possible. For example:

$$\begin{bmatrix} 1 & 2 & 1 & 1 \\ 3 & 1 & 4 & 2 \\ 5 & 1 & 6 & 3 \end{bmatrix} \begin{bmatrix} 1 & 0 \\ 2 & 1 \\ 1 & 0 \\ 0 & 1 \end{bmatrix}$$

would give a product matrix

$$\begin{bmatrix} 6 & 3 \\ 9 & 3 \\ 13 & 4 \end{bmatrix}$$

but

$$\begin{bmatrix} 1 & 3 & 5 \\ 2 & 1 & 1 \\ 1 & 4 & 6 \\ 1 & 2 & 3 \end{bmatrix} \begin{bmatrix} 1 & 0 \\ 2 & 1 \\ 1 & 0 \\ 0 & 1 \end{bmatrix}$$

has no product matrix. Try to multiply the matrices and you will see the system of rules breaks down. To obtain the element $r_1 c_1$ in the product matrix, we would have to multiply the elements in the first row of the first matrix by the elements in the first column of the second matrix, i.e. $1 \times 1 + 3 \times 2 + 5 \times 1 \ldots$ but there are insufficient elements in the row of the first matrix to complete the procedure. There is a simple rule for deciding whether matrix multiplication is possible – can you deduce it? There must be the same number of columns in the first matrix as there are rows in the second.

TUTORIAL ONE

1. To ensure that pigs receive a sufficient vitamin intake, a firm manufactures food additives in 10 kg. containers. The specification of the additives are as follows:

    Additive 1.   6 kg. of vitamin A and 4 kg. of vitamin B.
    Additive 2.   4 kg. of B, 3 kg. of C, 3 kg. of D.
    Additive 3.   3 kg. of A, 1 kg. of B, 3 kg. of C, 3 kg. of D.

    Put this information into matrix form.
    A farmer decides to add 6 containers of additive 1, 8 of additive 2, and 10 of additive 3 to a batch of feed. Put this into an appropriate vector, and deduce the total vitamin content of the food. If a kilo of each vitamin costs £3, £6, £7, and £10 respectively, use an

appropriate vector to obtain the cost of each additive, and hence by vector multiplication find the total cost of additives in the batch of feed.

2. Consider the matrices:

$$A = \begin{bmatrix} 2 & 1 \\ 2 & 1 \end{bmatrix} \qquad B = \begin{bmatrix} 2 & 1 \\ 3 & 1 \end{bmatrix}$$

Find the matrix AB and the matrix BA. What can you conclude?

3. Consider the matrices:

$$A = \begin{bmatrix} 3 & 0 & 1 \\ 1 & 0 & 2 \end{bmatrix} \qquad B = \begin{bmatrix} 1 & 0 & 1 \\ 3 & 0 & 0 \\ 0 & 1 & 0 \\ 2 & 1 & 0 \end{bmatrix} \qquad C = \begin{bmatrix} 1 & 0 & 1 \\ 0 & 0 & 1 \\ 1 & 0 & 0 \end{bmatrix}$$

$$D = \begin{bmatrix} 1 & 1 \\ 2 & 2 \\ 3 & 3 \end{bmatrix}$$

The dimensions are 2×3, 4×3, 3×3, and 3×2 respectively. What is the shape of:

$$AC, \quad DA, \quad AD, \quad BC, \quad CB.$$

Can you find a rule?

4.
$$\begin{bmatrix} 2 & 3 \\ 3 & 5 \end{bmatrix} \begin{bmatrix} X \\ Y \end{bmatrix} = \begin{bmatrix} 34 \\ 55 \end{bmatrix} \text{ find.} \begin{bmatrix} X \\ Y \end{bmatrix}$$

Operational Rules of Matrix Multiplication

In example 2 of the tutorial you were asked to find the matrix AB and the matrix BA given that $A = \begin{bmatrix} 2 & 1 \\ 2 & 1 \end{bmatrix}$ and $B = \begin{bmatrix} 2 & 1 \\ 3 & 1 \end{bmatrix}$

$$AB = \begin{bmatrix} 2 & 1 \\ 2 & 1 \end{bmatrix} \begin{bmatrix} 2 & 1 \\ 3 & 1 \end{bmatrix} = \begin{bmatrix} 7 & 3 \\ 7 & 3 \end{bmatrix}$$

$$BA = \begin{bmatrix} 2 & 1 \\ 3 & 1 \end{bmatrix} \begin{bmatrix} 2 & 1 \\ 2 & 1 \end{bmatrix} = \begin{bmatrix} 6 & 3 \\ 8 & 4 \end{bmatrix}$$

Thus the matrix AB is quite different from the matrix BA, and in this respect matrix multiplication differs fundamentally from algebriac multiplication. When multiplying matrices, it is vital that they are put in the right order, and to avoid confusion, mathematicians use the expressions 'pre-multiply' and 'post multiply' — rather than just say multiply.

For the example, the product matrix AB can be expressed as either pre-multiply B by A, or post-multiply A by B. If we have an operation (for example, multiplication) to perform on a number of variables, and it does not matter in which order the variables are placed, then the operation is called COMMUTATIVE. Thus, matrix multiplication is non-commutative.

Are there any exceptions to the rule that matrix multiplication is non-commutative?

$$\begin{bmatrix} 1 & 0 \\ 0 & 1 \end{bmatrix} \begin{bmatrix} 2 & 1 \\ 3 & 1 \end{bmatrix} = \begin{bmatrix} 2 & 1 \\ 3 & 1 \end{bmatrix}$$

$$\begin{bmatrix} 2 & 1 \\ 3 & 1 \end{bmatrix} \begin{bmatrix} 1 & 0 \\ 0 & 1 \end{bmatrix} = \begin{bmatrix} 2 & 1 \\ 3 & 1 \end{bmatrix}$$

The first thing to notice is that the operation is clearly commutative. You should notice that if any matrix A is either premultiplied or post-multiplied by the matrix

$$I = \begin{bmatrix} 1 & 0 \\ 0 & 1 \end{bmatrix}$$

then the product matrix is A. The matrix I acts in a similar fashion to unity in ordinary multiplication, and for this reason it is called the unit matrix.

Now let us consider the product matrix AA^{-1} where:

$$A = \begin{bmatrix} 2 & 1 \\ 3 & 2 \end{bmatrix} \text{ and } A^{-1} = \begin{bmatrix} 2 & -1 \\ -3 & 2 \end{bmatrix}$$

$$\begin{bmatrix} 2 & -1 \\ -3 & 2 \end{bmatrix} \begin{bmatrix} 2 & 1 \\ 3 & 2 \end{bmatrix} = \begin{bmatrix} 1 & 0 \\ 0 & 1 \end{bmatrix}$$

$$\begin{bmatrix} 2 & 1 \\ 3 & 2 \end{bmatrix} \begin{bmatrix} 2 & -1 \\ -3 & 2 \end{bmatrix} = \begin{bmatrix} 1 & 0 \\ 0 & 1 \end{bmatrix}$$

The operation is commutative. If the product matrix is the unit matrix I, the operation will always be commutative. If one of the matrices is called A, then the other, A^{-1}, is called the INVERSE of A. The unit matrix and inverse matrix have considerable analytical significance.

Suppose $A = \begin{bmatrix} 2 & 1 \\ 3 & 2 \end{bmatrix}$ $B = \begin{bmatrix} 1 & 2 \\ 3 & 1 \end{bmatrix}$ $C = \begin{bmatrix} 2 & 3 \\ 3 & 1 \end{bmatrix}$

If we wish to find the matrix $(AB)C$, then we must combine the normal rule of brackets (which specifies the order for an operation) with the non-commutative rule of matrices. First we must find AB, then post-multiply by C.

$$AB = \begin{bmatrix} 2 & 1 \\ 3 & 2 \end{bmatrix} \begin{bmatrix} 1 & 2 \\ 3 & 1 \end{bmatrix} = \begin{bmatrix} 5 & 5 \\ 9 & 8 \end{bmatrix}$$

$$(AB)C = \begin{bmatrix} 5 & 5 \\ 9 & 8 \end{bmatrix} \begin{bmatrix} 2 & 3 \\ 3 & 1 \end{bmatrix} = \begin{bmatrix} 25 & 20 \\ 42 & 35 \end{bmatrix}$$

Now let us find $A(BC)$.

$$BC = \begin{bmatrix} 1 & 2 \\ 3 & 1 \end{bmatrix} \begin{bmatrix} 2 & 3 \\ 3 & 1 \end{bmatrix} = \begin{bmatrix} 8 & 5 \\ 9 & 10 \end{bmatrix}$$

$$A(BC) = \begin{bmatrix} 2 & 1 \\ 3 & 2 \end{bmatrix} \begin{bmatrix} 8 & 5 \\ 9 & 10 \end{bmatrix} = \begin{bmatrix} 25 & 20 \\ 42 & 35 \end{bmatrix}$$

We can conclude that $A(BC) = (AB)C$, or the position of the bracket does not matter in matrix multiplication. Again using the jargon, we say that matrix multiplication is *associative*.

Solution of Simultaneous Linear Equations

Let us apply the operational rules derived so far. Using again the tailor problem of the last section, let us suppose that the objective was maximum use of resources. The tailor wishes to use all the material available to him, and keep his labour force fully occupied. Can you see that this is expressed mathematically by the statements

both $2y + 5x = 200$
and $3y + 5x = 225$

Putting the statements into matrix/vector form we have

$$\begin{bmatrix} 2 & 5 \\ 3 & 5 \end{bmatrix} \begin{bmatrix} y \\ x \end{bmatrix} = \begin{bmatrix} 200 \\ 225 \end{bmatrix}$$

For convenience sake, let

$$A = \begin{bmatrix} 2 & 5 \\ 3 & 5 \end{bmatrix} \quad B = \begin{bmatrix} y \\ x \end{bmatrix} \quad \text{and } C = \begin{bmatrix} 200 \\ 225 \end{bmatrix}$$

then,

$$AB = C$$

Now suppose we find A^{-1}, the inverse of A. Pre-multiplying both sides by A^{-1} gives

$$A^{-1}(AB) = A^{-1}C$$

and using the associative property of matrix multiplication

$$(A^{-1}A)B = A^{-1}C$$

But we already know that $A^{-1}A$ is the unit matrix I so

$$IB = A^{-1}C$$

and, we know that $IB = B$, hence

$$B = A^{-1}C$$

The vector B we require is obtained by premultiplying the vector C by the inverse of A. We shall see how to find the inverse in the next section, but for the moment let us take it as given that

$$A^{-1} = \begin{bmatrix} -1 & 1 \\ \frac{3}{5} & -\frac{2}{5} \end{bmatrix} \text{ so}$$

$$B = \begin{bmatrix} -1 & 1 \\ \frac{3}{5} & -\frac{2}{5} \end{bmatrix} \begin{bmatrix} 200 \\ 225 \end{bmatrix} = \begin{bmatrix} 25 \\ 30 \end{bmatrix}$$

He will maximise the use of his resources by producing 25 jackets and 30 overcoats.

TUTORIAL TWO

1. Consider the basic operations of 'normal' arithmetic: addition, subtraction, multiplication and division. Identify those which have a commutative property.
2. Why do you think that the inverse of A is called A^{-1}? (Hint: think back to the rules of indices.)
3. Give examples of operations in 'normal' arithmetic that have associative properties and non-associative properties.
4. Solve the pair of simultaneous equations

$$2x + y = 7$$
$$3x + 2y = 12$$

   Given that $A^{-1} = \begin{bmatrix} 2 & -1 \\ -3 & 2 \end{bmatrix}$

5. The tailor is experiencing fluctuations in the availability of labour and raw materials each week. He estimates the availabilities as follows:

Week	1	2	3	4
Labour	210	240	150	330
Material	240	300	200	370

Use the inverse given in the text to find his output in each of the four weeks (assuming maximum utilisation of resources).

Further Operational Rules: Row Transformations

Earlier we stated that a general rule for matrix multiplication was:

$$\begin{bmatrix} a & b \\ c & d \end{bmatrix} \begin{bmatrix} e & f \\ g & h \end{bmatrix} = \begin{bmatrix} ae + bg & af + bh \\ ce + dg & cf + dh \end{bmatrix}$$

Now suppose we decide to double the first row of the first matrix. If you remember the 'row by column' rule, you should realise that this will affect the first row, but not the second row of the product matrix. The first row would now become:

$$2ae + 2bg \qquad 2af + 2bh$$
i.e. $2(ae + bg) \qquad 2(af + bh)$

The first row of the product matrix would be doubled. If we divide the bottom row of the first matrix by 3, the first row of the product matrix remains unchanged, but the second row becomes:

$$\frac{ce}{3} + \frac{dg}{3} \qquad \frac{cf}{3} + \frac{dh}{3}$$

i.e. $\frac{1}{3}(ce + dg) \qquad \frac{1}{3}(cf + dh)$
i.e. $\frac{1}{3}$ of the second row of the previous produce matrix.

We have been performing *row transformations* and the rule is this: if any row of the first matrix is multiplied by n, then the corresponding row in the product matrix must be multiplied by n. If any row in the first matrix is divided by n, the corresponding row in the product matrix must be divided by n.

Now suppose we add the second row of the first matrix to the first row. The product matrix would become

$$\begin{bmatrix} a + c & b + d \\ c & d \end{bmatrix} \begin{bmatrix} e & f \\ g & h \end{bmatrix} = \begin{bmatrix} e(a + c) + g(b + d) & f(a + c) + h(b + d) \\ ce + bd & cf + dh \end{bmatrix}$$

We can write the top row of the product matrix like this

$$ac + ce + bg + db \qquad af + cf + bh + dh$$

Which is the sum of the rows of the first product matrix. In fact we can add or subtract any multiple of one row to another in the first matrix if we do the same to the product matrix. Consider the matrix multiplication.

$$\begin{bmatrix} 1 & 0 \\ 0 & 1 \end{bmatrix} \begin{bmatrix} 2 & 5 \\ 3 & 5 \end{bmatrix} = \begin{bmatrix} 2 & 5 \\ 3 & 5 \end{bmatrix}$$

Divide the top row by two

$$\begin{bmatrix} \frac{1}{2} & 0 \\ 0 & 1 \end{bmatrix} \begin{bmatrix} 2 & 5 \\ 3 & 5 \end{bmatrix} = \begin{bmatrix} 1 & \frac{5}{2} \\ 3 & 5 \end{bmatrix}$$

Subtract 3 times the top row from the bottom row

$$\begin{bmatrix} \frac{1}{2} & 0 \\ -\frac{3}{2} & 1 \end{bmatrix} \begin{bmatrix} 2 & 5 \\ 3 & 5 \end{bmatrix} = \begin{bmatrix} 1 & \frac{5}{2} \\ 0 & -\frac{5}{2} \end{bmatrix}$$

Add the bottom row to the top row —

$$\begin{bmatrix} -1 & 1 \\ -\frac{3}{2} & 1 \end{bmatrix} \begin{bmatrix} 2 & 5 \\ 3 & 5 \end{bmatrix} = \begin{bmatrix} 1 & 0 \\ 0 & -\frac{5}{2} \end{bmatrix}$$

Multiply the bottom row by $-\frac{2}{5}$

$$\begin{bmatrix} -1 & 1 \\ +\frac{3}{5} & -\frac{2}{5} \end{bmatrix} \begin{bmatrix} 2 & 5 \\ 3 & 5 \end{bmatrix} = \begin{bmatrix} 1 & 0 \\ 0 & 1 \end{bmatrix}$$

You should check the product matrix after each row transformation and satisfy yourself that the rules of multiplication still hold.

Can you see what we have done? We have found the inverse of $\begin{bmatrix} 2 & 5 \\ 3 & 5 \end{bmatrix}$, the input matrix for the tailor problem. We started with a system in the form

$$IA = A,$$

and by changing the product matrix A into I, we must also change I into A^{-1}, for

$$A^{-1}A = I$$

If we wish to find an inverse matrix, first write the system in the form

$$IA = A,$$

and by using the row transformation rules, change the product matrix into I.

You should notice that row transformations do not change the second matrix, and it seems a waste of effort to merely restate it after each transformation. To avoid this repetition, we can use a *partitioned matrix*. Suppose we wished to find the inverse of $\begin{bmatrix} 2 & 1 \\ 3 & 2 \end{bmatrix}$. Instead of an initial format like the following:

$$\begin{bmatrix} 1 & 0 \\ 0 & 1 \end{bmatrix} \begin{bmatrix} 2 & 1 \\ 3 & 2 \end{bmatrix} = \begin{bmatrix} 2 & 1 \\ 3 & 2 \end{bmatrix}$$

we can use the partitioned matrix

$$\left[\begin{array}{cc|cc} 1 & 0 & 2 & 1 \\ 0 & 1 & 3 & 2 \end{array} \right]$$

Can you see that if we transform the right-hand side of the partition into the unit matrix we must transform the left-hand side into the required inverse: Firstly, let us divide the bottom row by two, and subtract the transformed bottom row from the original top row:

$$\left[\begin{array}{cc|cc} 1 & -\frac{1}{2} & \frac{1}{2} & 0 \\ 0 & \frac{1}{2} & \frac{3}{2} & 1 \end{array} \right]$$

Now we can multiply the top row by 2, and subtract $\frac{3}{2}$ times the transformed top row from the bottom row:

$$\left[\begin{array}{cc|cc} 2 & -1 & 1 & 0 \\ -3 & 2 & 0 & 1 \end{array} \right]$$

You should check that $\begin{bmatrix} 2 & -1 \\ -3 & 2 \end{bmatrix}$ is the inverse required.

We now have all the necessary information required to solve simultaneous equations using matrices. Suppose we wish to solve:

$$x + y = 8$$
$$2x + 3y = 19$$

$$\begin{bmatrix} 1 & 1 \\ 2 & 3 \end{bmatrix} \begin{bmatrix} x \\ y \end{bmatrix} = \begin{bmatrix} 8 \\ 19 \end{bmatrix}$$

First, we find the inverse of $\begin{bmatrix} 1 & 1 \\ 2 & 3 \end{bmatrix}$

Setting up the partitioned matrix.

$$\left[\begin{array}{cc|cc} 1 & 0 & 1 & 1 \\ 0 & 1 & 2 & 3 \end{array} \right]$$

Subtracting twice the top row from the bottom row

$$\left[\begin{array}{cc|cc} 1 & 0 & 1 & 1 \\ -2 & 1 & 0 & 1 \end{array} \right]$$

Subtracting the bottom row from the top row.

$$\begin{bmatrix} 3 & -1 & | & 1 & 0 \\ -2 & 1 & | & 0 & 1 \end{bmatrix}$$

$$\begin{bmatrix} x \\ y \end{bmatrix} = \begin{bmatrix} 3 & -1 \\ -2 & 1 \end{bmatrix} \begin{bmatrix} 8 \\ 19 \end{bmatrix}$$

$$\begin{bmatrix} x \\ y \end{bmatrix} = \begin{bmatrix} 5 \\ 3 \end{bmatrix}$$

If you think carefully, you will realise that there is no need to find the inverse when solving simultaneous linear equations. Consider again the system.

$$\begin{bmatrix} 1 & 1 \\ 2 & 3 \end{bmatrix} \begin{bmatrix} x \\ y \end{bmatrix} = \begin{bmatrix} 8 \\ 19 \end{bmatrix}$$

The system will hold if we perform row transformations on the first matrix and the product vector. Suppose we transform the first matrix into the unit matrix. This is equivalent to pre-multiplying the first matrix by its inverse, and if we use the row transformation system we will simultaneously pre-multiply the product vector by this same inverse. In other words, we will solve the equations. Let us put the system into a partitioned matrix:

$$\begin{bmatrix} 1 & 1 & | & 8 \\ 2 & 3 & | & 19 \end{bmatrix}$$

Subtract twice the top row from the bottom row:

$$\begin{bmatrix} 1 & 1 & | & 8 \\ 0 & 1 & | & 3 \end{bmatrix}$$

Subtract the bottom row from the top row:

$$\begin{bmatrix} 1 & 0 & | & 5 \\ 0 & 1 & | & 3 \end{bmatrix}$$

Which gives the same result as before.

Now let us solve simultaneous linear equations with three variables.

$$x + y + Z = 5$$
$$2x + y + 2Z = 9$$
$$x + 2y + 2Z = 8$$

Modern Analytical Techniques

In matrix form this becomes:

$$\begin{bmatrix} 1 & 1 & 1 \\ 2 & 1 & 2 \\ 1 & 2 & 2 \end{bmatrix} \begin{bmatrix} x \\ y \\ z \end{bmatrix} = \begin{bmatrix} 5 \\ 9 \\ 8 \end{bmatrix}$$

We must now change the first matrix into the unit matrix, but clearly it cannot be changed into

$$\begin{bmatrix} 1 & 0 \\ 0 & 1 \end{bmatrix}$$

Now the unit matrix can also be written in a 3 × 3 form:

$$\begin{bmatrix} 1 & 0 & 0 \\ 0 & 1 & 0 \\ 0 & 0 & 1 \end{bmatrix}$$

You should check that this is a unit matrix. Multiply it by the first matrix in the system above. Again using the partitioned matrix

$$\left[\begin{array}{ccc|c} 1 & 1 & 1 & 5 \\ 2 & 1 & 2 & 9 \\ 1 & 2 & 2 & 8 \end{array}\right]$$

Subtracting twice the top row from the centre row, then subtract the top row from the bottom row:

$$\left[\begin{array}{ccc|c} 1 & 1 & 1 & 5 \\ 0 & -1 & 0 & -1 \\ 0 & 1 & 1 & 3 \end{array}\right]$$

Subtracting the bottom row from the top row, and multiplying the centre row by the minus one.

$$\left[\begin{array}{ccc|c} 1 & 0 & 0 & 2 \\ 0 & 1 & 0 & 1 \\ 0 & 1 & 1 & 3 \end{array}\right]$$

Subtracting the centre row from the bottom row completes the transformation.

$$\left[\begin{array}{ccc|c} 1 & 0 & 0 & 2 \\ 0 & 1 & 0 & 1 \\ 0 & 0 & 1 & 2 \end{array}\right]$$

The right-hand side of the partitioned matrix gives the column vector

$$\begin{bmatrix} x \\ y \\ Z \end{bmatrix} \text{ thus } \begin{bmatrix} x \\ y \\ Z \end{bmatrix} = \begin{bmatrix} 2 \\ 1 \\ 2 \end{bmatrix}$$

The Detached — Coefficient Method

Let us reconsider the equations

$$x + y + Z = 5$$
$$2x + y + 2Z = 9$$
$$x + 2y + 2Z = 8$$

If we set up the partitioned matrix, then the left-hand side is the co-efficient matrix, and the right-hand side the quantity vector. We can identify each column like this:

$$\begin{bmatrix} 1 & 1 & 1 & 5 \\ 2 & 1 & 2 & 9 \\ 1 & 2 & 2 & 8 \end{bmatrix}$$

↑	↑	↑	↑
x	y	Z	quantity
row	row	row	vector

The vertical partitioning line represents an equals sign. Reading the second row:

$$2x + y + 2Z = 9$$

We have already found the solution to this system. Let us perform row transformations, and use the solution to check the feasibility of each transformation. Firstly, we can subtract twice the top row from the centre row and then from the bottom row:

$$\begin{bmatrix} 1 & 1 & 1 & 5 \\ 0 & -1 & 0 & -1 \\ -1 & 0 & 0 & -2 \end{bmatrix}$$

Let us examine the matrix row by row. The first row is

$$x + y + Z = 5$$

We already know that $x = 2$, $y = 1$, $Z = 2$, so this equation is true. Likewise, the second and third rows.

$$-y = -1$$
$$-x = -2$$

are also true.

Now let us multiply the second and third rows by -1, and subtract the second row from the first.

$$\begin{bmatrix} 1 & 0 & 1 & | & 4 \\ 0 & 1 & 0 & | & 1 \\ 1 & 0 & 0 & | & 2 \end{bmatrix}$$

Clearly, the second and third rows are true. The first row is now

$$x + Z = 4$$

which is also true. Now let us subtract the bottom row from the top.

$$\begin{bmatrix} 0 & 0 & 1 & | & 2 \\ 0 & 1 & 0 & | & 1 \\ 1 & 0 & 0 & | & 2 \end{bmatrix} \quad \text{or} \quad \begin{matrix} Z = 2 \\ y = 1 \\ x = 2 \end{matrix}$$

This agrees with the result above, though we have not transformed the coefficient matrix into the unit matrix.

What can we conclude? We can represent any system of simultaneous equations by a partitioned matrix. We can perform row transformations without affecting the values of the variables in the equations. If we wish to solve the equations, we can select *any* coefficient in the coefficient matrix and by row transformation, transform it to unity. Then, again using row transformations, we *detach* the remaining coefficients in the same row or column. We then select any other coefficient *not* in the same row or column as the first, and repeat the operation. The solution is obtained when each row and column contains one unit coefficient, and the rest are detached.

TUTORIAL THREE

1.   Find the inverses of the following matrices:

$$\begin{bmatrix} 2 & 1 \\ 1 & 2 \end{bmatrix} \begin{bmatrix} 3 & 1 \\ 1 & 4 \end{bmatrix} \begin{bmatrix} 5 & 1 \\ 6 & 2 \end{bmatrix}$$

Check that each inverse is correct by matrix multiplication.

2.   Use the inverses found in question 1 to solve the equations:

$$\begin{matrix} 2x + y = 4. & 3x + y = 13. & 5x + y = 3. \\ x + 2y = -1. & x + 4y = 1. & 6x + 2y = 4. \end{matrix}$$

3.   Try to find the inverse of

$$\begin{bmatrix} 2 & 5 \\ 4 & 10 \end{bmatrix}$$

Why does this not have an inverse?

4. If a system of equations is to have a unique solution, what can you conclude about the shape of the coefficient matrix? Are any other conditions necessary?

5. In the last section we solved the partitioned matrix

$$\left[\begin{array}{ccc|c} 1 & 1 & 1 & 5 \\ 2 & 1 & 2 & 9 \\ 1 & 2 & 2 & 8 \end{array}\right]$$

Transform the coefficient matrix into:

$$\begin{bmatrix} 0 & 1 & 0 \\ 1 & 0 & 0 \\ 0 & 0 & 1 \end{bmatrix} \quad \text{and} \quad \begin{bmatrix} 1 & 0 & 0 \\ 0 & 0 & 1 \\ 0 & 1 & 0 \end{bmatrix}$$

showing that in each case the solution is the same.

6. Solve by the detached coefficient method.

$$2a + b + c + d = 11$$
$$a + b + 2c + 2d = 17$$
$$a + 2b + 3c + d = 18$$
$$2a + 3b + c + 2d = 19.$$

EXERCISES TO CHAPTER TWO

1. Suppose we have three matrices A, B and C such that

$$AB = C$$

Write an expression for the value of B in terms of A and C. Given that $A = \begin{bmatrix} 3 & 2 \\ 1 & 2 \end{bmatrix}$ and $C = \begin{bmatrix} 9 & 8 \\ 7 & 4 \end{bmatrix}$ find B.

Now write an expression for the value of A in terms of B and C. Given that $B = \begin{bmatrix} 3 & 0 \\ 0 & 2 \end{bmatrix}$ and $C = \begin{bmatrix} 6 & 2 \\ 3 & 6 \end{bmatrix}$ find A.

2. On a particular occasion a furniture manufacturer has 20 lengths of hardwood and 25 lengths of softwood. This enables him to make 115 chairs and 120 tables. On another occasion he has 30 lengths of hardwood and 20 lengths of softwood. This enables him to make 120 chairs and 110 coffee tables. If we wish to deduce the hardwood and softwood content of each chair and table, state the problem in matrix form and solve it.

3. On the piece of graph paper mark the points (1,1) and (1,2) and join them with a straight line. The line can be represented by the matrix

$$\begin{bmatrix} 1 & 1 \\ 1 & 2 \end{bmatrix}$$

Now imagine the line is rotated 90° clockwise with the origin as centre. Draw the new position of the line and deduce the matrix.

$$\begin{bmatrix} x_1 & x_2 \\ y_1 & y_2 \end{bmatrix}$$

Consider the matrix A such that

$$A \begin{bmatrix} 1 & 1 \\ 1 & 2 \end{bmatrix} = \begin{bmatrix} x_1 & x_2 \\ y_1 & y_2 \end{bmatrix}$$

The matrix A is the *transformation matrix* which maps $\begin{bmatrix} 1 & 1 \\ 1 & 2 \end{bmatrix}$ to its new position. Find this matrix and use it to map the line $\begin{bmatrix} 1 & 3 \\ 1 & 2 \end{bmatrix}$ to its new position if it is rotated in a similar fashion.

4. This question is concerned with the matrix A in the last example.

    Find the matrix $A \times A = A^2$
    $$A \times A \times A = A^3$$

    Give a precise meaning to the matrices A^2 and A^3.
    Premultiply the line $\begin{bmatrix} 1 & 1 \\ 1 & 2 \end{bmatrix}$ by the matrix A^2 and draw its new position on your graph.
    Without multiplying give the values of A^4, A^5 and A^6.

5. A man has a bottle of gin and a decanter of vermouth and the matrix

    $$P = \begin{bmatrix} a & b \\ c & d \end{bmatrix} \text{ is such that}$$

    a is the quantity of gin in the bottle.
    c is the quantity of vermouth in the bottle.
    b is the quantity of gin in the decanter.
    d is the quantity of vermouth in the decanter.

    He finds that the bottle contains 240 ml. of gin, and the decanter contains 300 ml. of vermouth. Write down the matrix P.
    He now decides to pour $\frac{1}{3}$ of the bottle into the decanter. Find the

new matrix P' that describes the quantities in each container. If Q is the matrix such that

$$QP = P'$$

Find the matrix Q.

He tastes the contents of the decanter and decides the mixture is not strong enough, so he pours half the contents of the decanter back into the bottle. Find a new matrix P'' that describes the quantities in each container (assume the quantity tasted is negligible). The contents of the bottle now suit him perfectly, and he realises that he would like to make more of this mixture in the future. Find the matrix R that describes how to do this i.e. the matrix R such that

$$RP = P''$$

6. Solve by the detached coefficient method

$$
\begin{aligned}
a + b + c &= 13 \\
2a + b + 2c &= 22 \\
a + 3b + 4c &= 39.
\end{aligned}
$$

Chapter Three

Linear Programming

Linear Programming: Simplex Method

Let us apply the detached coefficient method to linear programming, starting with a maximum profit, two product problem. Suppose a manufacturer is making two products, X and Y with the following input requirements:

Product	Raw materials	Machine time	Labour
X	4 tons	2 hrs.	1 hr.
Y	1 ton	1 hr.	1 hr.

The inputs are available in the following quantities per week.

90 tons of raw materials
50 hours of machine time
40 hours of labour

The profit is £4 on X and £3 on Y, and the objective is to maximise profit. Assuming the weekly output is x units of X and y units of Y, the following model would describe the problem.

$$\text{Maximise } P = 4x + 3y$$
$$\text{Subject to} \quad 4x + y \leqslant 90$$
$$2x + y \leqslant 50$$
$$x + y \leqslant 40$$
$$x \geqslant 0$$
$$y \geqslant 0$$

The graph shows the mapping of the inequalities and the feasibility polygon. The profit line shows that the solution is at point D, where $x = 10, y = 30$ and profit (maximum) is £130.

The detached coefficient method cannot cope with inequalities, so we must transform the model into a system of equations. Consider the first inequality $4x + y \leqslant 90$, i.e. the raw material inequality. Suppose 10 units of each is produced, then the raw material requirement would be $4 \times 10 + 10 = 50$ tons, and there would be a surplus of 40 tons. What can we deduce about the size of the surplus? If we produce

70

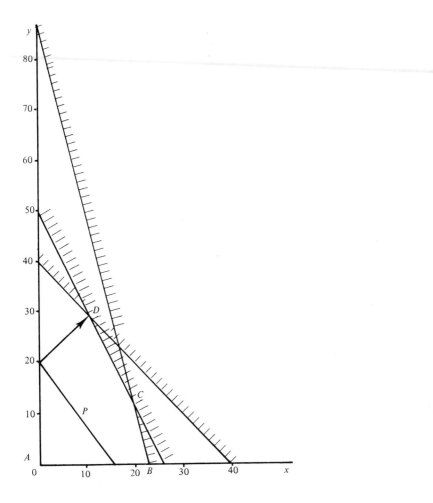

nothing, then the surplus is 90 tons, and if we use all of the raw material then the surplus would be zero. If the surplus is 'a' tons, then

$$0 < a < 90.$$

It follows from this that the amount used, plus the surplus, must equal 90 tons. Hence

$$4x + y + a = 90$$

The surplus 'a' is called a slack variable. If we call 'b' the surplus machine time and 'c' the surplus labour time, then we can rewrite the model as a system of equations

Maximise $P = 4x + 3y$

Subject to
$$4x + y + a = 90$$
$$2x + y + b = 50$$
$$x + y + c = 40$$
$$x \geqslant 0, y \geqslant 0$$
$$a \geqslant 0, b \geqslant 0, c \geqslant 0$$

Thus, we must satisfy the objective function subject to the restriction equations, and subject to the additional restriction that all variables must be non negative. We can represent the restriction equations in a partitioned matrix.

$$
\begin{bmatrix}
4 & 1 & 1 & 0 & 0 & \bigm| & 90 \\
2 & 1 & 0 & 1 & 0 & \bigm| & 50 \\
1 & 1 & 0 & 0 & 1 & \bigm| & 40
\end{bmatrix}
$$

↑	↑	↑	↑	↑	↑
x	y	a	b	c	Quantity
column	column	column	column	column	column

You will remember from the last chapter, that linear programming problems were solved by taking a basic feasible solution, and progressively improving it until the objective was achieved. Let us take as a basic solution that nothing is produced: this corresponds to point A on the graph. We now have $x = 0$, $y = 0$ and can concentrate on the last three columns of the partitioned matrix. Reading off as we did in the last section, we have $a = 90$, (i.e. 90 tons of raw material surplus), $b = 50$ (i.e. 50 machine hours are surplus) and $c = 40$ (i.e. 40 hours of labour are surplus). Thus the solution is feasible. If we wish to show the objective function in the matrix, we use a *double partitioned matrix.* In the first solution, the profit is zero, which we enter in the quantity column. We enter the coefficients of x and y as negative quantities: The reason for this will become apparent later. The double partitioned matrix (or Simplex tableau) now looks like this:

$$
\begin{bmatrix}
4 & 1 & 1 & 0 & 0 & \bigm| & 90 \\
2 & 1 & 0 & 1 & 0 & \bigm| & 50 \\
1 & 1 & 0 & 0 & 1 & \bigm| & 40 \\
\hline
-4 & -3 & 0 & 0 & 0 & \bigm| & 0
\end{bmatrix}
$$

x	y	a	b	c	Q
		↑	↑	↑	

Now let us deduce the effect of producing some units of one of the products. It would seem sensible to start with X, as it earns the greatest profit per unit. Suppose we decide to produce as much of X as possible — how many units can be produced? We can answer this problem by considering the x column and the Q column. The first row tells us it takes 4 tons of raw material to produce a unit of X, and 90 tons are available, i.e. there is sufficient raw material to produce $90 \div 4 = 22.5$ units. Reference to the second and third rows tell us there is sufficient machine time to produce $50 \div 2 = 25$ units, and sufficient labour time to produce $40 \div 1 = 40$ units. Raw material supplies restrict the output of x to a maximum of 22.5 units. But if we produce 22.5 units of X, we will use all the available raw material, and 'a' would be zero (remember, 'a' is the surplus raw material). Thus x is called the *entering variable*, and 'a' the *departing variable*. The non-zero variables (marked with an arrow in the matrix above) now become x, b and c.

We start as we did in the last section by dividing the top row by 4

| 1 | ¼ | ¼ | 0 | 0 | | 22½ |

Now we must detach the remaining coefficients in the x column. The appropriate row transformations would be

Subtract twice the transformed top row from the second row
Subtract the transformed top row from the third row
Add four times the transformed top row to the bottom row.

After completing the transformations, the new matrix would look like this

$$
\begin{bmatrix}
1 & ¼ & ¼ & 0 & 0 & 22½ \\
0 & ½ & -½ & 1 & 0 & 5 \\
0 & ¾ & -¼ & 0 & 1 & 17½ \\
\hline
0 & -2 & 1 & 0 & 0 & 90
\end{bmatrix}
$$

| x | y | a | b | c | Q |
| ↑ | | | ↑ | ↑ | |

Again, we are interested only in the columns containing non-zero variables (marked with an arrow). Reading — off this solution we have

$x = 22½$ (produce 22½ units of X)
$b = 5$ (5 hours machine time surplus)
$c = 17½$ (17½ hours of labour surplus)

You should check the feasibility of this solution, and notice that it corresponds to point B on the graph. Including the objective function in the matrix enables us to read-off the profit earned from this solution: £90 (which again you should check).

Now let us produce some units of Y as well as X — how much can we produce? We must use the same method as before i.e. divide the coefficients in the y column into the quantities in the Q column, and choose the smallest.

Why must we do this? If we do not choose the smallest value, then negative values will occur in the Q vector when we detach the remaining coefficients. You can verify this if you wish by multiplying the top row by 4 (or the third row by $\frac{4}{3}$) and detaching the remaining coefficients. We know that negative elements in the Q vector are not feasible. The second row gives the smallest total, so y is the entering variable, and b the departing variable. Multiplying the second row by 2 gives

$$0 \quad 1 \quad -1 \quad 2 \quad 0 \quad 10$$

To detach the remaining coefficients in the y column, we perform the following row transformations.

Subtract one quarter of the transformed second row from the top row

Subtract three quarters of the transformed second row from the third row

Add twice the transformed second row to the bottom row.

$$
\begin{bmatrix}
1 & 0 & \frac{1}{2} & -\frac{1}{2} & 0 & \Big| & 20 \\
0 & 1 & -1 & 2 & 0 & \Big| & 10 \\
0 & 0 & \frac{1}{2} & -\frac{3}{2} & 1 & \Big| & 10 \\
\hline
0 & 0 & -1 & 4 & 0 & \Big| & 110
\end{bmatrix}
$$

$$\underset{\uparrow}{x} \quad \underset{\uparrow}{y} \quad a \quad b \quad \underset{\uparrow}{c} \quad Q$$

$x = 20$ (produce 20 units of x)
$y = 10$ (produce 10 units of y)
$c = 10$ (10 hours of labour surplus)
Profit is £110.

This solution corresponds to position C on the graph.

Let us stop for a moment and consider exactly what we have been doing. We have selected an entering and a departing variable and performed row transformations to detach the remaining elements in the incoming variable column. To detach the element in the objective function row, we have added some multiple of the incoming variable row. As we have been adding, *this has caused profit to increase.* This gives us a signal as to when we have reached maximum profit. *If any column in the objective function is negative and we locate the entering variable in that column, then we must add to the objective row to detach the negative coefficient. Hence, profit must increase.* If, on the other hand, the coefficient in the objective row and entering variable column is positive, we will detach it by subtracting. This will cause profit to decrease. *Profit is at a maximum, then, when all the coefficients in the objective function cease to be negative.*

Now let us return to the matrix. If we look at the objective function row, we see that there is a negative coefficient in the 'a' column, and if we locate the entering variable in this row we will increase profit. In this column, the second row gives the smallest total, but this cannot be chosen (why?). Instead, we take the third row, and 'a' becomes the departing variable. Multiplying the third row by two gives

$$0 \quad\quad 0 \quad\quad 1 \quad\quad -3 \quad\quad 2 \quad\quad 20$$

To detach the remaining coefficients in the 'a' column we

Subtract half the transformed third row from the top row
Add the transformed row to the second and fourth row.

$$\begin{bmatrix} 1 & 0 & 0 & 1 & -1 & \bigm| & 10 \\ 0 & 1 & 0 & -1 & 2 & \bigm| & 30 \\ 0 & 0 & 1 & -3 & 2 & \bigm| & 20 \\ \hline 0 & 0 & 0 & 1 & 2 & \bigm| & 130 \end{bmatrix}$$

$$x \quad\quad y \quad\quad a \quad\quad b \quad\quad c \quad\quad\quad Q$$
$$\uparrow \quad\quad \uparrow \quad\quad \uparrow$$

$x = 10$ (produce 10 units of X)
$y = 30$ (produce 30 units of Y)
$a = 20$ (20 tons of raw material are surplus)
The profit is £130.

As there are no negative coefficients in the objective row, this is an optimum solution: profit is at a maximum. You should notice that this corresponds with position D on the graph, and agrees with the graphical method.

Let us check that profit is at a maximum. If we made b the entering variable, can you see that x would have to be the outgoing variable? To detach the last element in the b column, we would have to subtract the top row from the bottom. Thus the bottom row would become $[-1 \quad 0 \quad 0 \quad 0 \quad 3 \mid 120]$. Profit has declined by £10. In a similar fashion, we could show that if c was the incoming variable, then 'a' would be the departing variable, and after detaching the coefficients in the c column profit would decline again by £10.

Working in More Than Two Dimensions

At the beginning of this chapter, we stated that the purpose of introducing matrices was to solve linear — programming problems with more
than two unknown variables. Such problems are called *multi-
dimensional.* Let us consider a problem where more than two products
are produced.

A witch-doctor has a healthy business supplying a deadly poison, a
mother-in-law repellent spray, an elixir, and a love potion to a seemingly
insatiable market. The products are made from the same four ingredients in the following proportions:

Code	Product	Snake oil	Lizards' tongues	Frogs' legs	Bats' wings
X_1	Deadly poison	1 fluid oz.	2	1	3
X_2	Mother-in-law repellent	2 fluid ozs.	1	2	4
X_3	Elixir	2 fluid ozs.	2	1	2
X_4	Love potion	1 fluid oz.	1	2	1

He has been producing 10 units of X_1, 12 units of X_2, 6 units of X_3
and 2 units of X_4 per week at a profit of £4, £6, £4, £3, respectively,
using vector multiplication, his weekly profit is

$$[10 \quad 12 \quad 6 \quad 2] \begin{bmatrix} 4 \\ 6 \\ 4 \\ 3 \end{bmatrix} = £142.$$

The witch-doctor suspects that he is not maximising his weekly
profit, and engages our services as management consultants. We learn
that the ingredients can be obtained in the following weekly quantities.

70 fluid ozs. of snake oil.
56 lizards tongues.
70 frogs legs.
92 bats wings.

We realise that linear programming would give the product-mix that
would maximise profit. If we say X_1, X_2, X_3, and X_4 are the quantities
of each product made, and X_5, X_6, X_7, and X_8 are the slack variables,
then the mathematical model of his problem would be:

Maximise $P = 4X_1 + 6X_2 + 4X_3 + 3X_4$
Subject to $\quad X_1 + 2X_2 + 2X_3 + X_4 + X_5 = 70$
$\qquad\qquad 2X_1 + X_2 + 2X_3 + X_4 + X_6 = 56$

$$X_1 + 2X_2 + X_3 + 2X_4 + X_7 = 70$$
$$3X_1 + 4X_2 + 2X_3 + X_4 + X_8 = 92$$
$$X_1 \geq 0, \geq X_2 \geq 0, \geq X_3 \geq 0, X_4 \geq 0$$
$$X_5 \geq 0, \geq X_6 \geq 0, X_7 \geq 0, X_8 \geq 0$$

The first matrix would look like this:

$$\left[\begin{array}{cccccccc|c}
1 & 2 & 2 & 1 & 1 & 0 & 0 & 0 & 70 \\
2 & 1 & 2 & 1 & 0 & 1 & 0 & 0 & 56 \\
1 & 2 & 1 & 2 & 0 & 0 & 1 & 0 & 70 \\
3 & 4 & 2 & 1 & 0 & 0 & 0 & 1 & 92 \\
\hline
-4 & -6 & -4 & -3 & 0 & 0 & 0 & 0 & 0
\end{array}\right]$$

$$\quad X_1 \quad X_2 \quad X_3 \quad X_4 \quad X_5 \quad X_6 \quad X_7 \quad X_8 \quad\quad Q$$

If we choose X_2 as the entering variable (i.e. the product with the highest rate of profit) then X_8 becomes the departing variable. Performing the necessary row transformations, the second matrix is obtanied.

$$\left[\begin{array}{cccccccc|c}
-\frac{1}{2} & 0 & 1 & \frac{1}{2} & 1 & 0 & 0 & -\frac{1}{2} & 24 \\
\frac{5}{4} & 0 & \frac{3}{2} & \frac{3}{4} & 0 & 1 & 0 & -\frac{1}{4} & 33 \\
-\frac{1}{2} & 0 & 0 & \frac{3}{2} & 0 & 0 & 1 & -\frac{1}{2} & 24 \\
\frac{3}{4} & 1 & \frac{1}{2} & \frac{1}{4} & 0 & 0 & 0 & \frac{1}{4} & 23 \\
\hline
\frac{1}{2} & 0 & -1 & -\frac{3}{2} & 0 & 0 & 0 & \frac{3}{2} & 138
\end{array}\right]$$

$$X_1 \quad X_2 \quad X_3 \quad X_4 \quad X_5 \quad X_6 \quad X_7 \quad X_8 \quad\quad Q$$
$$\quad\quad \uparrow \quad\quad\quad\quad\quad \uparrow \quad\quad \uparrow \quad\quad \uparrow$$

We can continue to perform row transformations until all the negative components in the objective row are detached.

$$\left[\begin{array}{cccccccc|c}
-\frac{1}{3} & 0 & 1 & 0 & 1 & 0 & -\frac{1}{3} & -\frac{1}{3} & 16 \\
\frac{3}{2} & 0 & \frac{3}{2} & 0 & 0 & 1 & -\frac{1}{2} & 0 & 21 \\
-\frac{1}{3} & 0 & 0 & 1 & 0 & 0 & \frac{2}{3} & -\frac{1}{3} & 16 \\
\frac{5}{6} & 1 & \frac{1}{2} & 0 & 0 & 0 & -\frac{1}{6} & \frac{1}{3} & 19 \\
\hline
0 & 0 & -1 & 0 & 0 & 0 & 1 & 1 & 162
\end{array}\right]$$

$$X_1 \quad X_2 \quad X_3 \quad X_4 \quad X_5 \quad X_6 \quad X_7 \quad X_8 \quad\quad Q$$
$$\quad\quad \uparrow \quad\quad\quad\quad \uparrow \quad\quad \uparrow \quad\quad \uparrow$$

$$\left[\begin{array}{cccccccc|c}
-\frac{4}{3} & 0 & 0 & 0 & 1 & -\frac{2}{3} & 0 & -\frac{1}{3} & 2 \\
1 & 0 & 1 & 0 & 0 & \frac{2}{3} & -\frac{1}{3} & 0 & 14 \\
-\frac{1}{3} & 0 & 0 & 1 & 0 & 0 & \frac{2}{3} & -\frac{1}{3} & 16 \\
\frac{1}{3} & 1 & 0 & 0 & 0 & -\frac{1}{3} & 0 & \frac{1}{3} & 12 \\
\hline
1 & 0 & 0 & 0 & 0 & \frac{2}{3} & \frac{2}{3} & 1 & 176
\end{array}\right]$$

$$X_1 \quad X_2 \quad X_3 \quad X_4 \quad X_5 \quad X_6 \quad X_7 \quad X_8 \quad\quad Q$$
$$\quad\quad \uparrow \quad\quad \uparrow \quad\quad \uparrow \quad\quad \uparrow$$

This is the optimum solution. Reading from the matrix we have

X_2 = 12, make 12 mother-in-law repellent sprays.
X_3 = 14, make 14 units of elixir.
X_4 = 16, make 16 love potions.
X_5 = 2, 2 fluid ozs. of snake oil are surplus.
The profit is £176.

Let us check that the solution is feasible. If the above quantities were produced, the inputs required would be:

$(2 \times 12) + (2 \times 14) + 16$ = 68 fluid ozs. of snake oil.
$12 + (2 \times 14) + 16$ = 56 lizards tongues.
$(2 \times 12) + 14 + (2 \times 16)$ = 70 frogs legs.
$(4 \times 12) + (2 \times 14) + 16$ = 92 bats wings.

This agrees the 2 fluid ozs. surplus of snake oil yielded by the matrix.

A Mixed-Inequalities Example

We will state again the tailor problem dealt with in the last chapter.

Maximise $P = y + 2x$
Subject to $2y + 5x \leqslant 200$ (the labour restriction)
            $3y + 5x \leqslant 225$ (the material restriction)
            $x \geqslant 10$
            $y \geqslant 20$ the contract restrictions

Adding slack variables to the labour and material restrictions presents no problem.

$2y + 5x + a = 200, \quad a \geqslant 0$
$3y + 5x + b = 225, \quad b \geqslant 0$

How can we deal with the contract restrictions, which are 'greater than or equal to' ordering? Clearly, it would not make sense to add slack variables, we must subtract them.

$x - d = 10. \quad c \geqslant 0$
$y - c = 20. \quad d \geqslant 0$

It should be noted that c and d (which are called artificial variables) are also non-negative. The minus sign does not mean that the variable is negative, it means we must subtract the variable. If, for example $X = 50$, then c must be 40 to satisfy the equation

$x - c = 10$

The first matrix would look like this.

$$
\begin{bmatrix}
2 & 5 & 1 & 0 & 0 & 0 & 200 \\
3 & 5 & 0 & 1 & 0 & 0 & 225 \\
1 & 0 & 0 & 0 & -1 & 0 & 20 \\
0 & 1 & 0 & 0 & 0 & -1 & 10 \\
\hline
-1 & -2 & 0 & 0 & 0 & 0 & 0
\end{bmatrix}
$$

$$
\begin{array}{ccccccc}
y & x & a & b & c & d & Q \\
& & \uparrow & \uparrow & & &
\end{array}
$$

The first solution is to produce nothing, have 200 hours of labour surplus, and 225 yards of cloth surplus. But this is not a feasible solution, as it violates the contract restrictions. If we make first x then y the entering variables, we will obtain a feasible solution, but there will be no departing variables.

$$
\begin{bmatrix}
2 & 0 & 1 & 0 & 0 & 5 & 150 \\
3 & 0 & 0 & 1 & 0 & 5 & 175 \\
1 & 0 & 0 & 0 & -1 & 0 & 20 \\
0 & 1 & 0 & 0 & 0 & -1 & 10 \\
\hline
-1 & 0 & 0 & 0 & 0 & -2 & 20
\end{bmatrix}
$$

$$
\begin{array}{ccccccc}
y & x & a & b & c & d & Q \\
& \uparrow & \uparrow & \uparrow & & &
\end{array}
$$

$$
\begin{bmatrix}
0 & 0 & 1 & 0 & 2 & 5 & 110 \\
0 & 0 & 0 & 1 & 3 & 5 & 115 \\
1 & 0 & 0 & 0 & -1 & 0 & 20 \\
0 & 1 & 0 & 0 & 0 & -1 & 10 \\
\hline
0 & 0 & 0 & 0 & -1 & -2 & 40
\end{bmatrix}
$$

$$
\begin{array}{ccccccc}
y & x & a & b & c & d & Q \\
\uparrow & \uparrow & \uparrow & \uparrow & & &
\end{array}
$$

We now have a feasible solution: produce 20 of x, 10 of y and earn £40 profit. This gives us 110 hours of labour, and 115 yards of cloth surplus. Both c and d are zero. We can now proceed in the usual fashion to obtain an optimum solution.

$$
\begin{bmatrix}
0 & 0 & \frac{1}{5} & 0 & \frac{2}{5} & 1 & 22 \\
0 & 0 & -1 & 1 & 1 & 0 & 5 \\
1 & 0 & 0 & 0 & -1 & 0 & 20 \\
0 & 1 & \frac{1}{5} & 0 & \frac{2}{5} & 0 & 32 \\
\hline
0 & 0 & \frac{2}{5} & 0 & -\frac{1}{5} & 0 & 84
\end{bmatrix}
$$

$$
\begin{array}{ccccccc}
y & x & a & b & c & d & Q \\
\uparrow & \uparrow & & \uparrow & & \uparrow &
\end{array}
$$

$$\begin{bmatrix} 0 & 0 & \frac{3}{5} & -\frac{2}{5} & 0 & 1 & 20 \\ 0 & 0 & -1 & 1 & 1 & 0 & 5 \\ 1 & 0 & -1 & 1 & 0 & 0 & 25 \\ 0 & 1 & \frac{3}{5} & -\frac{2}{5} & 0 & 0 & 30 \\ 0 & 0 & \frac{1}{5} & \frac{1}{5} & 0 & 0 & 85 \end{bmatrix}$$

$$\begin{array}{ccccccc} y & x & a & b & c & d & Q \\ \uparrow & \uparrow & & & \uparrow & \uparrow & \end{array}$$

This is the optimum solution, and reading the variables we have.

$y = 25$ (make 25 overcoats)
$x = 30$ (make 30 jackets)
$c = 5$ (5 jackets made in excess of 20 required to meet the contract)
$d = 20$ (20 overcoats made in excess of the 10 required to meet the contract)

Degeneracy: A Blending Example

Let us suppose that by law a certain health food must contain at least 3% of vitamin A, and at least 7% of vitamin B. A manufacturer has stocks of two compounds, with the following vitamin content.

	Vitamin A	*Vitamin B*
Vito	2%	10%
Slam	5%	6%

He wished to blend Vito and Slam to produce a new product Wow, that meets the minimum vitamin requirements. The sales manager reckons that he can sell up to 60 lbs. of Wow per week. If Vito can be produced at a £4 profit and Slam at a £5 profit, in what proportions should they be mixed to produce the 60 lbs. of Wow?

As usual, we start by setting up the model. Assuming that x lbs. of Vito and y lbs. of Slam are blended, the objective function is

Maximise $P = 4x + 5y$

The marketing restriction gives

$x + y \leqslant 60$

The vitamin A restriction (i.e. at least 3%) gives

$2x + 5y \geqslant 3(x + y)$
or $2y - x \geqslant 0$

and the vitamin B restriction gives

$$10x + 6y \geqslant 7(x + y)$$
$$3x - y \geqslant 0$$

The mathematical model of the problem is

Maximise $P = 4x + 5y$
Subject to
$$x + y \leqslant 60$$
$$2y - x \geqslant 0$$
$$3x - y \geqslant 0$$
$$x \geqslant 0, y \geqslant 0$$

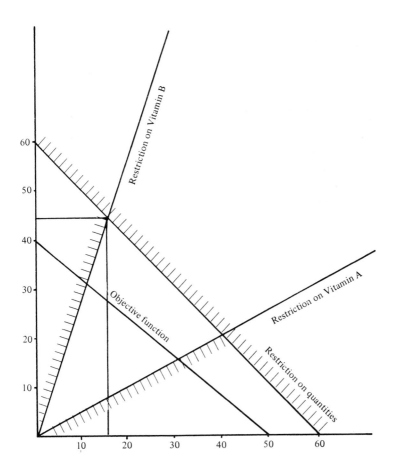

The objective function and restrictions are mapped in diagram 2. Using the method outlined in the last chapter, we read off that profit would be maximised when $x = 15$ and $y = 45$, or 15 lbs. of Vito is blended with 45 lbs. of Slam. The profit would be £285. The vitamin A content is

$$\frac{2 \times 15}{100} + \frac{5 \times 45}{100} = \frac{51}{20} \text{ lbs.}$$

or, as a percentage,

$$\frac{51}{20} \times \frac{1}{60} \times \frac{100}{1} = \frac{17}{4} = 4.25\%$$

This is in excess of the minimum requirements. The vitamin B content is

$$\frac{10 \times 15}{100} + \frac{6 \times 45}{100} = 4.2 \text{ lbs.}$$

or, as a percentage,

$$\frac{21}{5} \times \frac{1}{60} \times \frac{100}{1} = 7\%$$

which just meets the minimum requirements.

Now let us solve the same problem using the Simplex method. Notice that two of the inequalities are in a 'greater than or equal to' form, which suggests that an artificial variable should be added. However, if you think back to the last chapter, you will remember that the inequality sign can be reversed if we multiply throughout by minus one. Hence, instead of writing the vitamin restrictions like this:

$$2y - x \geqslant 0 \text{ and } 3x - y \geqslant 0,$$

we can write them like this

$$x - 2y \leqslant 0, \text{ and } y - 3x \leqslant 0.$$

We could not do this in the last example – why?

Adding slack variables in the usual way, we can obtain the first matrix

$$\begin{bmatrix} 1 & 1 & 1 & 0 & 0 & 60 \\ 1 & -2 & 0 & 1 & 0 & 0 \\ -3 & 1 & 0 & 0 & 1 & 0 \\ \hline -4 & -5 & 0 & 0 & 0 & 0 \end{bmatrix}$$

x	y	a	b	c	Q
		↑	↑	↑	

We select y as the incoming variable, but which is to be the outgoing variable? The problem is the presence of the zeros in the quantity column. Such matrices are called *degenerate* and we overcome this problem by adding a small positive amount E to the zeros. The quantity E is so small that it does not affect our final solution.

$$
\begin{bmatrix}
1 & 1 & 1 & 0 & 0 & \bigm| & 60 \\
1 & -2 & 0 & 1 & 0 & \bigm| & E \\
-3 & 1 & 0 & 0 & 1 & \bigm| & E \\
\hline
-4 & -5 & 0 & 0 & 0 & \bigm| & 0
\end{bmatrix}
$$

x	y	a	b	c	Q
		↑	↑	↑	

It is now easy to see that if y is the incoming variable, then c must be the outgoing variable.

$$
\begin{bmatrix}
4 & 0 & 1 & 0 & -1 & \bigm| & 60 - E \\
-5 & 0 & 0 & 1 & 2 & \bigm| & 3E \\
-3 & 1 & 0 & 0 & 1 & \bigm| & E \\
\hline
-19 & 0 & 0 & 0 & 5 & \bigm| & 5E
\end{bmatrix}
$$

x	y	a	b	c	Q
	↑	↑	↑		

Now we make x the incoming and a the outgoing variable

$$
\begin{bmatrix}
1 & 0 & \frac{1}{4} & 0 & -\frac{1}{4} & \bigm| & 15 - \frac{E}{4} \\
0 & 0 & \frac{5}{4} & 1 & \frac{3}{4} & \bigm| & 75 + \frac{7E}{4} \\
0 & 1 & \frac{3}{4} & 0 & \frac{1}{4} & \bigm| & 45 + \frac{E}{4} \\
\hline
0 & 0 & \frac{19}{4} & 0 & \frac{1}{4} & \bigm| & 285 + \frac{E}{4}
\end{bmatrix}
$$

x	y	a	b	c	Q
↑	↑		↑		

We now drop the E's from the solution and obtain that £285 profit can be earned by blending 15 lbs. of x with 45 lbs. of y. What do you think the slack variable $b = 75$ means? The slacks a and c are both zero — what do you conclude from this?

TUTORIAL ONE
N.B. Before attempting the questions that follow, you might prefer to read the appendix to this chapter which gives an easier method of performing the row transformations.

1. Derive a solution matrix for the tailor problem if the supply of cloth was 245 yards per week. Store your result carefully as you will need it for the next tutorial.

2. Let us consider an extension to the tailor problem. A second tailor is producing four products, the input requirement of each being

	Labour	Material	Lining
Coats	5 hrs.	6 yds.	4 yds.
Jackets	4 hrs.	4 yds.	3 yds.
Slacks	2 hrs.	3 yds.	1 yd.
Waistcoats	2 hrs.	1 yd.	1 yd.

His labour force is sufficient to provide 750 hours per week. A wholesaler can supply him with up to 950 yds. of cloth and 560 yds. of lining per week. Labour costs 50p per hour, cloth £1 per yard and lining 25p per yard.

After carefully assessing the prices charged by competitors, he decides to publish the following trade prices.

Coats	£10.10
Jackets	£ 7.25
Slacks	£ 4.65
Waistcoats	£ 2.50

If his objective is to maximise profit, what is the objective function? A department store buying manager sees potential in the tailor's designs, and a contract is arranged to supply 30 coats, 40 jackets, 50 pairs of slacks and 10 waistcoats per week. What are the restrictions on the objective function?

Find a solution matrix to satisfy the objective. Store the result carefully.

Duality

Let us yet again consider the tailor problem, and ask how does the tailor earn his profit. You may think that there is an obvious answer to this question — by producing jackets and overcoats. But we could give a second answer to this question. He earns his profit by combining hours of labour with yards of cloth. If we consider profit from this angle, then we can deduce a considerable amount of information useful to the tailor in addition to the product — mix that maximises profit.

Now suppose the tailor, realising his profit is derived from his inputs of cloth and labour, attempts to assess the relative values of these resources: he is *imputing* his profits to his inputs. He imputes an amount V_1 to each hour of labour, and an amount V_2 to each yard of cloth. Now, we know that it takes 2 hours of labour and 3 yards of cloth to produce a jacket, so the value of inputs in each jacket is

$$2V_1 + 3V_2$$

However, he will want to impute values V_1 and V_2 sufficiently great to account for the profit on jackets (£1). In other words

$$2V_1 + 3V_2 \geqslant 1 \quad \ldots\ldots\ldots \quad (1)$$

Likewise, as it requires 5 hours labour and 5 yards of cloth to make an overcoat, the values of the inputs would be

$$5V_1 + 5V_2,$$

and as the profit is £2 on each overcoat

$$5V_1 + 5V_2 \geqslant 2 \quad \ldots\ldots\ldots \quad (2)$$

It has been stated that the values V_1 and V_2 must be sufficient to account for the profit on jackets and overcoats. Why, then, state that

$$2V_1 + 3V_2 \geqslant 1$$

rather than that

$$2V_1 + 3V_2 = 1 \quad \ldots\ldots \quad ?$$

Well, the reason for this will become apparent later, but for the moment consider this point: if we had included a third input (lining perhaps) we would have two equations and three unknown variables. Any schoolboy will tell you that it is usually impossible to satisfy such a condition!

Think back to the last chapter, and you will realise that there is no unique solution to the simultaneous inequalities (1) and (2). Are there any limits to the values we can place on V_1 and V_2? Well, we know that the tailor has 200 hours of labour and 225 yards of cloth available per week, so the total value of all inputs is

$$A = 200V_1 + 225V_2$$

It might seem reasonable to find the smallest valuation of the *scarce* inputs that would account for all the profits of the outputs. If we did this, then the problem would become

Minimise $A =$ $200V_1 + 225V_2$
Subject to $5V_1 + 5V_2 \geqslant 2$
              $2V_1 + 3V_2 \geqslant 1$
              $V_1 \geqslant 0, V_2 \geqslant 0$

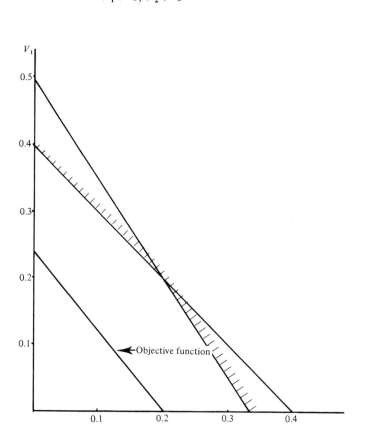

Figure 3 shows the mapping of this model. The objective function it-self has not been mapped: this would not be convenient. Instead, the objective function divided by 1000 has been mapped (you should realise that this would not affect its slope). Reading from the graph we see that A is minimised when $V_1 = V_2 = \frac{1}{5}$. Thus 20p is the minimum value we can impute to each hour of labour and each yard of cloth, and still account for all the profit. Now the total imputed value of the inputs is

$$200 \times \tfrac{1}{5} + 225 \times \tfrac{1}{5} = £85.$$

Does this ring a bell? The total imputed value is the same as the maximum profit the tailor could earn. *If we choose values for V_1 $V_2 \ldots V_n$ that would minimise the total imputed value, we would simultaneously find maximum profit.*

Let us place the linear programming model to maximise profit alongside the previous model.

Maximise $P = y + 2x$	Minimise $A = 200V_1 + 225V_2$
Subject to $2y + 5x \leqslant 200$	Subject to $5V_1 + 5V_2 \geqslant 2$
$3y + 5x \leqslant 225$	$2V_1 + 3V_2 \geqslant 2$
$y \geqslant 0. \ x \geqslant 0.$	$V_1 \geqslant 0. \ V_2 \geqslant 0.$

It is apparent that there is a strong connection between the two models. Seemingly, if we turn the maximising problem 'inside out' we obtain the minimising problem. What is a column in one becomes a row in the other, the inequality signs are reversed, the coefficients in the quantity column of the first problem become the coefficients of the objective function in the second, and vice versa. Only the non-negative condition for the variables remains unchanged. If the first model is called the *primal problem*, then the second is called the *dual*. Let us again examine the solution matrix for the primal problem.

$$
\begin{bmatrix}
0 & 0 & \frac{3}{5} & -\frac{2}{5} & 0 & 1 & 20 \\
0 & 0 & -1 & 1 & 1 & 0 & 5 \\
1 & 0 & -1 & 1 & 0 & 0 & 25 \\
0 & 1 & \frac{3}{5} & -\frac{2}{5} & 0 & 0 & 30 \\
\hline
0 & 0 & \frac{1}{5} & \frac{1}{5} & 0 & 0 & 85
\end{bmatrix}
$$

$$
\begin{array}{cccccc}
y & x & a & b & c & d \\
\uparrow & \uparrow & & & \uparrow & \uparrow
\end{array}
$$

The variables a and b were slacks attached to labour and cloth. If you examine the coefficients in the objective function row, you will see that a and b both have a value of $\frac{1}{5}$. This agrees with the values for V_1 and V_2 obtained from the graph. If we solve a primal problem by the Simplex method we simultaneously solve its dual. The solution to the dual is given in the objective row where there will be an imputed value each scarce resource.

Marginal Revenue Product: Scarcity Values

The objective function in the last problem demands that we find the smallest valuation of the scarce inputs that would account for all the profits of the outputs. You may be wondering why we used the word 'scarce'. If we impute values to scarce resources, then resources that are not scarce will have a zero imputed value. Now of course, businessmen are more interested in resources that are scarce, for they limit the possible output and total profit. Let us again examine the solution matrix for the witch doctor problem.

$$
\begin{bmatrix}
-\frac{4}{3} & 0 & 0 & 0 & 1 & -\frac{2}{3} & 0 & -\frac{1}{3} & 2 \\
1 & 0 & 1 & 0 & 0 & \frac{2}{3} & -\frac{1}{3} & 0 & 14 \\
-\frac{1}{3} & 0 & 0 & 1 & 0 & 0 & \frac{2}{3} & -\frac{1}{3} & 16 \\
\frac{1}{3} & 1 & 0 & 0 & 0 & -\frac{1}{3} & 0 & \frac{1}{3} & 12 \\
\hline
1 & 0 & 0 & 0 & 0 & \frac{2}{3} & \frac{2}{3} & 1 & 176
\end{bmatrix}
$$

$$x_1 \quad x_2 \quad x_3 \quad x_4 \quad x_5 \quad x_6 \quad x_7 \quad x_8 \quad Q$$
$$\uparrow \quad \uparrow \quad \uparrow \quad \uparrow$$

Suppose we want the imputed values $V_1 . V_2 . V_3$ and V_4 that satisfy the dual problem. The values are given in that part of the objective function row that is concerned with the slack variables. Reading from the matrix we obtain

$V_2 = \frac{2}{3}$, or the imputed value of lizards tongues is £0.66.
$V_3 = \frac{2}{3}$, or the imputed value of frogs legs is £0.66.
$V_4 = 1$, or the imputed value of bats wings is £1.

Notice that $V_1 = 0$, i.e. there is a zero imputed value on snake oil. This is because snake oil is not scarce (there are 2 fluid ozs. surplus).

When we solve the dual problem, we are finding *Scarcity values* for the inputs. Such values are of considerable operational significance. Consider the scarcity value of bats wings: if the restriction on bats wings was lifted so that an extra one was available, then profit would rise by £1. Thus, the scarcity value of an input shows by how much profit would increase if the supply of that resource was increased by one unit. Economists call the scarcity value of an input its *marginal revenue product*.

From any matrix giving an optimal solution to a primal problem we can read values for the dual. This enriches our knowledge of the nature and importance of scarce resources. In the example above, we can deduce that if the supply of lizards tongues or frogs legs was increased by one unit, profit would rise by £0.66, and if the supply of bats wings was increased by one unit, profit would rise by £1. Now suppose the witch-doctor finds an additional, but more expensive, source of supply of bats' wings. How much more can he afford to pay for them? As each extra bats' wing adds £1 to total profit, he could afford to pay anything up to £1 extra for additional supplies, and still increase his profit.

If in a particular week the witch-doctor could obtain an extra 3 bats' wings, we would expect his profit to rise by £3. We could check this by restating the primal problem with the bats' wings restriction as

$$3x_1 + 4x_2 + 2x_3 + x_4 + x_8 = 95$$

rather than

$$3x_1 + 4x_2 + 2x_3 + x_4 + x_8 = 92$$

The simplex routine could then be reworked. However, exactly the same row transformations as before would be used, and the new solution matrix would be identical with the previous one but for the components in the last column. In fact, there is no need to rework the solution. Let us extract the colum vector x_8 from the solution matrix, and identify each row.

$$
\begin{bmatrix}
-\frac{1}{3} \\
0 \\
-\frac{1}{3} \\
\frac{1}{3} \\
1
\end{bmatrix}
\quad
\begin{array}{l}
\text{Snake oil surplus row.} \\
\text{Elixir production row.} \\
\text{Love potion production row.} \\
\text{Mother-in-law repellent production row.} \\
\text{Objective function row.}
\end{array}
$$

This column vector tells us precisely what would happen if an extra bats' wing was made available, i.e. the snake oil surplus would fall by $\frac{1}{3}$ of a fluid ounce, love potion production would decline by $\frac{1}{3}$ of a unit, and mother-in-law repellent production would increase by $\frac{1}{3}$ of a unit. Also, profit would increase by £1. In fact, for any input that has a scarcity value we can say that

Column vector X change in availability of input = change in Q vector.

We are considering an increase of 3 bats' wings

$$
\begin{bmatrix}
-\frac{1}{3} \\
0 \\
-\frac{1}{3} \\
\frac{1}{3} \\
1
\end{bmatrix}
\begin{bmatrix} 3 \end{bmatrix}
=
\begin{bmatrix}
-1 \\
0 \\
-1 \\
1 \\
3
\end{bmatrix}
$$

The snake oil surplus would decline by 1 fluid oz.
The production of elixir would remain unchanged.
One love potion less would be produced.
One extra mother-in-law repellent would be produced.
Profit would rise by £3.

It would seem desirable to check that this new solution is feasible.

Producing one love potion less would release 1 fluid oz. of snake oil, 1 lizards tongue, 2 frogs legs and one bats wing. Producing one extra mother-in-law repellent would use 2 fluid ozs. of snake oil, 1 lizards tongue, 2 frogs legs and 4 bats wings. The change in product mix requires an extra fluid oz. of snake oil (i.e. the surplus declines by one) and 3 bats' wings extra (which we assumed were available). Also, as the profit on love potions was £3 each, and £6 each on mother-in-law repellents, the change in product mix would cause profit to rise by £3.

In the original solution, the snake oil surplus was two fluid ozs. An additional 3 bats wings would give a new product mix with a reduction in the snake oil surplus to one fluid oz. If an additional six bats wings were available, then the new product mix would give a zero snake oil surplus. If we considered any increase above six in the availability of bats wings, then the column vector would give a non-feasible solution. For example, 9 extra bats wings would give a change in snake oil surplus of $-\frac{1}{3} \times 9 = -3$, i.e. the new snake oil surplus would be $2 - 3 = -1$. But one cannot have a negative surplus! What can we conclude? Bats' wings have a scarcity value of £1 as long as we do not consider increases in the availability of greater than 6. Why? Because if we increase the availability by greater than 6, snake oil becomes scarce and cannot itself have a zero scarcity value. Can you see that the scarcity value of lizards tongues (£0.66) holds as long as supplies do not increase by more than 3? What is the maximum increase in frogs legs if their scarcity value is to remain at £0.66?

So far, we have considered increases in the availability of scarce resources, but the method holds equally well for decreases. Suppose the availability of bats wings was reduced by 36. Then the new situation would be

$$\begin{bmatrix} -\frac{1}{3} \\ 0 \\ -\frac{1}{3} \\ \frac{1}{3} \\ 1 \end{bmatrix} [-36] = \begin{bmatrix} 12 \\ 0 \\ 12 \\ -12 \\ -36 \end{bmatrix}$$

The snake oil surplus would rise by 12 fluid ozs. (to 14).
Elixir production would remain unchanged (at 14).
Love potion production would rise by 12 (to 28).
Mother-in-law repellent production would fall by 12 (to zero).
Profit would fall by £36 (to £140).

Can you see that if the availability of bats wings fell by more than 36, the vector would not give a feasible solution? The scarcity value of bats wings holds as long as its availability does not decline by more than 36, nor increase by more than 6. Finding such limits is called *ranging*. There are two simple rules for ranging a scarce resource – can you find them? What are the ranges for frogs legs and lizards tongues?

Accounting Losses: Opportunity Costs

It was stated earlier that the restrictions in the dual problem were by nature of a 'greater than or equal to' form. The time has now come to justify this. First, let us state the dual of the witch-doctors' problem.

$$\text{Minimise } A = 70V_1 + 56V_2 + 70V_3 + 92V_4$$

$$\text{Subject to} \quad V_1 + 2V_2 + V_3 + 3V_4 \geqslant 4$$
$$2V_1 + V_2 + 2V_3 + 4V_4 \geqslant 6$$
$$2V_1 + 2V_2 + V_3 + 2V_4 \geqslant 4$$
$$V_1 + V_2 + 2V_3 + V_4 \geqslant 3$$
$$V_1 \geqslant 0, V_2 \geqslant 0, V_3 \geqslant 0, V_4 \geqslant 0$$

Now we know that to solve linear programming problems, we must change inequalities into equations, so artificial variables must be subtracted from the left hand side of each restriction. The first restriction now becomes

$$V_1 + 2V_2 + V_3 + 3V_4 - V_5 = 4$$

Suppose that when we solve this model we find that

$$V_1 + 2V_2 + V_3 + 3V_4 > 4.$$

What does this imply? *It implies that the value of scarce resources used in producing this product exceeds the profit it earns.* Surely, this indicates that the producer would be advised to employ his resources producing something else. Also, there is a second implication: the artificial variable, V_5, will be non-negative. Let us examine both of these implications more clearly.

We know that if the objective function is to be satisfied subject to the restrictions, then

$$V_1 = 0. \quad V_2 = \tfrac{2}{3}. \quad V_3 = \tfrac{2}{3}. \quad V_4 = 1.$$

Now we will substitute these values in the first restriction (i.e. the one for deadly poison)

$$1 \times 0 + 2 \times \tfrac{2}{3} + \tfrac{2}{3} + 3 \times 1 = 5, \text{ hence } V_5 = 1.$$

The value of scarce resources used in the production of deadly poison exceeds the profit it can earn. We would not expect the witch-doctor to produce this product (note that the solution to the primal problem tells him not to). What meaning can we attach to the artificial variable? The value of the scarce resources used producing deadly poison exceeds the profit it can earn by £1. Hence $V_5 = 1$ is a measure of the *accounting loss* made on producing this product. In fact, each unit of deadly poison produced costs the witch-doctor £1 in lost profit, and for this reason, economists call the artificial variable the *opportunity cost* of producing the product. If the opportunity cost of producing a good is zero, then its production is worthwhile. However, if a good has an opportunity cost, then the producer will switch resources away from its production.

Look again at the matrix which gives the solution to the primal problem. The first column refers to x_1, deadly poison. The value of the component in the objective function row of this column is one − the same as the opportunity cost. The column vector containing the opportunity cost demonstrates what would happen to total profit if we could reduce the production of deadly poison by one unit. It also assumes that it is possible to make fractional units of the products

$$\begin{bmatrix} -\frac{1}{3} \\ 1 \\ -\frac{1}{3} \\ \frac{1}{3} \\ 1 \end{bmatrix}$$

snake oil surplus row.
elixir row.
love potion row.
mother-in-law repellent row.
artificial variable row.

Thus, if it was possible to reduce the production of deadly poison by one unit, this would leave sufficient resources to produce an extra unit of elixir, and $\frac{1}{3}$ of a unit of the mother-in-law repellent, provided that love potion production was also reduced by $\frac{1}{3}$ of a unit. The surplus snake oil would be reduced by $\frac{4}{3}$ fluid ounces. We shall now check if this is feasible.

	Snake oil	Lizards tongues	Frogs legs	Bats wings
Resources released by producing:				
1 unit less of deadly poison	1	2	1	3
$\frac{1}{3}$ unit less of love potion	$\frac{1}{3}$	$\frac{1}{3}$	$\frac{2}{3}$	$\frac{1}{3}$
Resources used producing:				
1 extra unit of elixir	2	2	1	2

	Snake oil	Lizards tongues	Frogs legs	Bats wings
Resources used producing:				
$\frac{1}{3}$ unit extra of mother-in-law repellent	$\frac{2}{3}$	$\frac{1}{3}$	$\frac{2}{3}$	$\frac{4}{3}$
Difference	$\frac{4}{3}$	0	0	0

Thus the quantities of resources involved are feasible. Now let us check the feasibility of the profit change. The value of the artificial variable tells us that the change in production would cause profit to *increase* by £1

Profit reduction from	
one unit less of deadly poison	£4
$\frac{1}{3}$ unit less of love potion	£1
Profit increase from	
one extra elixir	£4
$\frac{1}{3}$ unit extra of mother-in-law repellent	£2
Net change	£1

The profit change is feasible.

Summary of the Pay-off of the Simplex Method

What has linear programming done for our witch-doctor? It has given him the quantities of each product that must be produced to maximise his profit. It isolates his scarce resources and gives them scarcity values. Hence, the witch-doctor can see how much more he can afford to pay for additional supplies of scarce factors. By using an appropriate column vector, he can deduce his product mix if the availability of one of his scarce factors changes. Finally, he can obtain any accounting losses on his products. For example, let us suppose that the local chief urgently requires a bottle of deadly poison (his mother-in-law is immune to the repellent). Now the accounting loss on deadly poison is £1 — if the chiefs' order is satisfied profit for that week would decline by £1. Hence, the price of deadly poison must be increased by £1 if profit is to be maintained. The pay-off of the Simplex Method is considerable — is it any wonder that linear programming has made such a considerable impact on industry?

TUTORIAL TWO

1. Examine the solution matrix for the two product tailor problem. Select an appropriate vector, and deduce the product mix and profit if the supply of cloth is increased by 20 yards in a particular week. Compare your result with the solution matrix you obtained in the first question of the last tutorial.

2. Find your solution matrix for the second question of the last tutorial. Which inputs are scarce? What are their scarcity values? Over what ranges do the scarcity values operate?
   What operational significance can you attach to the scarcity value and ranging for labour? (Hint: consider overtime working.)
   Which products have opportunity costs? How does this affect the validity of the pricing policy? Comment on the validity of the contract.

3. Refer to the solution matrix for the tailor problem.
   Write a report to the directors on the state of the firm, including any recommendations you think necessary.

A Minimising Example

We can now examine how the Simplex Method can be used to solve minimising problems. The analysis will be confined to a two dimensional problem so that the Simplex results can be compared with a graphical method. We will use the fact that by maximising a primal problem we simultaneously minimise the dual.

An oil company put additives to petrol to give improved performance and to reduce engine wear. Ideally, each 1000 gallons should contain at least 40 mgs. of additive p, 14 mgs. of additive q and 18 mgs. of additive r. The company can obtain stocks of ingredients which have the following weights of additive per litre.

Ingredient	mgs. p	mgs. q	mgs. r
y	4	2	3
x	5	1	1

Both ingredients cost £1 per litre. The company wishes to know the quantities of each ingredient which will satisfy the conditions at minimum cost i.e.

Minimise $C = y + x$
Subject to $\quad 4y + 5x \geqslant 40$
$\qquad\qquad 2y + \ x \geqslant 14$
$\qquad\qquad 3y + \ x \geqslant 18$
$\qquad\qquad y \geqslant 0$
$\qquad\qquad x \geqslant 0$

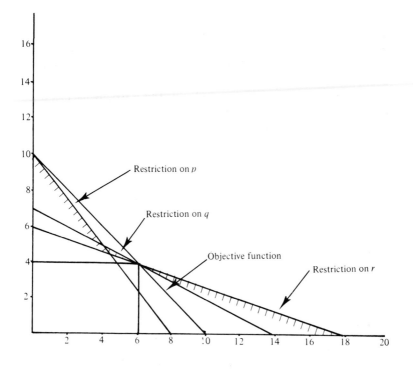

The mapping of the inequalities and objective function is shown in diagram 4. We find that $y = 5$ and $x = 4$ and the cost is £9 per 1000 gallons.

Now let us solve this problem by the Simplex Method. We can solve it by considering the dual problem rather than the primal problem. We shall impute a cost V_1 to each mg. of p, V_2 to each mg. of q and V_3 to each mg. of r. Now if we add 1 litre of ingredient y to 1000 gallons of petrol, then the total imputed cost would be

$$4V_1 + 2V_2 + 3V_3$$

Now we use this ingredient solely because it contains additives. Hence, we do not wish the total imputed cost of additives in a litre of ingredient y to exceed its cost to the company. In other words

$$4V_1 + 2V_2 + 3V_3 \leqslant 1$$

Likewise for each litre of x

$$5V_1 + V_2 + V_3 \leqslant 1$$

It would be useful to compare these inequalities with those obtained in previous examples of dual problems. Previously, we were considering only scarce resources — what are we considering here? We will impute costs only to those additives that just meet the requirements. If any additive exceeds the requirements, then so much the better. We will give any such additive a zero imputed cost. This is strictly parallel to imputing values to scarce resources only.

If we add slack variables to the left hand side of the inequalities we will transform them into equations.

$$4V_1 + 2V_2 + 3V_3 + V_4 = 1$$
$$5V_1 + V_2 + V_3 + V_5 = 1$$

What meaning can we attach to the slack variables? Now we want the cost of the additives to account for the cost of the ingredient. If this is not the case the slack variable will be non-negative and will again be an accounting loss. It is the opportunity cost of using this ingredient as opposed to another.

If the requirements are to be met, then the total imputed cost of additives will be:

$$40V_1 + 14V_2 + 18V_3$$

Again, comparing with previous examples, we wish to find the very largest valuation on additives that will account for the cost of the ingredients i.e.

Maximise $40V_1 + 14V_2 + 18V_3$
Subject to $2V_1 + 2V_2 + 3V_3 \leqslant 1$
$\qquad\qquad 5V_1 + V_2 + V_3 \leqslant 1$
$\qquad\qquad V_1 \geqslant 0. \quad V_2 \geqslant 0. \quad V_3 \geqslant 0.$

Setting up the matrix in the usual fashion, and performing row transformations we have:

$$
\begin{bmatrix}
4 & 2 & 3 & 1 & 0 & 1 \\
5 & 1 & 1 & 0 & 1 & 1 \\
-40 & -14 & -18 & 0 & 0 & 0
\end{bmatrix}
$$

$\quad V_1 \qquad V_2 \qquad V_3 \qquad V_4 \qquad V_5 \qquad Q$
$\qquad\qquad\qquad\qquad\qquad\;\uparrow\qquad\;\uparrow$

$$
\begin{bmatrix}
0 & \frac{6}{5} & \frac{11}{5} & 1 & -\frac{4}{5} & \frac{1}{5} \\
1 & \frac{1}{5} & \frac{1}{5} & 0 & \frac{1}{5} & \frac{1}{5} \\
0 & -6 & -10 & 0 & 8 & 8
\end{bmatrix}
$$

$\quad V_1 \qquad V_2 \qquad V_3 \qquad V_4 \qquad V_5 \qquad Q$
$\;\;\uparrow\qquad\qquad\qquad\qquad\uparrow$

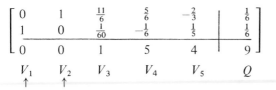

$$
\begin{bmatrix}
0 & 1 & \frac{11}{6} & \frac{5}{6} & -\frac{2}{3} & \bigg| & \frac{1}{6} \\
1 & 0 & \frac{1}{60} & -\frac{1}{6} & \frac{1}{5} & \bigg| & \frac{1}{6} \\
\hline
0 & 0 & 1 & 5 & 4 & \bigg| & 9
\end{bmatrix}
$$

$$V_1 \quad V_2 \quad V_3 \quad V_4 \quad V_5 \qquad Q$$

$$\uparrow \qquad \uparrow$$

Firstly, let us read off the solution to the maximising problem. The imputed cost of the additives p and q can account entirely for the cost of the ingredients if we impute a cost of £$\frac{1}{6}$ to each. The very largest valuation on additives that will account for the cost of the ingredients is £9. By analogy with the previous analysis, the largest valuation on additives that will account for the cost of the ingredients will also be the smallest cost for which the ingredients can satisfy the requirements of additive content. Thus, we cannot obtain the desired additive content for less than £9. This agrees with the minimum cost obtained by graphical methods.

Now think back to the last dual problem we examined. The solution to the maximisation problem was given in the quantity column, and the solution to the minimising problem was given in the objective function row. Hence, the values in the V_3, V_4, and V_5 columns of the objective function row must give the solution we require. Let us examine each of these columns in turn. The V_5 column was obtained because a slack variable was added to the coefficients obtained from the quantities of y in the primal problem. Thus V_5 is related to y and we can conclude in our minimising problem $y = 5$. In the same way, we can argue that V_4 is related to x, and $x = 4$. These values agree with the graphical solution. Finally, we notice that $V_3 = 1$. Now the V_3 column was obtained from the coefficients of r, and the value gives the artificial variable subtracted from the inequality giving the desired content of r in 1000 gallons of petrol. Hence, if 5 litres of y and 4 litres of x are added to 1000 gallons of petrol,

the quantity of additive p is $4 \times 5 + 5 \times 4 = 40$ grms.
the quantity of additive q is $2 \times 5 + 4$ $= 14$ grms.
the quantity of additive r is $3 \times 5 + 4$ $= 19$ grms. (1 grm. in
                excess of requirement)
the cost is $5 + 4 = £9$, which is minimum

Appendix: The Use of a Pivot in the Simplex Method

It will be small consolation to you to be told that the linear programming problems we have considered are extremely simple compared with those met in the real world. You have probably found that row transformations are extremely tedious, and sustained arithmetic accuracy is difficult to achieve. However, the point of this chapter has not been to teach you to solve linear programming problems. It is intended to show you that many problems can be solved by such methods. If you can recognise when linear programming is appropriate and if you can interpret a solution matrix, then this chapter will have achieved its purpose. We can with relief leave the actual computation to a computer.

However, many of you will want to work through the stages of the Simplex Method, and may prefer to use a pivot rather than perform row transformations. Let us see how a pivot works.

$$
\begin{bmatrix}
4^* & 1 & 1 & 0 & 0 & 90 \\
2 & 1 & 0 & 1 & 0 & 50 \\
1 & 1 & 0 & 0 & 1 & 40 \\
\hline
-4 & -3 & 0 & 0 & 0 & 0
\end{bmatrix}
$$

Firstly, we select the column with the largest digit preceded by a minus sign in the objective row and call this the pivot column. In this case, then, we select the first column. We divide the non-negative components in this column into the components in the quantity column thus

$90 \div 4 = 22.5$
$50 \div 2 = 25$
$40 \div 1 = 40$

The row which gives the smallest result to the division is selected as the pivot row. In this case, the first row is the pivot row. The component in the pivot row *and* pivot column is called the pivot, and this is clearly marked in the matrix by a star.

Next we divide the pivot row by the pivot, and replace the other components in the pivot column with zero. The matrix would now look like this.

1	¼	¼	0	0	22½
0					
0					
0					

We replace the remaining components like this. We imagine a rectangle with the pivot in one corner, and the component we wish to replace in the diagonally opposite corner. We then find the product of the components in the other corners, divide by the pivot, and subtract from the component we are replacing.

Thus, if we wish to replace the third component in the quantity column, the rectangle would be

4^* 90
1 40

The product we derive is $1 \times 90 = 90$. Dividing by the pivot we have $90 \div 4 = 22\frac{1}{2}$. Hence, the new value of the component we are replacing is $40 - 22\frac{1}{2} = 17\frac{1}{2}$. The remaining components are treated in exactly the same way.

EXERCISES TO CHAPTER THREE

1. A manufacturer of two commodities has to use three separate materials in their manufacture. The relevant input data is:

    | | | Product | |
Material	A	B	Material available
x	3	2	84 units
y	4	5	140 units
z	1	3	63 units

    He makes £1 profit on a unit of *A*, and £2 on a unit of *B*. What product mix will maximise his profit?

2. A firm producing three products has a bottleneck in production at a particular point of the productive process involving two machines. It realises there is something wrong with its product mix and wishes to feed into the bottleneck that combination which utilises the available machine time to its fullest extent and is also the most profitable. The relevant data showing the time spent on the production of each product is —

		Product		Time available
	x	y	z	
Machine 1	24 mins.	60 mins.	72 mins.	240 hrs.
Machine 2	60 mins.	72 mins.	48 mins.	240 hrs.
Profit/unit	£1	£2	£2	

    State the objective function and the restrictions and deduce the optimum product mix. Check that capacity is fully utilised.

3. Space travellers colonising Mars discover that it can be made fertile by using a compost containing at least 26% phosphates and 17% nitrates. The compost comes from soils found on Saturn and Venus and is made up in bales of 100 tons by mixing these two soils with inactive soil from Mars. Soil from Saturn contains 60% phosphates and 20% nitrates and costs £8 a ton, while soil from Venus contains 20% phosphates and 40% nitrates and costs £5 per ton.

    Set up the initial matrix and hence calculate the quantities of the two soils which must be mixed into each 100 ton bale so as to minimise cost of production. What is the minimum price that must be charged for the compost if the project is to break even?

4. A manufacturer wishes to make a certain mixture that contains not more than 5% of sulphur and at least 8% of phosphorus. He intends to mix two compounds with the following specifications

Compound	Sulphur content	Phosphorus content	Cost
A	6%	13%	£1
B	1%	4%	£2

The manufacturer has a budget of £600 to produce this mixture, and he wants to make as much as possible. Formulate an appropriate model and solve it with a matrix.

5. A firm makes three products A B and C. Inputs per unit are

	Machine time	Labour time	Raw materials	Profit
a.	3 hrs.	1 hr.	2 tons	£4
b.	2 hrs.	2 hrs.	1 ton	£6
c.	1 hr.	2 hrs.	3 tons	£5
Available per week	218 hrs.	134 hrs.	150 tons	

He has a contract to supply 10 units of each product per week. Derive a solution matrix which will give maximum profit.

6. Suppose that ideally a batch of animal food should contain at least 270 units of vitamin A, 100 units of vitamin B and 190 units of vitamin C. The vitamin content is put into the food by adding certain additives. Additive X contains 2 units of each vitamin per gram and costs 5p per gram. Additive Y contains 3 units of vitamin A, and one unit each of vitamins B and C per gram and costs 6p. Additive Z contains 3 of vitamin A, 1 of B and 6 of C, costing 8p per gram. A manufacturer wishes to make a batch of this feed at a minimum cost, how many grams of X, Y and Z should be added?

7. Find the example on wine blending in the chapter on inequalities. Write down the objective function and restrictions and then derive the first matrix. Using the column for wine A as the entering variable perform the first iteration.

8. Find your solution matrix for question 4. If your solution is correct, you will notice an amount 1600 appears in the quantity vector. What is its significance?

9. Find your solution matrix for question 1. Interpret the dual solution to this problem.

10. Using the solutions to the primal and dual, write a report to the directors of the firm in question 5.

Chapter Four

Special Techniques for Linear Programming:
Transportation and Assignment Models

When considering linear programming methods, we have so far confined our attention to production problems — maximum profit product — mix, or minimum cost combination for inputs, and blending problems. The time has now come to widen the horizons of linear programming techniques, and introduce various techniques which will simplify the method under certain conditions. Firstly, let us see how linear programming can help the transport manager to solve routing problems. Suppose a firm is producing a certain product in two separate factories. Factory A produces 75 units and factory B 25 units per week. The whole output is sold to three distributors, who take 45, 35 and 20 units per week. The transport manager has calculated the unit cost of moving the output to the distributors as follows:

		to distributor		
		a	b	c
from factory	A	£3	£4	£6
	B	£1	£4	£3

Thus it costs £3 to move a unit from factory B to distributor c, £4 to move a unit from Factory A to distributor b etc.

The Simplex Form of the Transportation Problem

Let us suppose amounts X_{11} X_{12} and X_{13} are distributed from factory A to distributors a, b and c respectively. Also that amounts X_{21}, X_{22} and X_{23} are distributed from factory B to each of the three distributors. The TRANSPORTATION MATRIX would look like this.

		to distributor		
		a	b	c
from factory	A	X_{11}	X_{12}	X_{13}
	B	X_{21}	X_{22}	X_{23}

Notice the convenience of using suffixes to X when stating a general solution to such problems: X_{13} means the quantity in the first row, third column. Thus the first digit of the suffix locates the row, and the second locates the column.

We can now state that the quantity distributed from Factory A is

$$X_{11} + X_{12} + X_{13}$$

and as we wish the quantity distributed to be the same as the quantity produced

$$X_{11} + X_{12} + X_{13} = 75$$

Likewise, for factory B

$$X_{21} + X_{22} + X_{23} = 25$$

The quantity received by distributor 'a' is

$$X_{11} + X_{21},$$

and as we wish distributor a to receive 45 units

$$X_{11} + X_{21} = 45$$

Finally, from the restriction placed on distributors b and c

$$X_{12} + X_{22} = 35$$
$$\text{and } X_{13} + X_{23} = 20$$

To find the cost of the allocation we note that for each unit sent from factory A to warehouse 'a', a cost of £3 is incurred, so the cost of sending X_{11} units is $3X_{11}$. Using similar arguments for the other quantities, the total allocation cost is

$$3X_{11} + 4X_{12} + 6X_{13} + X_{21} + 4X_{22} + 3X_{23}$$

As the objective is to minimise cost, the problem becomes

Minimise $C = 3X_{11} + 4X_{12} + 6X_{13} + X_{21} + 4X_{22} + 3X_{23}$
Subject to
$$X_{11} + X_{12} + X_{13} = 75 \ldots \ldots \ldots \ldots \quad (1)$$
$$X_{21} + X_{22} + X_{23} = 25 \ldots \ldots \ldots \ldots \quad (2)$$
$$X_{11} + X_{21} = 45 \ldots \ldots \ldots \ldots \ldots \quad (3)$$
$$X_{12} + X_{22} = 35 \ldots \ldots \ldots \ldots \ldots \quad (4)$$
$$X_{13} + X_{23} = 20 \ldots \ldots \ldots \ldots \ldots \quad (5)$$
All variables $\geqslant 0$

A First Feasible Solution: the 'Northwest Corner' Method

Of course, it would be perfectly possible to solve this problem using methods outlined in the previous chapter, but there is a better way. Firstly, we combine the transportation matrix and the cost matrix like this.

3	4	6
X_{11}	X_{12}	X_{13}
1	4	3
X_{21}	X_{22}	X_{23}

The quantities X are the amounts allocated, and the amounts in the top right-hand corner of each cell represent the allocation costs. Now all linear programming problems that we have investigated involved finding a first feasible solution, and improving on it until an optimum is reached. This method (called the Transportation Method) is no exception. A first feasible solution can be obtained by the 'Northwest Corner' method, so called because we consider the cell in the top left-hand corner first. We then make the amount allocated to this cell (X_{11}) as large as possible. Now we know that

$$X_{11} + X_{12} + X_{13} = 75 \quad \ldots \ldots \ldots \ldots \ldots \quad (1)$$
$$\text{and } X_{11} + X_{21} \quad\quad = 45 \quad \ldots \ldots \ldots \ldots \ldots \quad (3)$$

As all variables must be non-negative, it follows that it is equation (3) that restrict the size of X_{11}, i.e. $X_{11} \leqslant 45$. Now if X_{11} is to be as large as possible, then $X_{11} = 45$, $X_{21} = 0$, and the restrictions now become

$$X_{12} + X_{13} = 30 \quad \ldots \ldots \ldots \ldots \quad (1)$$
$$X_{22} + X_{23} = 25 \quad \ldots \ldots \ldots \ldots \quad (2)$$
$$X_{12} + X_{22} = 35 \quad \ldots \ldots \ldots \ldots \quad (4)$$
$$X_{13} + X_{23} = 20 \quad \ldots \ldots \ldots \ldots \quad (5)$$

We now turn to the second cell in the first row, (X_{12}) and make this quantity as large as possible. From equation (1) we obtain that $X_{12} = 30$, and $X_{13} = 0$ (equation (4) does not restrict the size of X_{12}) and the restrictions now become

$$X_{22} + X_{23} = 25 \quad \ldots \ldots \ldots \ldots \quad (2)$$
$$X_{22} \quad\quad = 5 \quad \ldots \ldots \ldots \ldots \quad (4)$$
$$X_{23} \quad\quad = 20 \quad \ldots \ldots \ldots \ldots \quad (5)$$

As the three restrictions are consistent with each other, we now have a feasible solution. The matrix now looks like this

x_{11} 3	x_{12} 4	x_{13} 6
45	30	0
x_{21} 1	x_{22} 4	x_{23} 3
0	5	20

The cost of this solution is $3 \times 45 + 4 \times 30 + 4 \times 5 + 3 \times 20 =$ £335.

TUTORIAL ONE
A firm produces a certain product at three factories, and sells the output to four distributors. Output and demand are

Factory	*A*	*B*	*C*	
Output (weekly)	65	25	10	
Distributor	*a*	*b*	*c*	*d*
Demand (weekly)	45	25	20	10

The costs of distribution are

	a	*b*	*c*	*d*
A	£2	£3	£4	£6
B	£1	£4	£3	£2
C	£5	£1	£3	£2

If the objective is to minimise cost, form the linear programming model of the problem. Derive a feasible solution using the 'Northwest Corner' method. Find the cost of this allocation.

Finding an Improved Feasible Solution

Returning to our earlier example, we know that the output can be allocated to the distributors at a cost of £335. Let us now see if it is possible to find another allocation at a lower cost. As the system now stands, factory A does not supply any units to distributor c, nor factory B to distributor a. In other words, all the cells are *occupied* except X_{13} and X_{21}. Now X_{21} is the cell with the lowest transport cost: it costs only £1 to supply distributor 'a' from factory B. It would seem sensible, then, to let factory B supply distributor a.

Let us investigate the effect on cost of supplying distributor '*a*' with one unit from factory *B*. How will this affect the solution? Cell X_{21} will now become occupied, having one unit. However, factory *A* will no longer supply distributor '*a*' with 45 units, as this would mean that distributor '*a*' would be receiving a total of 46 units — one more than he required. Hence cell X_{12} would contain 44 units. This now means that factory *A* now has a surplus unit, which we can distribute to cell X_{12} (notice that we do not put it in cell X_{13}, the other unoccupied cell — for the moment we are interested only in unoccupied cell X_{21}). If we remove one unit from cell X_{22}, the allocation would be feasible.

Thus we have

X_{11} = 44 (saving £3 in transportation costs)
X_{21} = 1 (costing £1 more in transportation costs)
X_{12} = 31 (costing £4 more in transportation costs)

and

X_{22} = 4 (saving £4 in transportation costs)

The net effect on cost, of allocating a unit to cell X_{21} is

$$-3 + 1 + 4 - 4 = -£2$$

i.e. such a move would save £2. We can conclude that every unit we could allocate to cell X_{21} would reduce transport costs by £2, and hence, it would pay us to move as much as possible to that cell.

How much can we allocate to cell X_{21}? For every unit that is allocated to X_{21}, we must remove a unit from both X_{11} and X_{22}. We can remove only 5 units from X_{22} — if we removed more than this we would have a negative (and hence non-feasible) solution. Hence we can move a maximum of 5 units into X_{21}. This would reduce transport costs by £10. (£2 per unit.) The solution would now look like this:

x_{11} 3	x_{12} 4	x_{13} 6
40	35	0
x_{21} 1	x_{22} 4	x_{23} 3
5	0	20

You should satisfy yourself that this solution is feasible. The cost of this allocation is $3 \times 40 + 4 \times 35 + 1 \times 5 + 3 \times 20 = £325$. This agrees with the predicted cost reduction.

Reallocation Routes

Let us look a little more closely at the problem of allocating into a chosen unoccupied cell, because it can be tricky. Suppose the following allocation had been selected.

x_{11} 77 $-$	x_{12}	x_{13}	x_{14} 38	x_{15} $+$
x_{21} 9 $+$	x_{22} 35	x_{23}	x_{24}	x_{25} 51 $-$
x_{31}	x_{32} 39	x_{33} 14	x_{34}	x_{35}
x_{41}	x_{42}	x_{43} 48	x_{44}	x_{45}

Thus there are four factories supplying five distributors. Suppose it was decided that in future factory (1) should supply warehouse 5 i.e. cell X_{15} should become occupied – how could this be achieved? If distributor 5 is to be supplied from factory 1, he must receive less from factory 2. We mark cell X_{15} with a plus sign, and X_{25}, with a minus. Now if factory (1) is to supply distributor (5) it must reduce its supply to other distributors – otherwise it will be asked to supply more than it can produce. It cannot supply less to distributor (4), as it is distributor (4)'s only supplier, and if we continue to use occupied cells only there would be no method of sympathetically increasing the amount allocated to distributor (4). Hence, it must supply less to distributor (1) and cell X_{11} is marked with a minus. Distributor (1) must now receive more from another factory, and as factory (2) now has surplus output cell X_{21} is marked with a plus. A reallocation route has now been obtained.

Using this route, we see that less is to be allocated to X_{11} and X_{25}. As negative allocations are not permitted, we can remove up to 51 units from both these cells. Thus 51 units is the most that can be allocated into X_{15}. Adding and subtracting 51 units as appropriate we obtain the new allocation. You can check the feasibility of the new solution: the row totals and column totals are the same as the initial allocation

x_{11} 26	x_{12}	x_{13}	x_{14} 38	x_{15} 51
x_{21} 60	x_{22} 35	x_{23}	x_{24}	x_{25}
x_{31}	x_{32} 39	x_{33} 14	x_{34}	x_{35}
x_{41}	x_{42}	x_{43} 48	x_{44}	x_{45}

Let us now state the rules for finding a reallocation route.
1. Mark the unoccupied cell chosen for allocation with a plus.
2. Examine the row containing the chosen cell, and mark the other occupied cell with a minus. If there are two occupied cells, mark the one which contains another cell within its column.
3. Scan the column containing this cell, and mark the cell that contains another occupied cell within its row with a plus.
4. Continue the process until each row that contains a plus also contains a minus (and vice versa) and until the same conditions apply to columns. A unique reallocation route will then have been found.
5. The most that can be allocated into the chosen unoccupied cell is found by examining those cells containing minus signs. The smallest quantity in these cells is the most that can be moved.

Sometimes the reallocation route is quite involved. If we wanted to allocate into cell X_{45}, then the route would be

x_{11} 26 +	x_{12}	x_{13}	x_{14} 38	x_{15} 51 −
x_{21} 60 −	x_{22} 35 +	x_{23}	x_{24}	x_{25}
x_{31}	x_{32} 39 −	x_{33} 14 +	x_{34}	x_{35}
x_{41}	x_{42}	x_{43} 48 −	x_{44}	x_{45} +

A maximum of 39 units could be moved into cell X_{45}.

If we reallocate using occupied cells only (and we have seen that it is logical to do so as we are interested in one unoccupied cell only) then the Northwest Corner method gives a unique reallocation route. Sometimes it is not possible to use occupied cells only and we shall investigate this later. Finding the reallocation route is of cardinal importance in solving the transportation problem, and the following tutorial is designed to help you to master the technique.

TUTORIAL TWO

x_{11} 77	x_{12}	x_{13}	x_{14} 38	x_{15}
x_{21} 9	x_{22} 35	x_{23}	x_{24}	x_{25} 51
x_{31}	x_{32} 39	x_{33} 14	x_{34}	x_{35}
x_{41}	x_{42}	x_{43} 48	x_{44}	x_{45}

Reallocate to the following cells, performing each operation successively i.e. work reallocation (2) on the matrix resulting from reallocation (1)
(1) X_{42}, (2) X_{41}, (3) X_{23}, (4) X_{32}, (5) X_{35}, (6) X_{34}, (7) X_{43}, (8) X_{41}, (9) X_{15}, (10) X_{22}, (11) X_{43}, (12) X_{34}.

Your final matrix should be:

x_{11} 86	x_{12}	x_{13}	x_{14} 29	x_{15}
x_{21}	x_{22} 30	x_{23} 14	x_{24}	x_{25} 51
x_{31}	x_{32} 44	x_{33}	x_{34} 9	x_{35}
x_{41}	x_{42}	x_{43} 48	x_{44}	x_{45}

The Optimum Solution

We left our original problem looking like this:

x_{11} 3 40 —	x_{12} 4 35	x_{13} 6 +
x_{21} 1 5 +	x_{22} 4	x_{23} 3 20 —

The 'Northwest Corner' solution was improved upon by allocating into cell X_{21} and clearing cell X_{22}. Can we get an even better solution? The only other solution possible would be to allocate into cell X_{13}. The route for this allocation is marked on the matrix, and we can see that the cost of moving one unit into cell X_{13} would be $+6 +1 -3 -3 = +£1$. Thus, each unit moved into cell X_{13} would increase total cost by £1. Satisfy yourself that this is the case by moving the 5 units into X_{13}, and work out the new total cost. You will find that total cost has increased by £5. As we have now investigated all possible solutions we can see that the second solution was indeed optimal.

Let us now summarise the Transportation Method of Linear Programming. The Transportation Matrix is set up and the 'Northwest Corner' technique used to obtain a first feasible solution. A low cost unoccupied cell is chosen and a route is found for allocating into that cell. The route is costed to see whether reallocation to that cell is worthwhile. The solution is optimal when no further reallocations are worthwhile.

TUTORIAL THREE

1. Look again at your first feasible solution to the question in tutorial 1. How many other solutions are possible? What is the optimal solution?
2. The weekly output of three factories is:

Factory	A	B	C
	139	74	32

The weekly demand from 5 distributors is:

Distributor	a	b	c	d	e
Demand	75	65	55	40	10

and the unit allocation costs are

	a	b	c	d	e
A	£2	£3	£5	£6	£7
B	£3	£6	£5	£2	£1
C	£1	£7	£5	£3	£2

Allocate the output of the factories to the distributors at minimum cost.

Shadow Costs

You are probably thinking that the method is tedious. It becomes most time consuming to examine the cost of all the alternative solutions, especially as the dimensions of the transportation matrix increase. Can you imagine trying to solve a ten factory, twenty distributor problem? We want a system that will identify immediately which of the alternative solutions will increase and which will decrease the cost of allocation. Let us examine a general 3×3 matrix. The 'Northwest Corner' solution is rewritten below, but the zero's have been omitted from the unoccupied cells. Each cell is identified in the usual fashion. The quantities A are the amounts allocated, and the quantities Y the allocation costs for each cell.

Dispatch costs
↓

x_{11} y_{11} A_1 $-$	x_{12} y_{12} A_2	x_{13} y_{13} A_3 $+$	y_{11}
x_{21} y_{21} $+$ $y_{23}-y_{13}$ $+y_{11}-y_{21}$	x_{22} y_{22} $y_{23}-y_{13}$ $+y_{12}-y_{22}$	x_{23} y_{23} $-A_4$	$y_{23}-y_{13}$ $+y_{11}$
x_{31} y_{31} $y_{32}-y_{12}$ $+y_{11}-y_{31}$	x_{32} y_{32} A_5	x_{33} y_{33} $y_{32}-y_{12}$ $+y_{13}-y_{33}$	$y_{32}-y_{12}$ $+y_{11}$
reception costs → 0	$y_{12}-y_{11}$	$y_{13}-y_{11}$	

There are 5 possible solutions to this matrix, and with the method used so far it would be necessary to check the cost of each. Suppose we decided to check the solution using cell X_{21}, then the reallocation route would be the one shown. The change in cost resulting from allocating a unit to X_{21} would be

$$Y_{21} + Y_{13} - (Y_{11} + Y_{23})$$
$$= Y_{21} + Y_{13} - Y_{11} - Y_{23}$$

A better way of putting this would be to state that the cost of *not* allocating a unit to X_{21} is

$$Y_{11} + Y_{23} - Y_{13} - Y_{21}$$

We can calculate the cost of not using any unoccupied cell more directly using *shadow costs*. We start by supposing that the allocation cost of each occupied cell is made up from two components — the dispatch cost from each factory and the reception cost of each distributor. We arbitrarily assign a zero reception cost to distributor (1) and enter this at the foot of column (1). (You should remember that column (1) shows distributor (1)'s sources of supply.) Now the allocation cost of cell X_{11} is Y_{11}, and for any occupied cell

Allocation Cost = Dispatch cost + reception cost.

If reception cost is zero, it must follow that the dispatch cost of factory 1 is Y_{11} minus zero, i.e. Y_{11}. This is entered at the end of row (1).

Using a similar reasoning for cells X_{12} and X_{13}, we obtain

Reception cost of distributor (2) is $Y_{12} - Y_{11}$
and
Reception cost of distributor (3) is $Y_{13} - Y_{11}$

If we use cell X_{23}, we can obtain the dispatch cost of factory 2.

i.e. $Y_{23} - (Y_{13} - Y_{11})$
or $Y_{23} - Y_{13} + Y_{11}$

and using cell X_{32} we obtain the dispatch cost of factory 3.

i.e. $Y_{32} - (Y_{12} - Y_{11})$
or $Y_{32} - Y_{12} + Y_{11}$

Now let us examine an unoccupied cell. If we add the reception cost and the dispatch cost, we obtain the *shadow cost* of that cell. For X_{21}

the shadow cost is $Y_{23} - Y_{13} + Y_{11}$, and if we subtract the allocation cost from the shadow cost, we have

$$Y_{23} - Y_{13} + Y_{11} - Y_{21},$$

which is identical to the cost of not using that cell. This process has been done for all the unoccupied cells and entered in the bottom right-hand corner of each cell. You should satisfy yourselves using both methods that each entry is correct.

Summarising, the cost of not using an unoccupied cell is

Reception cost plus dispatch cost minus allocation cost

Now we shall re-examine the 'Northwest Corner' solution to our original problem.

Dispatch costs ↓

x_{11} 3 45	x_{12} 4 30	x_{13} 6 −2	3
x_{21} 1 2	x_{22} 4 5	x_{23} 3 20	3
Reception costs → 0	1	0	

Again we start by giving distributor (1) a zero reception cost, and using cell X_{11} we obtain that factory (1) has a dispatch cost of $3 - 0 = 3$. Using cell X_{12} we see that distributor (2) has a reception cost of $4 - 3 = 1$. Now we use cell X_{22} to find the dispatch cost of factory (2) i.e. $4 - 1 = 3$. Finally, cell X_{23} gives distributor 3 a reception cost of $3 - 3 = 0$. The shadow cost of unoccupied cell X_{21} is $3 + 0 = 3$, and hence the cost of not using cell X_{21}, is $3 - 1 = 2$. Likewise, the shadow cost of X_{13} is $3 + 0 = 3$, and the cost of not using that cell is $3 - 5 = -2$.

What can we conclude? The cost of not using cell X_{21} is £2 — we would reduce total cost by £2 for every unit we allocated to that cell X_{13} is − £2, − we would increase total cost by £2 for every unit we allocated to that cell. *Thus we will allocate to cells where the cost of not using them is positive, and avoid those cells where the cost of not using them is negative. Also we will concern ourselves first with the cell that has the greatest positive cost of remaining unused.* The shadow cost method has the great advantage of finding such costs without the necessity of identifying all the reallocation routes first. In fact we find only one allocation route: the one for the unoccupied cell that reduces cost most. Let us now examine a matrix with greater dimensions

x_{11} 4 90	x_{12} 5 36 −	x_{13} 5	x_{14} 1 + 1	x_{15} 2	4
x_{21} 4 +	x_{22} 4 22	x_{23} 2 46 −	x_{24} 1 0	x_{25} 2	3
x_{31} 5	x_{32} 6	x_{33} 3 27 +	x_{34} 2 38 −	x_{35} 3	4
x_{41} 5 13	x_{42} 1 5	x_{43} 5	x_{44} 4	x_{45} 2 22	5
0	1	−1	−2	−3	

Reception costs and dispatch costs have been calculated in the usual way, only the positive costs of not using the unoccupied cells have been entered, so that we can see at a glance the cells we should concentrate on. It is easy to decide whether such costs will be positive − if shadow cost exceeds allocation costs, then such costs will indeed be positive. The matrix shows that by allocating into either X_{14} or X_{42} we could reduce costs. It also tells us to concentrate first on X_{42}, as this gives the greatest saving per unit allocated. Incidently, the shadow cost method gives the correct cost of not allocating to an empty cell even when the route is not straight forward. The route for reallocating into X_{14} is marked on the matrix, and the change in cost per unit moved is $+1 -2 + 3 - 2 + 4 - 5 = -£1$. Hence the cost of not using this route is £1, which agrees with the result obtained by shadow costs.

This enlarged example clearly illustrates the labour saved using the shadow cost method. The method outlined earlier would have involved finding the 12 other routes, and calculating the cost of using each!

TUTORIAL FOUR
1. Solve the examples in tutorial 1 and tutorial 2 using the shadow cost method.
2. Complete the 4 X 5 matrix introduced in the last section.

Degeneracy

Earlier it was stated that a unique re-allocation route for each unoccupied cell could be found. In fact, this is true only if the number of occupied cells is $n + m - 1$, where n is the number of rows, and m is the number of columns. If the number of occupied cells is less than this, then it may not be possible to find a re-allocation route using occupied cells only. Moreover, it will be impossible to find all the shadow costs. Such matrices are called *degenerate*, and we require a method for dealing with them.

The advantage of the 'Northwest Corner' method is that it avoids degeneracy. Why then, you may ask, should we bother to investigate a method for dealing with it? Well the 'Northwest Corner' method is not a very efficient method of obtaining a first feasible solution — we would be nearer the optimal solution if we allocated to the low allocation cost cells first. But this system of allocation often results in a degenerate matrix. Secondly, it sometimes happens that a non-degenerate matrix goes degenerate after reallocation. Consider the following matrix.

x_{11}	x_{12}
20	30
x_{21}	x_{22}
	20

In this case, $n + m - 1 = 3$, the same as the number of occupied cells. Now suppose it was found to be worthwhile to allocate into cell X_{21}, the resultant matrix would be

x_{11}	x_{12}
	50
x_{21}	x_{22}
20	

which is degenerate.

Let us suppose we were presented with the following feasible solution to a transportation problem. We want to know whether there is a better solution.

x_{11} 1 15 +	x_{12} 2 ϵ 	x_{13} 3 30 −	1
x_{21} 2 2	x_{22} 5 35 −	x_{23} 4 + 2	4
x_{31} 1 20	x_{32} 3	x_{33} 2 1	1
0	1	2	

We begin as usual by assigning a zero reception cost to column 1. This gives dispatch costs of 1 to rows 1 and 3. Using cell X_{13}, we can obtain a reception cost of 2 for column 3. It is not possible to calculate any other costs as there are insufficient occupied cells. The way out of this problem is to treat one of the unoccupied cells as if it were occupied — it doesn't matter which cell we choose, though it is best to choose a cell with a low allocation cost. Can you see why? We allocate a very small quantity epsilon to one of the cells. Epsilon is so small that it does not affect the feasibility of the solution, nor does it affect the total cost. We will assign epsilon (ϵ) to cell X_{12}, treat this cell as occupied, and calculate the remaining costs. Allocating into X_{21} and X_{23} gives the same saving per unit, and X_{23} is arbitrarily chosen.

x_{11} 1 15 −	x_{12} 2 $\epsilon + 30$ +	x_{13} 3 	1
x_{21} 2 + 2	x_{22} 5 5 −	x_{23} 4 30	4
x_{31} 1 20	x_{32} 3	x_{33} 2	1
0	1	0	

x_{11} 1 10	x_{12} 2 $\epsilon + 35$	x_{13} 3 	1
x_{21} 2 5 +	x_{22} 5	x_{23} 4 30 −	2
x_{31} 1 20 −	x_{32} 3	x_{33} 2 + 1	1
0	1	2	

x_{11} 1 10 −	x_{12} 2 $\epsilon + 35$	x_{13} 3 + 0	1
x_{21} 2 25 +	x_{22} 5	x_{23} 4 10 −	2
x_{31} 1	x_{32} 3	x_{33} 2 20	0
0	1	2	

This is an optimal solution, though there is a second solution obtained by allocating to X_{13} (the cost of not using X_{13} is zero). Notice that the second optimal solution is degenerate.

x_{11} 1 ϵ	x_{12} 2 $\epsilon + 35$	x_{13} 3 10	1
x_{21} 2 35	x_{22} 5	x_{23} 4 0	2
x_{31} 1	x_{32} 3	x_{33} 2 20	0
0	1	2	

As an optimal solution can be degenerate, this gives a third reason for mastering the technique of handling degeneracy. In the solutions above, we can drop the epsilon as it is so insignificantly small that it affects neither feasibility nor cost. In practice epsilon is dropped when the matrix ceases to be degenerate.

A Better Method of Obtaining a First Feasible Solution

We noted earlier that we can improve on the Northwest Corner method of finding a first feasible solution. This method involves finding the low cost cells and allocating to them first. This will give a solution nearer to the optimum. Let us examine the following example to see how the method works.

> Output of four factories: 110, 80, 60 and 50 units.
> Demand of five distributors: 140, 70, 40, 30 and 20 units.
> Allocation costs:

	1	2	3	4	5
1	£4	£5	£5	£6	£1
2	£2	£4	£4	£3	£5
3	£3	£2	£4	£5	£5
4	£3	£5	£2	£3	£4

The transportation matrix is presented in the usual form, except for a new first row which gives distributor's demand, and a new first column which gives factory output.

The cell with the lowest allocation cost is X_{15}, and we begin by allocating as much as possible to that cell. Now although factory (1) can supply 110 units, distributor (5) can only take 20 units, hence 20 is the most we can allocate to X_{15}. Now we will allow the first row to indicate how much is to be allocated to each distributor, and the column to indicate the amount of output undistributed. Distributor (5) has his full quota, and factory (1) has $110 - 20 = 90$ units undistributed. We show this information at the beginning of column 5 and row 1. We now examine cells with £2 allocation costs (X_{21}, X_{32}, and X_{43}). We allocate as much as possible to these cells, and adjust the row and column headings accordingly. Now we examine cells with £3 allocation costs. We cannot allocate to X_{24} (factory 2's output has all been allocated) nor X_{31} (factory 3's output has all been allocated). We can allocate to either X_{41} or X_{44}. If we do not allocate to X_{41}, then we must allocate to some other cell in row 1. Likewise, if we do not allocate to X_{44} we must allocate to some other cell in column (4). Now as the cells

	1̶4̶0̶ 6̶0̶	7̶0̶ 1̶0̶	4̶0̶	3̶0̶ 2̶0̶	2̶0̶	
1̶1̶0̶ 9̶0̶ 3̶0̶ 2̶0̶	x_{11} 4 60 +	x_{12} 5 10	x_{13} 5	x_{14} 6 20 −	x_{15} 1 20	4
8̶0̶	x_{21} 2 80 −	x_{22} 4	x_{23} 4	x_{24} 3 + 1	x_{25} 5	2
6̶0̶	x_{31} 3	x_{32} 2 60	x_{33} 4	x_{34} 5	x_{35} 5	1
5̶0̶ 1̶0̶	x_{41} 3	x_{42} 5	x_{43} 2 40	x_{44} 3 10	x_{45} 4	1
	0	1	1	2	−3	

in column (4) have higher allocation costs than column (1), it would seem more logical to allocate to X_{44}. The allocation is now fixed and we have no further choices. The remaining 90 units produced by factory 1 must be allocated as shown. Shadow costs are calculated in the usual fashion, and the only way total cost can be reduced is to allocate into X_{24}. The reallocation route is marked, and allocating 20 units into X_{24} gives:

x_{11} 4 80	x_{12} 5 10	x_{13} 5	x_{14} 6	x_{15} 1 20	4
x_{21} 2 60	x_{22} 4	x_{23} 4	x_{24} 3 20	x_{25} 5	2
x_{31} 3	x_{32} 2 60	x_{33} 4	x_{34} 5	x_{35} 5	1
x_{41} 3	x_{42} 5	x_{43} 2 40	x_{44} 3 10	x_{45} 4	2
0	1	0	1	−3	

This is the optimum allocation.

TUTORIAL FIVE

Here are three questions to practice the technique of transportation.
Find the least – cost allocation in each case.

1.

	30	20	10	5
35	x_{11} 1	x_{12} 1	x_{13} 3	x_{14} 4
15	x_{21} 6	x_{22} 3	x_{23} 4	x_{24} 2
15	x_{31} 3	x_{32} 2	x_{33} 4	x_{34} 5

2.

	50	40	30	20	10
70	x_{11} 1	x_{12} 10	x_{13} 20	x_{14} 5	x_{15} 1
50	x_{21} 20	x_{22} 30	x_{23} 1	x_{24} 40	x_{25} 10
30	x_{31} 40	x_{32} 1	x_{33} 20	x_{34} 10	x_{35} 1

3.

	190	80	55	60	40
220	x_{11} 1	x_{12} 3	x_{13} 2	x_{14} 4	x_{15} 5
100	x_{21} 4	x_{22} 1	x_{23} 2	x_{24} 3	x_{25} 5
65	x_{31} 5	x_{32} 4	x_{33} 1	x_{34} 3	x_{35} 2
40	x_{41} 2	x_{42} 4	x_{43} 5	x_{44} 6	x_{45} 6

Solution of Transportation Matrices When Supply and Demand Are Not Equal

So far, we have considered examples where supply and demand are
matched exactly. This situation is seldom found in practice – shortages
of supply or excess capacity are most common. Consider the case where

a good is a by-product of some other good — it would be almost impossible to match supply to demand. Let us see how we can deal with such
situations.

Output	65,	25,	10
Demand	45,	25,	20

Thus, supply exceeds demand by 10 units. The allocation costs are

£2	£3	£4
£1	£4	£2
£5	£1	£3

We deal with this problem by supposing that the surplus output is
put into store; and that the cost of moving the output into store is zero.
We introduce an extra column to represent the store and assign to it
zero allocation costs. The amount to be allocated to this column is ten
units. Allocating according to the 'least cost cell' first method, and calculating shadow costs in the usual fashion we have:

	~~45~~ ~~20~~	~~25~~ ~~15~~	~~20~~	~~10~~	
~~65~~ ~~45~~ ~~30~~ ~~10~~	x_{11} 2 20 +	x_{12} 3 15	x_{13} 4 20 −	x_{14} 0 10	2
~~25~~	x_{21} 1 25 −	x_{22} 4	x_{23} 2 + 1	x_{24} 0 2	1
~~10~~	x_{31} 5	x_{32} 1 10	x_{33} 3	x_{34} 0	0
	0	1	2	−2	

Suppose we had decided not to store the surplus output, but to
destroy it. We could then have assigned a very high cost X to each cell
of a 'disposal' column — i.e. we would have assumed the high allocation
costs represent the loss of revenue. Would this have made any difference
to our solution? Try it for yourself — what can you conclude?

If demand exceeds supply, then we use exactly the same method, but
this time introducing a false row.

Allocating into cell X_{23} gives

x_{11} 2	x_{12} 3	x_{13} 4	x_{14} 0	
40	15		10	2
x_{21} 1	x_{22} 4	x_{23} 2	x_{24} 0	
5		20		1
x_{31} 5	x_{32} 1	x_{33} 3	x_{34} 0	
	10			0
0	1	1	−2	

— the optimum solution.

A Problem Involving Maximisation

Suppose that rather than calculating the cost of allocating the output of a given number of factories to a given number of distributors, a transport manager calculates what would be the profit earned on all the possible allocations. For example, suppose the output of three factories could be sold to four distributors at the following rates of profit

£2	£3	£4	£6
£1	£4	£3	£2
£5	£1	£3	£2

The output of the factories are 65, 25, and 10 units, and the demand of the distributors are 45, 25, 20 and 10 units. The problem is to allocate the output so as to maximise the profit.

Think back to the Simplex Method — it was used to find solutions to objective functions which are to be maximised. If we wished to minimise, we solved the dual problem. Now the transportation method is completely the reverse; if we wish to minimise we solve the primal problem, and if we wish to maximise we solve its dual. Can you remember that we find the dual by turning the primal problem 'inside out'? How can we turn the primal transportation problem inside out? We select the greatest unit profit (in this case it is £6) and subtract all the other profits from this.

4	3	2	0
5	2	3	4
1	5	3	4

Compare the two matrices carefully — the component which had the greatest value in the profit matrix has the least value in the transformed matrix. Also, the component which had the least value in the profit matrix has the greatest value in the transformed matrix. In fact, the order of magnitude of the components has been completely reversed. Now if we minimise the transportation matrix using the transformed components, surely we will be finding that allocation which *maximised* profit. Setting up this matrix in the usual fashion we have:

	~~45~~ 35	~~25~~	~~20~~	~~10~~	
~~65~~ ~~55~~ ~~35~~	x_{11} 4 35	x_{12} 3	x_{13} 2 20	x_{14} 0 10	4
~~25~~	x_{21} 5 ϵ	x_{22} 2 25	x_{23} 3	x_{24} 4	5
~~10~~	x_{31} 1 10	x_{32} 5	x_{33} 3	x_{34} 4	1
	0	-3	-2	-4	

This is the optimum alloctaion, and to find the total profit earned by this allocation, we must use the actual profit figures, not the transformed components i.e. $2 \times 35 + 4 \times 20 + 6 \times 10$ etc.

'Reduced Costs' Method

Suppose an oil company has its refineries sited at Southhampton, the Thames estuary, Merseyside and Milford Haven. One of the by-products of refining oil is phenol, which is used by five divisions of the company: Plastics Division of Birmingham, Paints Division of Manchester, Fibres Division of Leeds, Adhesives Division at Bristol, and Drugs Division at Nottingham. The company has its own fleet of vehicles, and it has been calculated that the cost of transporting phenol is 1p per unit per mile. Other relevant information is as follows

Phenol Production
Southampton	115,000 units
Thames Estuary	95,000 units
Merseyside	53,000 units
Milford Haven	48,000 units
	311,000 units

Phenol Consumption

Plastics Division	86,000 units
Paints Division	74,000 units
Fibres Division	62,000 units
Adhesives Division	38,000 units
Drugs Division	51,000 units
	311,000 units

Road Distances

	Birmingham	Manchester	Leeds	Bristol	Nottingham
Southampton	128	206	224	75	158
Thames Estuary	110	184	190	116	122
Merseyside	90	35	73	160	97
Milford Haven	167	208	248	146	216

The problem is to allocate phenol to the divisions at minimum cost. It will be convenient to deal in thousand units of phenol. You will realise that the calculation of shadow costs will also be rather cumbersome owing to the nature of the mileages. Obviously, it will be convenient to simplify the mileages as much as possible. This can be done by calculating the minimum distance that the entire output of phenol could move, irrespective of feasibility. The nearest division to Southampton is at Bristol. If the entire Southampton output could be shipped to Bristol, then the cost would be

$$75 \times 115 \times 1000 = £86,250$$

This is the minimum cost of moving the entire output of Southampton. Now if we add £86,250 to the total cost of moving Southampton's phenol, then the mileages from Southampton can be reduced by 75 miles thus

	Birmingham	Manchester	Leeds	Bristol	Nottingham
Southampton	53	131	149	0	83

An arithmetic example may clarify this point. Suppose the entire output of Southampton was equally distributed to the five divisions. The cost using the true mileages would be

To Birmingham	23×128
To Manchester	23×206
To Leeds	23×224
To Bristol	23×75
To Nottingham	23×158

Total cost = 23 × 1000 (128 + 206 + 224 + 75 + 158)
            = 23,000 × 791
            = £181,930

And using the reduced mileages.

To Birmingham   23 × 53
To Manchester   23 × 131
To Leeds   23 × 149
To Bristol   23 × 0
To Nottingham   23 × 83

Total cost = 23 + 1000 (53 + 131 + 149 + 0 + 83)
            = 23,000 × 416
            = £95,680

Add minimum
cost   £86,250

                       £181,930

Reducing the mileages in this way, then, is a legitimate process. It is strictly comparable to the method we used for calculating the arithmetic mean i.e. using a false origin.

The minimum cost of moving the phenol produced on the Thames Estuary is

110 × 95 × 1000 = £104,500,

and the mileages from the Thames Estuary can be reduced by 110 miles if £104,500 is added to total cost. Likewise, the distances from Merseyside can be reduced by 35 miles if

35 × 53 × 1000 = £18,550

is added to total cost, and the distances from Milford Haven can be reduced by 146 miles if

146 × 48 × 1000 = £70,080

is added to total cost.

The minimum cost of moving phenol is

£86,250 + £104,500 + £18,550 + £70,080 = £279,380

You should realise that the calculations above have completely ignored the feasibility of the allocation. If this amount is added to total cost. Then the mileages can be represented like this:

	Birmingham	Manchester	Leeds	Bristol	Nottingham
Southampton	53	131	149	0	83
Thames Estuary	0	74	80	6	12
Merseyside	55	0	38	125	62
Milford Haven	21	62	102	0	70

All the rows in the reduced matrix now contain a zero, as do all the columns except columns three and five. The lowest mileage in the Leeds column is 38, and, as 62,000 units are delivered, the mileages from Leeds can be reduced by 38 if

$$38 \times 62 \times 1000 = £23,560$$

is added to total cost. The mileages from Nottingham can be reduced by 12,

$$12 \times 51 \times 1000 = £6120$$

is added to total cost. The final form of the reduced mileage matrix is

	Birmingham	Manchester	Leeds	Bristol	Nottingham
Southampton	53	131	111	0	71
Thames Estuary	0	74	42	6	0
Merseyside	55	0	0	125	50
Milford Haven	21	62	64	0	58

and if we are to use this matrix, we must add

$$£279,380 + £23,560 + £6120 = £309,060 \quad \text{to total cost.}$$

	86	74	62	38	51	
115	x_{11} 53 / 21 / 15	x_{12} 131 / —	x_{13} 111 / 56 / +	x_{14} 0 / 38	x_{15} 71	68
95	x_{21} 0 / 44	x_{22} 74	x_{23} 42 / 1	x_{24} 6	x_{25} 0 / 51	0
53	x_{31} 55	x_{32} 0 / 53	x_{33} 0	x_{34} 125	x_{35} 50	−63
48	x_{41} 21 / 42 / +	x_{42} 62 / 22	x_{43} 64 / 6 / —	x_{44} 0	x_{45} 58	21
	0	63	43	−68	0	

Reducing the mileage in this way has two uses. Firstly, it considerably aids the calculation of shadow costs, and secondly the zero's help to obtain a more rational first feasible solution i.e. we allocate to cells with zero cost first. Our first feasible solution might look like the matrix on page 125.

x_{11} 53 + 37	x_{12} 131 15 −	x_{13} 111 62	x_{14} 0 38	x_{15} 71 29	90
x_{21} 0 44	x_{22} 74	x_{23} 42	x_{24} 6	x_{25} 0 51	0
x_{31} 55	x_{32} 0 53	x_{33} 0	x_{34} 125	x_{35} 50	−41
x_{41} 21 42 −	x_{42} 62 6 +	x_{43} 64	x_{44} 0	x_{45} 58	21
0	41	21	−90	0	

x_{11} 53 15 +	x_{12} 131	x_{13} 111 62 −	x_{14} 0 38	x_{15} 71	53
x_{21} 0 44	x_{22} 74	x_{23} 42 16	x_{24} 6	x_{25} 0 51	0
x_{31} 55	x_{32} 0 53 −	x_{33} 0 + 19	x_{34} 125	x_{35} 50	−41
x_{41} 21 27 −	x_{42} 62 21 +	x_{43} 64 15	x_{44} 0	x_{45} 58	21
0	41	58	−53	0	

x_{11} 53 42 +	x_{12} 131	x_{13} 111 35 −	x_{14} 0 38	x_{15} 71	53
x_{21} 0 44	x_{22} 74	x_{23} 42 58	x_{24} 6	x_{25} 0 51	0
x_{31} 55	x_{32} 0 26	x_{33} 0 27	x_{34} 125	x_{35} 50	−58
x_{41} 21	x_{42} 62 48	x_{43} 64	x_{44} 0	x_{45} 58	4
0	58	58	−53	0	

x_{11} 53 77	x_{12} 131	x_{13} 111	x_{14} 0 38	x_{15} 71	53
x_{21} 0 9	x_{22} 74	x_{23} 42 35	x_{24} 6	x_{25} 0 51	0
x_{31} 55	x_{32} 0 26	x_{33} 0 27	$x3_4$ 125	x_{35} 50	−42
x_{41} 21	x_{42} 62 48	x_{43} 64	x_{44} 0	x_{45} 58	20
0	42	42	−53	0	

This is the least − cost allocation, and the cost is:

£309,060 + 1000 $[(77 \times 53) + (35 \times 42) + (48 \times 62)]$ = £394,330

TUTORIAL SIX

1. Find the optimum allocation assuming that Drugs Division is closed.
2. The Thames refinery is put out of action by a strike, and it is decided to buy phenol from abroad. Assuming it costs more to import phenol than to produce it, find the optimum allocation.
3. Assume that the allocation costs given in examples (2) and (3) of Tutorial 5 are in fact profits. Find the allocation that maximises profit in each case.

4. Rework Examples (2) and (3) of tutorial 5 using the 'reduced costs' method. (Although it is hardly worth while using this method in these cases, the method will be used extensively later, and it is as well to practice it now!

5. Under what circumstances can the Transportation Method be used to solve linear programming problems? (Hint: carefully examine the restriction equations.)

Other Examples of the Transportation Method

The quarterly production of a certain chemical compound is a constant 50 units, but sales vary seasonally. The estimated sales for a particular year are:

Quarter	1	2	3	4
Sales	45	65	60	30

Although output matches demand, there is a marked difference between quarterly production and quarterly sales. The product costs £100 to produce. If, in a particular quarter there is excess output, then the surplus is put into store at a cost of £5 per unit. If a unit is held in store for two quarters, then the storage cost will be £10, but in addition to this it will need filtering at a cost of £10 per unit. If a unit is held for three quarters, then in addition to the £15 storage cost, it will need distilling at a cost of £25 per unit.

The customers of the firm require that the goods be delivered on time, and impose penalties on the firm for late deliveries. If a unit is delivered one quarter late, the price is reduced by £10. If it is delivered two quarters late, the price is reduced by £15, and the price is reduced by £17 if delivery is delayed by three quarters.

Suppose we let the first suffix of X represent the quarter in which the output was produced, and the second suffix represent the quarter in which it was sold. Thus X_{23} represents any units of output that were produced in the second quarter but sold in the third. The total amount produced in the first quarter will be:

$$X_{11} + X_{12} + X_{13} + X_{14}$$

Now, as output matches demand *for the year as a whole*, it follows that the output of the first quarter must be sold at some time during the year, i.e.

$$X_{11} + X_{12} + X_{13} + X_{14} = 50$$

and using the same reasoning

$$X_{21} + X_{22} + X_{23} + X_{24} = 50$$
$$X_{31} + X_{32} + X_{33} + X_{34} = 50$$
$$X_{41} + X_{42} + X_{43} + X_{44} = 50$$

The output that is used for satisfying demand in the first quarter is

$$X_{11} + X_{21} + X_{31} + X_{41}$$

and as the first quarters' orders must be satisfied at some time during the year,

$$X_{11} + X_{21} + X_{31} + X_{41} = 45$$

Similarly

$$X_{12} + X_{22} + X_{32} + X_{42} = 65$$
$$X_{13} + X_{23} + X_{33} + X_{43} = 60$$
$$X_{14} + X_{24} + X_{34} + X_{44} = 30$$

Having obtained the restriction equations, let us find the objective function. Now if both digits in the suffix are the same then this shows that the output was sold in the same quarter that it was produced. The only costs incurred in this case are the production costs of £100 per unit. Thus part of the objective function is

$$100X_{11} + 100X_{22} + 100X_{33} + 100X_{44}$$

If the second suffix of X is one greater than the first then the output of a particular quarter is stored at a cost of £5 i.e.

$$105X_{12} + 105X_{23} + 105X_{34}$$

and using similar arguments, we can obtain

$$120X_{13} + 120X_{24} \text{ and } 140X_{14}$$

If the second suffix of X is one less than the first, then demand in a particular quarter is satisfied from the next quarters output. This involves a price penalty of £10 per unit or, which is the same thing, costs could be fixed at £110. Thus we have

$$110X_{21} + 110X_{32} + 110X_{43}$$

and again using similar arguments, we can obtain

$$115X_{31} + 115X_{42}$$
$$\text{and } 117X_{41}$$

Solving this problem is the same as

$$\begin{aligned}
\text{Minimise } C = {}& 100X_{11} + 100X_{22} + 100X_{33} + 100X_{44} + 105X_{12} \\
& + 105X_{23} + 105X_{34} + 120X_{13} + 120X_{24} + 140X_{14} \\
& + 110X_{21} + 110X_{32} + 110X_{43} + 115X_{31} + 115X_{42} \\
& + 117X_{41}
\end{aligned}$$

Subject to the above restrictions.

If you managed the last question of Tutorial 6, you will realise that this problem can be solved using the Transportation Technique.

	~~45~~	~~65~~ ~~15~~ 10	~~60~~ 10	~~30~~	
~~50~~ ~~5~~	x_{11} 100 45	x_{12} 105 5	x_{13} 120	X_{14} 140	100
~~50~~	x_{21} 110	x_{22} 100 50	x_{23} 105	x_{24} 120	95
~~50~~	x_{31} 115	x_{32} 110	x_{33} 100 50	x_{34} 105	100
~~50~~ ~~20~~ ~~10~~	x_{41} 117	x_{42} 115 10	x_{43} 110 10	x_{44} 100 30	110
	0	5	0	−10	

This is the least − cost allocation.

TUTORIAL SEVEN

1. Suppose that in the above example the estimated demand for the year was

Quarter	1	2	3	4
Quantity	55	65	59	40

Demand exceeds supply by 19 units. In order to keep the customers satisfied, the buying manager is instructed to buy the extra 19 units from an outside firm. He contacts four firms, who quote the following terms:

When goods can be supplied	Price
1st quarter	£140
2nd quarter	£135
3rd quarter	£130
4th quarter	£125

Find the least cost allocation.

2. A manufacturer wishes to make shirts from a large consignment of cloth. He can offer two types of shirt: semi-fitted and fully-fitted in four colours: red, blue, green and yellow. He has sufficient dye to produce 55 red, 40 blue, 30 green and 20 yellow. His salesman thinks he can sell 85 semi-fitted and 60 fully-fitted shirts of this type. The rate of profit per shirt varies because of differing production costs and differing levels of demand, and the relevant profit figures are:

	Red	Blue	Green	Yellow
Semi-fitted	10p	30p	20p	40p
Fully-fitted	20p	40p	20p	50p

How many shirts of each type should be made to maximise profit?

3. A recruitment officer has five different vacancies to fill, and five men available to fill them. He devised an aptitude test for each vacancy, and administers each test to the applicants. Assessment is on a ten point scale, and a low mark indicates a high aptitude. Thus a score of one would mean the candidate is highly suitable, and a score of ten would indicate that the candidate was most unsuitable. The candidates were assessed as follows

	Vacancy				
	1	2	3	4	5
Candidate A	8	7	8	4	1
B	3	6	2	3	4
C	7	2	1	2	3
D	4	6	5	5	2
E	8	8	5	7	1

The objective of the recruitment officer is to allocate men to the tasks in such a fashion that their effectiveness is maximised.
What can you conclude about the amount that can be allocated to each cell? i.e. the size of the variables X_{11}, X_{12} etc?
Set up the objective function and restrictions, also set up the basic transportation matrix with a first feasible solution. Into how many

cells would you have to allocate epsilon to make the matrix non-degenerate? You can try to solve this problem by the Transportation Method if you wish, but it is quite a task!

The Assignment Method

Solving question 3 of the last tutorial is best achieved by the Assignment Method, which uses a technique extremely similar to the reduced cost matrix technique. We start by copying-out the matrix again, and inserting the row minima.

					Row Minima
8	7	8	4	1	1
3	6	2	3	4	2
7	2	1	2	3	1
4	6	5	5	2	2
8	8	5	7	1	1

Now we subtract the row minima from each element in that row

7	6	7	3	0
1	4	0	1	2
6	1	0	1	2
2	4	3	3	0
7	7	4	6	0

We now examine the matrix, and draw in as few straight lines as possible (horizontally or vertically) to cover all the zeros. In this case, two lines would cover them. Now this is a 5 × 5 matrix, and when we have arrived at the situation where a minimum of 5 lines covers all the zeros, we have an optimum solution. Can you see why? It would imply that each row and each column contains at least one zero. Clearly we do not yet have an optimum solution, and the next step is to find the column minima:

1   1   0   1   0

We now subtract the column minima from the other elements in that row. Again, we draw in the minimum number of lines to cover all the zeros.

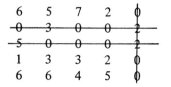

This is still not optimal. Next, we select the smallest uncovered element (in this case it is one) and subtract it from all the uncovered elements. Also, we add the smallest uncovered element to any component that has two lines running through it. The other covered elements remain unchanged.

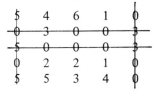

Still this is not optimal. We repeat the last stage until an optimum is achieved.

5	3	5	0	0
1	3	0	0	4
6	0	0	0	4
0	1	1	0	0
5	4	2	3	0

Now we know that we have reached an optimum and all that remains is to extract the allocation. Firstly, we recopy the matrix inserting only the zero components. The position of the zeros indicates the required allocation.

	Vacancy				
	1	2	3	4	5
Candidate A				0	⊗
B			0	⊗	
C		0	⊗	⊗	
D	0			⊗	⊗
E					0

Column (1) contains only one zero, so candidate *D* is allocated to vacancy (1). Now row *D* indicates that candidate *D* could also fill vacancies (4) and (5), but he cannot do this as we have already allocated him to vacancy (1). Hence, we delete the other zeros in row *D* as they are non feasible. Column (2) also contains one zero, so we allocate candidate *C* to vacancy 2, and elimate the two other zeros in row *C*. Column (3). Now column (3) has one zero, so we allocate man *B* to task 3, and delete the other zero in row *B*. Finally, we allocate candidate *A* to vacancy 4 and delete the other zero in row *A*. The position of the zeros now indicate the optimum allocation.

The example worked above is a minimising example. If we wished to maximise, we would subtract each component from the largest component, and thus reverse their order of magnitude. We could then proceed as before. In fact, we are finding the dual in the same way as in the previous model.

TUTORIAL EIGHT

1. Suppose that in the example above, a mark of 10 would represent the greatest degree of aptitude, and a mark of 1 the least. The problem would then be to maximise. Find the optimum under such circumstances.
2. If we use the assignment technique, then we will have an optimum allocation for an $M \times M$ matrix when at least M lines are required in order to cover all the zeros. Let us return to the example worked in the text, and suppose that it was decided to scrap the last two vacancies. However five candidates are still available. What would be the optimum allocation in this case? We now have a 5 × 3 matrix, and the problem is what is the minimum number of lines necessary to obtain an optimum, and indeed how can the allocation take place?

   We solve the problem by introducing two imaginery vacancies, and assuming that each candidate has a zero rating. Now we have a 5 × 5 matrix and we can proceed as before. Notice that there is no point in subtracting the row minima (they are all zero), so proceed directly to the column minima. You can now find the optimum allocation.
3. Now suppose two of the candidates withdraw. Which vacancies should be filled and by whom?

EXERCISES TO CHAPTER FOUR

1. A firm makes a product in five different factories, the weekly output from each being:

Factory	A	B	C	D	E	
	71	26	50	133	76	units

   The output is sold to five distributors, with the following weekly demands.

Distributor	a	b	c	d	e
	110	38	52	130	26

The distribution costs per unit (£) are

Distributor	a	b	c	d	e
Factory	7	5	9	4	17
	12	13	9	10	6
	4	5	6	5	6
	6	6	3	5	8
	4	5	5	7	9

If the objective is to minimise cost, find the objective function and restrictions.

2. Draw up a transportation matrix for example 1, and find the least cost allocation. What is the cost of this allocation?

3. Now suppose that factory *D* must be closed down for 3 weeks for modernisation. How much should each distributor receive during this period, and which factory should supply them? If the profit per unit (ignoring transport costs) is £100 find the total loss of profit during this period.

4. Suppose the figures given in example 1 were profits and not allocation costs. Write down the objective function *to be solved by the transportation method* and the restrictions. Find the most profitable allocation.

5. A vehicle manufacturer has divided production in the body shop to 45% saloons, 45% vans and 10% convertibles. Each vehicle is capable of taking three different engines – 1000 cc., 1250 cc. and 1500 cc. The sales manager supplies suitable selling prices, and the accountant calculates the profit per vehicle to be as follows

	1000 cc.	1250 cc.	1500 cc.
Saloon	£100	£115	£130
Van	£160	£170	£180
Convertible	£ 60	£ 75	£ 90

If 70% of engines produced are 1000 cc., 10% are 1250 cc. and 20% are 1500 cc., decide which engine should be fitted into which vehicle, and find the proportion of output for each variant. What is the total profit per 100 vehicles sold?

6. A national plant-hire firm receives orders for cranes from firms in Preston, Taunton, Northampton, Oxford and York. It has a crane

available in Southampton, Bristol, Birmingham, Nottingham and Shrewsbury. The road distances are

	Preston	Taunton	Northampton	Oxford	York
Southampton	233	87	106	65	236
Bristol	184	43	102	69	214
Birmingham	105	130	50	64	127
Nottingham	100	180	57	94	78
Shrewsbury	82	146	93	104	130

Decide on the allocation of cranes to meet the orders.

7. It is decided to train six astronauts for a flight to Mars. They all go through a general training and then receive aptitude tests for their roles on the mission. The marks awarded in the aptitude test were:

	Astronaut					
Test for	*A*	*B*	*C*	*D*	*E*	*F*
Flight Commander	21	5	21	15	15	28
Pilot	30	11	16	8	16	4
Navigator	28	2	11	16	25	25
Engineer	19	16	17	15	19	3
Back-up man	26	21	22	28	29	24
Communications	3	21	21	11	26	26

Decide upon the composition of the crews.

8. Suppose it was decided to dispense with the back-up man. What would be the composition of the crew?

Chapter Five

Probability Theory

How often during the last week have you used expressions such as 'Probably', 'It is likely', or 'almost certainly'? If you have there is little doubt that you know exactly what you mean, but suppose that someone were to ask you, 'How probable?', 'How likely?', or 'How certain?'. What would you reply?

A common answer would be, 'I am 90% certain'. Would you really mean this, or would you be using this expression merely as a means of indicating that only in exceptional circumstances can you see yourself as being wrong in your assertion? Even this latter raises the secondary question of how probable is the occurrence of exceptional circumstances.

Such vagueness is acceptable in everyday life but if large scale expenditure depends on your decision, and on your degree of assurance, do you think that such vagueness is good enough? Every business decision that is taken involves some degree of risk, but with his livelihood at stake and quite possibly the jobs of many workers dependent on the accuracy of his decision, the businessman must know the odds against him when he follows a particular line of action. This is precisely why the theory of probability is important – it provides a technique of calculating with mathematical precision the probability of a given event occurring or not occurring.

The Probability of an Event

Suppose that you had to trace a letter from a customer which is in a file containing 200 letters. What do you think the odds are against your opening the file at random and there finding the letter you require? The answer should be obvious. You can open the file in 200 different places but only one of them would give you the result you want. The probability of success is then 1 in 200, or $\frac{1}{200}$, or 0.005.

We call the probability of an event occurring $P(E)$ and in this case we would say $P(E) = 0.005$. If an action can have n equally likely results and *r* of these produce a given event we may assume the probability of a single event occurring is

$$\frac{\text{the number of times the desired event can occur}}{\text{the total number of events that can occur}}$$

This concept enables us to give a precise meaning to what we mean by certainty. The only situations in which we could be absolutely certain of choosing the right letter with a random choice would be if the file contained only one letter, (the one we want), or if it contained 200 letters all of which we wanted. In the first case $P(E) = 1$ in 1 and in the second case $P(E) = 200$ in 200. In both cases then $P(E) = 1.0 =$ absolute certainty of the event occurring. The opposite case is that which arises when there is no chance of the event occurring. This we can look at in two ways. Suppose that the letter we want is not in the file. The probability of our finding it is now $\frac{0}{200} = 0.0.$ $P(E) = 0.$

Alternatively we could say that we are absolutely certain that the event will not occur i.e.

$P(\text{non } E) = 1.0$

Consider these two statements carefully. There are only two things that can happen, we find the letter or we do not. This is the one thing of which we can be absolutely certain.

Thus we may come to the conclusion that certainty is the sum of the probability that an event will happen and the probability that it will not happen.

If we let $P(E) = p$ and $P(\text{non } E) = q$
we may say $p + q = 1.0$

Probability of Two Events – 'Either' 'Or' Choices

Let us extend our customer's correspondence. Suppose that we wrote two separate letters to him, both containing the details we wish to know, (possibly his house number). Both are in our file and either will do.

The probability of choosing either one or the other is now increased. The number of occasions on which the desired event will occur is now 2 and so the probability is $\frac{2}{200} = .01.$

This probability is compounded of two events. The probability of choosing the first letter is .005 and the probability of choosing the second letter is also .005.

Either will satisfy us and so we may say that

$$P(E_1 \text{ or } E_2) = P(E_1) + P(E_2)$$

It is important to remember that this will hold good only when E_1 does not depend on E_2 having previously occurred or vice versa. Choosing letter one does not depend on our previously having chosen letter two or vice versa. In probability jargon the two events, since they cannot occur together, are mutually exclusive i.e. if we choose letter 1 we cannot also, on the same choice, choose letter 2.

Probability of Two Events – 'Both' 'And' Choices

Extending our analogy further we may now visualise circumstances in which we require both letters to give us the information we need. What is the probability of our opening the file in two different places, and on each occasion choosing the letter we desire? Common sense will tell you that the chances are very much less than in the previous example. But why? It must be obvious that the first choice involves 200 distinct outcomes, but on the second choice each of these could be combined with any one of the remaining 199 letters. Thus there are 200×199 possible outcomes, only one of which is desirable. Thus the probability would be

$$P(E_1 \text{ and } E_2) = \frac{1}{39,800}$$

We could obtain this by saying,

$$P(E_1) = \frac{1}{200},$$

but having obtained the desired outcome there remain only 199 letters in the file so the probability of successfully choosing the second letter is $\frac{1}{199}$.

Since in this case we require both events to occur,

$$P(E_1 + E_2) = P(E_1) \times P(E_2) = \frac{1}{200} \times \frac{1}{199} = \frac{1}{39,800}$$

But there is more than this to this apparently simple problem. What we have calculated is the probability of both E_1 and E_2 occurring *when E_1 is the first event to occur*. Of course there is a possibility that E_2

might occur first. In other words, there are two ways in which we might choose at random our two letters:

(a) E_1 first, followed by E_2.
(b) E_2 first, followed by E_1.

In this case you can see that the probability of choosing E_2 first followed by E_1, is exactly the same as the probability of E_1, followed by $E_2 \left[\dfrac{1}{39,800} \right]$

Thus we have

(a) $P(E_1 \text{ and } E_2) = \dfrac{1}{39,800}$

(b) $P(E_2 \text{ and } E_1) = \dfrac{1}{39,800}$

Now choices *a* and *b* are 'either, or' choices, they are mutually exclusive, and the probability (you will remember) of either occurring is,

$P(A \text{ or } B) = P(A) + P(B)$

The problem now boils down to this,

$P(E_1 \text{ and } E_2) = P(E_1 E_2) + P(E_2 E_1)$

Thus the probability of our selecting the two required letters is,

$$\frac{1}{39,800} + \frac{1}{39,800} = \frac{1}{19,900}$$

This characteristic of multiple choice can be seen clearly in what we know as a 'Tree diagram'.

Consider the choice of food in a restaurant. The menu offers us a choice of tomato or mushroom soup, followed by chicken, steak or salad, with either ice cream or apple pie as sweet, each choice being mutually exclusive. The situation may be illustrated as follows

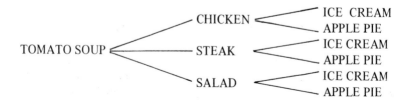

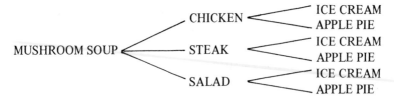

Thus there are 12 distinct meals served and the probability of any one of them (say mushroom soup, steak, and ice cream) being asked for is $\frac{1}{12}$.

The probability of a customer demanding mushroom soup is $\frac{1}{2}$, of his demanding steak $\frac{1}{3}$, and of demanding ice cream $\frac{1}{2}$.

But the probability of his demanding all three is the product of these probabilities, not the sum i.e.

$$P(E_1, E_2 \text{ and } E_3) = \frac{1}{2} \times \frac{1}{3} \times \frac{1}{2} = \frac{1}{12}$$

Permutations and Combinations

The problems considered so far are quite easy to solve using the simple rules given, but often it is more convenient to solve probability problems using combinations. Suppose a recruitment officer was attempting to fill 4 different vacancies in a factory, and 10 applicants were available. Suppose also that after testing the applicants, he found that they were all equally suitable for each vacancy. He decides to make the appointments by drawing 4 names from a hat — how many different allocations are possible?

The first name drawn out of the hat can be any one of the 10 applicants: so there are 10 ways (all different) of drawing the first name. This leaves 9 applicants names in the hat, so there are 9 different ways of selecting the second name. How many ways are there of drawing the first two names? Suppose we assigned the letters A to J to each applicant, and drew two names at random. There would be 9 ways of selecting A first $(AB, AC, AD \ldots \ldots AJ)$, 9 ways of selecting B first $(BA, BC, BD \ldots \ldots BJ)$, 9 ways of selecting C first, and so on. Thus there will be $9 \times 10 = 90$ ways of selecting the first two names.

Using similar reasoning, it should be obvious to you that there are $10 \times 9 \times 8 \times 7 = 5040$ ways of allocating the applicants to the vacancies.

A mathematician would tell you that this problem involves finding the number of permutations size 4 from a group of items size 10, and would write the problem like this,

$$^{10}P_4 = 10 \times 9 \times 8 \times 7 = 5040$$

If we write the problem this way,

$$^{10}P_4 = 10(10 - 1)(10 - 2)(10 - 3) = 5040$$

it is easy to see that the number of permutations size r from a group of items size n is

$$^{n}P_r = n(n - 1)(n - 2)\ldots\ldots(n - r + 1)$$

It was stated in the problem that all the vacancies were different — this is important. One allocation might be:

vacancy	1	2	3	4
applicant	A	B	C	D

here is another:

vacancy	1	2	3	4
applicant	B	A	C	D

Suppose we just consider the men ABC and D. In how many different ways can they be allocated to the four vacancies? Using the permutation formula

$$^4P_4 = 4 \times 3 \times 2 \times 1 = 24$$

Now suppose the tasks were all the same; then the 24 arrangements $ABCD$, $BACD$, $DABC$ etc., would all be the same, and any distinction between them would be meaningless. All we wish to know is that the four men ABC and D are allocated to the tasks, the order in which they were drawn does not matter. The collective name for all permutations which are the same is a combination. Suppose we wished to find the number of combinations of 4 applicants that could be made from a group of 10. If we arranged each combination in as many different ways as possible, we would have the total number of combinations. Hence if $^{10}C_4$ is the number of combinations possible, as we know each combination can be arranged in 24 different ways.

$$24 \ ^{10}C_4 = 5040$$
$$^{10}C_4 = \frac{5040}{24}$$
$$= 210$$

If combinations of 8 items were chosen from a group of 10, then each item could be arranged in

$$8 \times 7 \times 6 \times 5 \times 4 \times 3 \times 2 \times 1 \text{ ways}$$

Mathematicians call this quantity 'factorial 8' and write it like this

$$8! = 8 \times 7 \times 6 \times 5 \times 4 \times 3 \times 2 \times 1$$

Likewise, a group of 6 items could be arranged in

$$6! = 6 \times 5 \times 4 \times 3 \times 2 \times 1 \text{ ways}$$

Now suppose combinations of r items were chosen from a group of n items. Each combination could be arranged in $r!$ ways. Using similar arguments as in the example above, as

$$^nP_r = n(n-1)(n-2)\ldots\ldots(n-r+1)$$
$$^nC_r = \frac{n(n-1)(n-2)\ldots\ldots(n-r+1)}{r!}$$

Problems involving permutations and combinations are called problems of random selection. When considering random selection it is most important to consider which is the appropriate selection to use: permutations or combinations. Most of the problems you will meet in this book involve combinations, i.e. selections in which the order is immaterial.

EXAMPLE
In a fashion competition, it is necessary to list the 10 most attractive dresses in order from a group of 12. How many entries are required to guarantee a correct result? Here order matters, so we use permutations

$$^{12}P_{10} = 12 \times 11 \times 10 \times 9 \times 8 \times 7 \times 6 \times 5 \times 4 \times 3 = 239,500,800$$

EXAMPLE
It is known that 10% of a very large consignment of components is defective. If six components are drawn, what is the probability that the sample contains two defectives and four non-defectives?

Suppose the first two components were defective and the rest non-defective. The probability of this event would be

$$0.1 \times 0.1 \times 0.9 \times 0.9 \times 0.9 \times 0.9 = \frac{9^4}{10^6}$$

(Can you see why the consignment was made very large?) Clearly, there is more than one way of drawing *this sample:* in fact there are . . .

$$^6C_2 = \frac{6 \times 5}{2!} = 15 \text{ ways}$$

Check it by writing out all the ways. Hence, the required probability is

$$\frac{15 \times 9^4}{10^6} = 0.0984$$

TUTORIAL ONE

1. If a coin is tossed what is the probability of the result being (a) a head, (b) a tail, (c) either a head or a tail, (d) both a head and a tail?

2. If on the previous experiment heads occur on 17 consecutive occasions, what is the probability of obtaining a head on the eighteenth toss?

3. If three coins are tossed simultaneously what is the probability that all three will fall alike?

4. In a certain town 25% of girls are natural blonde, 12½% are between the ages of 18 and 25, and 30% speak French. What is the probability that the first girl you meet on your way to work should be,

    (a) a French speaking blonde between the ages of 18 and 25
    (b) a brunette who does not speak French?

5. A lunar landing craft has a completely duplicated system in case the primary system should fail. The probability of failure of the primary communications system is .001 and of the secondary system .005, what is the probability of a lunar landing being aborted due to communications failure?

6. A box contains 10 packets of soap powder 3 of which are Brand X. Two packets are picked out at random and found to be Brand X. What is the probability of this occurring?

7. In a T.V. quiz programme prizes are locked in boxes. There are 10 boxes, three of which contain booby prizes. Once the box is opened it is eliminated. What is the probability that none of four contestants will select a booby prize?

8. A new born baby may be either a boy or a girl. Why does it not follow that probability of the baby being a boy is 0.5?

9. Show that $^nC_r = {}^nC_{n-r}$.

10. At a race meeting, eight horses competed for the major prize. What is the probability that a punter who knows nothing about horse racing will,

(a) pick the winner.

(b) be able to place the first three in order.

(c) be able to forecast the first three, irrespective of order?

11. You wish to win a fortune by picking eight draws on a football coupon, but choose 12 selections and write the instructions:

Perm any 8 from 12 = 495 lines at 1p = £4.95

What is wrong, mathematically, with this statement?

Probability Distributions

Suppose some fairground operator devised a new 'game' of chance. From a certain number of balls in a bag you are asked to pick blindfold two red ones, in which case you will be paid. The picking of a blue ball counts as a loss. After each ball has been picked it is put back in the bag. Your choice, believe it or not, can result in four different combinations of balls.

1. Red   Red
2. Red   Blue
3. Blue   Red
4. Blue   Red

If we call the probability of picking a red ball p, and of picking a blue ball q, the probability of

Red	Red	is p^2
Red	Blue	is pq
Blue	Red	is pq
Blue	Blue	is q^2

Thus our total probability is $p^2 + 2pq + q^2$, telling us the probability of choosing all red balls, one of each colour, and all blue balls.

It should be apparent that $p^2 + 2pq + q^2 = (p + q)^2$ and we may say that the expansion of $(p + q)^2$ gives us the distribution of probabilities over all possible choices in this example.

If the bag contains 5 red and 15 blue balls, the probability of picking a red ball is $\frac{5}{20} = \frac{1}{4}$ and of a blue ball $\frac{3}{4}$. (Can you see why these probabilities remain constant? Remember what happens to each ball when it has been chosen.)

Expanding the function . . . . . $(q + p)^2 = q^2 + 2pq + p^2$ we find that the probability of total failure $= q^2 = \dfrac{9}{16}$, the probability of partial failure $= 2qp = 2 \times \dfrac{1}{4} \times \dfrac{3}{4} = \dfrac{6}{16}$, and the probability of complete success $= p^2 = \dfrac{1}{16}$.

So long as the showman pays odds of 15 to 1 or less, he is bound to win in the long run.

This does not, of course, mean that if the showman has exactly 16 clients in a given period of time he will win. It may well be that every one of those customers will succeed in choosing two red balls, (although it is most unlikely). We mean merely that if the experiment is performed a sufficiently large number of times we would expect only one-sixteenth of those who participate to be successful.

Let us now consider the same situation but having to pick three balls instead of two. We still retain p for the probability of a red ball and q for the probability of a blue ball. Examine the possible results.

Red	Red	Red	Probability $= p^3$
Red	Red	Blue	$p^2 q$
Red	Blue	Red	$p^2 q$
Blue	Red	Red	$p^2 q$
Red	Blue	Blue	$p q^2$
Blue	Red	Blue	$p q^2$
Blue	Blue	Red	$p q^2$
Blue	Blue	Blue	q^3

and our total probability is $(p^3 + 3p^2 q + 3pq^2 + q^3)$ which you ought to recognise as $(p + q)^3$.

Again we will write the function as $(q + p)^3$. Expanding into $q^3 + 3q^2 p + 3qp^2 + p^3$.

We get the probability of 3 blue balls as $\left(\dfrac{3}{4}\right)^3 = \dfrac{27}{64}$.

of 2 blue and one red as $3 \times \dfrac{9}{16} \times \dfrac{1}{4} = \dfrac{27}{64}$

of 2 red and one blue as $3 \times \dfrac{3}{4} \times \dfrac{1}{16} = \dfrac{9}{64}$

of 3 red balls as $\left(\dfrac{1}{4}\right)^3 = \dfrac{1}{64}$

and odds of less than 64 to 1 would pay the stallholder.

Consider now the two functions and their expansions ...

1. $(q + p)^2 = q^2 + 2qp + p^2$
2. $(q + p)^3 = q^3 + 3q^2p + 3qp^2 + p^3$

It should be obvious that the index of each function represents the number of items (balls) chosen. Thus if we had to pick 20 balls the function giving the probability distribution would be $(q + p)^{20}$ and for n items $(q + p)^n$.

This is so similar to the basic expression for the so called Binomial Expansion that the Probability Distribution represented by this type of expansion is called the Binomial Distribution.

Expansions of the form $(q + p)^n$ are easy if n is small, but what if n is large?

Let us consider the expansion of $(q + p)^5$. This represents the probability of various combinations of successes and failures when we select 5 items from a number of items $A + B$ the combinations will be ...

$5A$	$4A + 1B$	$3A + 2B$	$2A + 3B$	$1A + 4B$	$5B$
q^5	q^4p	q^3p^2	q^2p^3	qp^4	p^5

If all 5 items selected are A there is no problem involved, but consider the selection of $4A$ and $1B$. Any one of the items we select may be B, the first, second, third fourth or fifth. Thus there are five ways in which the combination $4A$ and $1B$ may arise each of which has a probability q^4p and since each combination is mutually exclusive the total probability will be $5q^4p$. You will readily see that we can express this term as $^5c_1q^4p$.

Similarly when we consider the combination $3A + 2B$, the total probability of this choice is the probability of $3A + 2B$ i.e. (q^3p^2) multiplied by the number of ways in which this selection can occur in a choice of 5 items, i.e. $^5c_2q^3p^2$.

Knowing this we can write down the expansion of $(q + p)^5$ as follows,

$$(q + p)^5 = q^5 + {}^5C_1q^4p + {}^5C_2q^3p^2 + {}^5C_3q^2p^3 + {}^5C_4qp^4 + p^5$$

$$= q^5 + nq^4p + \frac{n(n-1)}{2!} q^3p^2 + \frac{n(n-1)(n-2)}{3!} q^2p^3$$

$$+ \frac{n(n-1)(n-2)(n-3)}{4!} qp^4 + p^5 = q^5 + 5q^4p + 10q^3p^2$$

$$+ 10q^2p^3 + 5qp^4 + p^5$$

It remains only to know the values of q and p to enable us to state the probability of any given combination of A and B occurring.

We may now state a general form of the Binomial Expansion $(q + p)^n$

$$(q + p)^n = q^n + {}^nC_1q^{n-1}p + {}^nC_2q^{n-2}p^2 + {}^nC_3q^{n-3}p^3$$
$$+ {}^nC_4q^{n-4}p^4 \ldots p^n$$

You will realise of course that this expansion is valid only if n is selected from a very large number of items. Why? Simply because if the number of items is small the probability p will change as we select consecutive items. If for example we select items from a group of 10 containing 4 of the type required the probability of choosing a required item is $\frac{4}{10}$.

But the probability of choosing a second one is $\frac{3}{9}$ and so on. But if we select from a group of 20,000 the resultant probabilities vary so little that they can be taken as being constant.

EXAMPLE

A manufacturer knows that he produces 5% defectives in the manufacture of his commodity. He sells the commodity in boxes of 20 and is satisfied if each box contains 2 defectives or less. What is the probability of a customer buying a box with more than 2 defective items?

Here $q = .95$ and $p = .05$
$p(\text{defects} \geqslant 3) \quad = 1 - p(\text{defects} \leqslant 2)$
Probability of 2 defects or less $= p(0 \text{ defects}) + p(1 \text{ defect})$
$$+ p(2 \text{ defects})$$

Thus with a binomial expression $(.95 + .05)^{20}$ the probability of two defects or less is the sum of the first three terms of the expression

$p(\text{defects} \leqslant 2) = .95^{20} + {}^{20}c_1 (.95)^{19} (.05) + {}^{20}c_2 (.95)^{18} (.05)^2$

$= .95^{20} + 20 (.95)^{19} (.05) + 190 (.95)^{18} (.05)^2$

$= .36 + .38 + .18 = .92$

It follows that the probability of a customer not having three or more defects is $(1 - .92) = .08$.

We may interpret this as meaning that if three defects represent a dissatisfied customer it is probable that 8% of customers will be dissatisfied.

It will be useful to graph the binomial distribution. Remember what the distribution gives you — the probability of obtaining a given number of 'successes' when you pick a given number of items from a larger number.

Let us take the case when $p = 0.1$ and we are choosing 3 items. We know that the probability of

3 successes is $p^3 = .001$
2 successes is $3p^2q = 3 \times .01 \times .9 = .027$
1 success is $3pq^2 = 3 \times .1 \times .81 = .243$
0 successes is $q^3 = .729$

This would be graphed as in diagram 1.

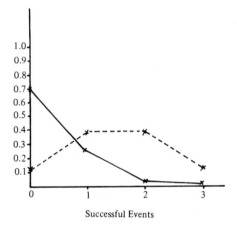

Successful Events

A rather special case of this distribution is when $p = 0.5$. This is also graphed. We know

$p(3 \text{ successes}) = p^3 = (.5)^3 = .125$
$p(2 \text{ successes}) = 3p^2q = 3 \times (.5)^2 \times .5 = .375$
$p(1 \text{ success}) = 3pq^2 = 3 \times .5 \times (.5)^2 = .375$
$p(0 \text{ successes}) = q^3 = (.5)^3 = .125$

You will notice that this curve is symmetrical — a special case we will take up later.

A weakness of the Binomial Distribution is that we are required to know the probability of an event occurring, and this we can know only if we know the possible total events. Thus if we record the average number of gusts of wind of over 40 m.p.h. per day we cannot obtain a probability for such a gust since we cannot ascertain the numbers of occasions on which the wind did not reach 40 m.p.h. Certainly in such a case, n is very large, and p small.

Moreover, even if we knew the value of n and p, you will readily appreciate that calculation becomes cumbersome. Imagine, for example having to expand an expression such as $(.99 + .01)^{4000}$! Fortunately, however, we do not have to perform such calculations. We shall see in the next section that when n is large and p is small the expansion of a binomial expression with a given arithmetic mean, gives us a distribution which approximates very closely indeed to that given by the Poisson distribution with the same arithmetic mean: and this latter is not only easier to calculate, but can also be expanded without our having to ascertain either n or p.

Since most statistical data are such that neither n nor p can be determined beyond the fact that n is very large a p small, the Poisson distribution will probably be used far more than the binomial.

TUTORIAL TWO

1.   Evaluate $^{10}c_2 \cdot {}^{25}c_{23}$. What is the value of n if $^{n}c_2 = 10$?

2.   In a class test you are required to answer 10 questions. The only answers you can give are 'yes' and 'no'. What is the probability that you will get the answer to question 1 right by guessing?
If the pass mark is 5 out of 10 what is the probability of passing the exam by guessing?

3.   A manufacturer guarantees that he will repair, free of charge any of his products that prove to be defective within one year of purchase. He estimates that at the worst 5% of his output will be defective. What is the probability that in a batch of 20 he will be required to repair free of charge 3 or more items?
If the average cost of repair is £30 per unit and the accountant estimates that not more than £210 a week should be spent on this guarantee, what is the probability of the firm overspending if 20 units a week are sold.

4.   A sales manager will accept delivery of an order if a random sample of 10 items contains no defective articles. If in fact 3% of total production is defective what is the probability that he will accept delivery?

5.   Graph the probability distribution derived from the data in question 3 when the size of the sample is

   (a)   5 items
   (b)   10 items
   (c)   20 items

6.   Graph the probability distribution in question 3, when batches of 10 are taken as a sample but when the probability of defects is

   (a)   .1
   (b)   .3
   (c)   .5

7. From your results in questions 4 and 5 what can you deduce about the effect on the probability distribution of varying p and n?
8. Interpret the expression $1000(.9 + .1)^{10}$.

The Poisson Distribution

If you were to be told that over the last 10 years an average of 20 flashes of lightning a month had been seen by people standing on top of Snowdon. Could you caclulate the probability of, say 12 flashes being seen next month? Of course not — not using the Binomial expansion anyway. You could not assess either p or q simply because you do not know the number of occasions on which lightning did not flash.

The Poisson distribution is a distribution enabling us to assess probability when we know only the recorded average number of occurrences. But first we must look at one or two facts about probability distributions. You will remember that when you studied frequency distributions, one of the important exercises you undertook was to calculate the arithmetic mean $\frac{\Sigma fx}{\Sigma f}$. Now the Binomial distribution may be regarded as a type of frequency distribution. X is the number of times a desired event occurs, and f the probability of that result occurring. The justification for this is that if we perform the experiment sufficiently often the relative frequency of the results achieved will approach closer and closer to the proportions indicated by the probability. Thus as $n \to \infty$, $p \to$ relative frequency, and the arithmetic mean of the binomial distribution becomes $\frac{\Sigma px}{\Sigma p}$.

Suppose we expand the expression $(.8 + .2)^4$ we get

$$.8^4 + 4(.8^3)(.2) + 6(.8^2)(.2^2) + 4(.8)(.2^3) + .2^4$$
$$= .4096 + .4096 + .1536 + .0256 + .0016$$

This gives the following frequency distribution

x No. of times the event happens	f Probability	fx
0	.4096	.0
1	.4096	.4096
2	.1536	.3072
3	.0256	.0768
4	.0016	.0064
	1.0000	.8000

$$\bar{x} = \frac{\Sigma fx}{\Sigma f} = \frac{.8}{1.0} = .8$$

This means the average number of times the desired event occurs when 4 items are selected is 0.8, or better, when 40 items are selected is 8.

You will see that $\bar{x}$ in this case is $n \times p$ i.e. $4 \times .2$.

If we now do the same calculation for the expression $(.9 + .1)^{20}$ we find that $\Sigma fx = 2.0$, $\Sigma f = 1.0000$, and so $\bar{x} = 2.0$ again $n \times p$ ($20 \times .1$).

This is in fact a characteristic of the binomial distribution. The average number of times the desired event will occur is always np.

Now let us turn our attention to a very useful algebraic constant that we will use frequently – 'e'. You may have heard of logarithms to base e. This is the same constant. The value of e is given as

$$e = 1 + \frac{1}{1!} + \frac{1}{2!} + \frac{1}{3!} + \frac{1}{4!} + \frac{1}{5!} + \ldots \ldots \frac{1}{n!}$$

The value of 'e' can be found with any degree of accuracy by taking additional terms. It is calculated as follows

$$1 \quad = 1.00000$$

$$\frac{1}{1!} = 1.00000$$

$$\frac{1}{2!} = .50000 \quad \text{(Dividing 1 by 2)}$$

$$\frac{1}{3!} = .16667 \quad \text{(Dividing 0.5 by 3)}$$

$$\frac{1}{4!} = .04167 \quad \text{(Dividing .16667 by 4)}$$

$$\frac{1}{5!} = .00833$$

$$\overline{ 2.71667 }$$

If we take the series far enough we will find that the value of e correct to 4 places of decimals is 2.7183.

This series is the exponential series and the exponential theorem states

$$e^x = 1 + \frac{x}{1!} + \frac{x^2}{2!} + \frac{x^3}{3!} + \ldots \ldots \frac{x^n}{n!}$$

where x is an index.

Thus

$$e^{2a} = 1 + \frac{2a}{1!} + \frac{(2a)^2}{2!} + \frac{(2a)^3}{3!} + \ldots \ldots \frac{(2a)^n}{n!}$$

Any book of tables will give you values of e^x and e^{-x} so you need not worry about having to work the value out.

In considering the Poisson distribution it is useful to consider another meaning of e^x.

The expression $1 + \dfrac{x}{1!} + \dfrac{x^2}{2!} + \dfrac{x^3}{3!} \ldots\ldots\ldots$ is the expansion of $\left(1 + \dfrac{x}{n}\right)^n$ Try it!

$$\left(1 + \frac{x}{n}\right)^n = 1 + \frac{nx}{n} + \frac{n(n-1)}{2!}\left(\frac{x}{n}\right)^2 + \frac{n(n-1)(n-2)}{3!}\left(\frac{x}{n}\right)^3 + \ldots$$

$$= 1 + x + \frac{\left(1 - \dfrac{1}{n}\right)x^2}{2!} + \frac{\left(1 - \dfrac{3}{n} + \dfrac{2}{n}\right)x^3}{3!} + \ldots\ldots$$

But as n approaches infinity the terms containing n become infinitely small and the expression becomes

$$1 + \frac{x}{1!} + \frac{x^2}{2!} + \frac{x^3}{3!} + \frac{x^4}{4!} + \ldots\ldots$$

So $e^x = \left(1 + \dfrac{x}{n}\right)^n$ when n is very large.

Similarly $e^{-x} = \left(1 - \dfrac{x}{n}\right)^n$

Finally before turning at last to the Poisson distribution let us examine an amazing characteristic of the binomial distribution when n is very large. Providing that we keep the arithmetic mean (np) constant, when n is large, to increase it gives us exactly the same terms correct to several decimal places. And the larger we make n the greater the number of decimal places to which the series is the same.

As an example let us expand the two series.

$(.999 + .001)^{1000}$ ($np = 1.0$) and $(.9999 + .0001)^{10,000}$ ($np = 1.0$)

$(.999 + .001)^{1000} = .368 + .368 + .184 + .061 + .015 + .003 + .001 + \ldots\ldots$

$(.9999 + .0001)^{10,000} = .368 + .368 + .184 + .061 + .015 + .003 + .001 + \ldots\ldots$

This result was discovered by Poisson and gives rise to the Poisson distribution which is a limiting case of the binomial distribution when n is very large and p small. Even if we do not know the exact value of n it makes no difference since so long as it is large variations will not affect the distribution (so long as np is held constant). If p exceeds about 0.1 the Poisson distribution ceases to be a satisfactory approximation to the Binomial.

To derive the Poisson Distribution let us expand the Binomial expression $(q + p)^n$. Since $np = a$, $p = \dfrac{a}{n}$ and $(q + p)^n = \left(q + \dfrac{a}{n}\right)^n$

$$\left(q + \frac{a}{n}\right)^n = q^n + {}^nc_1 \, q^{n-1}\frac{a}{n} + {}^nc_2 \, q^{n-2}\left(\frac{a}{n}\right)^2 + {}^nc_3 \, q^{n-3}\left(\frac{a}{n}\right)^3$$

$$+ \ldots = q^n + \frac{nq^{n-1}a}{n} + \frac{n(n-1)q^{n-2}a^2}{2! \, n^2} + \frac{n(n-1)(n-2)q^{n-3}a^3}{3! \, n^3}$$

$$+ \ldots$$

We have already discovered that as n approaches infinity all terms in n tend to become unity and the expression becomes

$$q^n + q^{n-1}a + \frac{q^{n-2}a^2}{2!} + \frac{q^{n-3}a^3}{3!} + \ldots$$

$$= q^n \left(1 + \frac{a}{q} + \frac{a^2}{2! \, q^2} + \frac{a^3}{3! \, q^3} + \ldots \right)$$

$$= (1 - p)^n \left(1 + \frac{a}{q} + \frac{a^2}{2! \, q} + \frac{a^3}{3! \, q} + \ldots \right)$$

As p approaches zero q tends to equal 1 and the series approximates to

$$(1 - p)^n \left(1 + a + \frac{a^2}{2!} + \frac{a^3}{3!} + \ldots \right)$$

Now $(1 - p)^n = \left(1 - \dfrac{a}{n}\right)^n = e^{-a}$

So the Poisson distribution becomes

$$e^{-a}\left(1 + \frac{a}{1!} + \frac{a^2}{2!} + \frac{a^3}{3!} + \frac{a^4}{4!} + \ldots \ldots \frac{a^n}{n!}\right)$$

You will readily see that this series approximates to the Binomial Expansion only if n is very large　otherwise $\dfrac{n(n-1)(n-2)}{3!}$ (etc.) $\neq 1$ and if p is very small [otherwise q, q^2, q^3, etc. do not tend to equal 1]. In other words, the larger is n, and the smaller is p, the more closely do the Binomial and the Poisson Distributions coincide.

Let us see how closely the Poisson Distribution approximates to the Binomial by considering a specific example, namely a distribution with an arithmetic mean of 0.5.

Poisson Distribution

$$e^{-0.5}\left(1 + \frac{.5}{1!} + \frac{(.5)^2}{2!} + \frac{(.5)^3}{3!} + \frac{(.5)^4}{4!} + \dots\right)$$

$$= .6065 + .304 + .076 + .013 + .002 + \dots .$$

Binomial Distribution

$$(.99 + .01)^{50} = (.99)^{50} + {}^{50}c_1 (.99^{49})(.01) + {}^{50}c_2 (.99^{48})(.01^2)$$
$$+ {}^{50}c_3 (.99^{47})(.01^3) + {}^{50}c_4 (.99^{46})(.01^4) + \dots .$$

$$= (99)^{50} + 50(.99)^{49} (.01) + \frac{50.49}{2!} (.99^{48})(.01^2) + \frac{50.49.48}{3!}$$

$$(.99^{47})(.01^3) + \frac{50.49.48.47}{4!} (.99^{46})(.01^4)$$

$$= .606 + .305 + .076 + .012 + .001$$

Thus even with n as small as 50, there is a remarkably good approximation of the Poisson to the Binomial Distribution. If you were to expand $(.999 + .001)^{500}$ the two series would be identical.

If you compare the series you will readily agree that the simplicity of calculation of the Poisson distribution more than makes up for any approximation of the magnitude that is likely to occur – so long as n is not small.

EXAMPLE

A telephone exchange receives an average of 5 calls in each 3 minute period. It is capable of handling 6 incoming calls at any one time. What is the probability that the switchboard will not be able to handle all incoming calls? (Assume all calls last for three minutes.) Using the Poisson distribution we may say that the probability of

0 incoming calls is e^{-5}			=	.00674
1 ,,	,, ,,	$e^{-5} \times 5$	=	.03370
2 ,,	,, ,,	$e^{-5} \times \frac{(5)^2}{2!}$	=	.08425
3 ,,	,, ,,	$e^{-5} \times \frac{(5)^3}{3!}$	=	.14042

$$4 \quad , , \quad , , \; , , \; e^{-5} \times \frac{(5)^4}{4!} \quad = \quad .17552$$

$$5 \quad , , \quad , , \; , , \; e^{-5} \times \frac{(5)^5}{5!} \quad = \quad .17552$$

$$6 \quad , , \quad , , \; , , \; e^{-6} \times \frac{(5)^6}{6!} \quad = \quad .14627$$

$$\text{Probability of 6 incoming calls} \quad = \quad \overline{.76242} \\ \text{of fewer} \ldots$$

The probability of more than six incoming calls being received in any one period is $(1 - .762) = 0.238$. There is a 23.8% chance of the switchboard being overloaded and the caller failing to make a connection

TUTORIAL THREE

1. In calculating the arithmetic mean of a probability distribution you will find that Σf is always equal to 1.0. Why should this be so?
2. In a particular house on average fuses blow 4 times a year. The owner wishes to have on hand sufficient fuses to ensure that he can replace any number that may blow. How many must he buy to ensure that the probability of his running short of fuses is less than .01?
3. During the peak period a firm receives an average of 120 calls an hour. The maximum number of connections it can make is 4 per minute. What is the probability that the board will be overtaxed during any given minute? Is the probability of its being overloaded at some time during the hour 60 times this figure?
4. The absentee rate in a typing pool is shown in the following table.

No. of absentees	0	1	2	3	4	5
No. of days	1	9	7	5	3	0

   If more than 5 typists are absent a girl must be hired from an appointments bureau. What is the probability that such an appointment would have to be made? Is it justifiable, in the light of your result, to suggest the appointment of an additional typist?
5. How could you use the poisson distribution to test that a particular phenomenon is randomly distributed?

   A manufacturer takes samples of 200 components produced by a particular process. He finds that these samples contain defective items as follows.

No. of defectives:	0	1	2	3	4	5
No. of samples:	109	65	22	3	1	0

Do you think that such a distribution of defective items is to be expected?

6. A shopkeeper sells, on average, three refrigerators per week. He obtains deliveries from the manufacturer each Monday. How many should he order to ensure that he has less than 3% chance of running out of stock during the week?

EXERCISES TO CHAPTER FIVE

1. Excluding the final letter indicating the year of registration, car registration numbers consist of three letters followed by three numbers. What is the probability that a car coming off the production line at Dagenham will have the registration PTB 739?

2. A library shelf contains 12 books written by Hammond Innes and 30 books written by Agatha Christie. What is the probability that a reader picking books at random will select one book by the former and one by the latter writer?

3. A retailer sells electric light bulbs in cartons of ten, and always selects three bulbs at random to check that the carton is not damaged. If a particular carton contains three defective bulbs, in how many ways can a sample be selected which includes at least one defective? In how many ways can the sample be selected so as to exclude defective bulbs?

4. If, in example three, the sample consisted of only two bulbs, what would be the probability that both were (a) defective, (b) non-defective?

5. If the probability of a child being a male is ½, what is the probability that in a family of 6 there will be not less than 2 and not more than 4 girls?

6. It is known that a salesman can make a sale on average on 2 out of every five visits he makes. What is the probability that on any particular day when he makes five calls he will do better than average?

7. It is known that in a particularly competitive industry there is a 25% chance of senior executives dying of a heart attack before they reach the age of 60. What is the probability that in a firm with six such executives 3 or more will have a heart attack before reaching this age?

8. In a bombing attack it is known that there is a 50% chance that any one bomb will hit the target. If it takes 3 direct hits to destroy the target and 10 bombs are dropped, is there an 80% chance of destroying the target?

9. A manufacturer finds that his output contains 2% defectives. What is the probability that any particular sample of 100 items will contain 6 or more defectives?

He now modifies his production techniques and finds that the first two batches he checks contain 6 defectives each. Would you say that the modification has adversely affected his output?

10. At a particular airport it is observed that over a period of time an average of three aircraft take off in any interval of 10 minutes. What is the probability that no aircraft will take off in the next ten minutes? What is the probability that 2 will take off? It it is desired to make repairs to the runway that will take 30 minutes to complete, what is the probability that it can be done with no interruption to scheduled take offs? (e^{-9} = .00012).

11. The average number of new strikes in heavy industry in Britain has been shown by Kendall to be about 0.9 per week for the period 1948–59.

What is the probability that during any one week (a) there will be no new strikes and (b) there will be 4 new strikes or more?

12. A doctor finds that he can cope adequately with three patients an hour in normal circumstances. If over a period of time he finds that his flow of patients is as follows, what is the probability of his being unable to cope satisfactorily during any period of one hour.

No. of patients per hour:	0	1	2	3	4	5	6
No. of occasions:	8	39	55	71	26	7	4

Chapter Six

Sampling Theory

Without doubt, we are entering the age of the numerate manager. Vast armies of clerks are engaged in collecting information, and computers work 24 hours a day to process it. Is such information truly representative or can it give a misleading picture? We can all quote examples of numerical information that is grossly misleading – especially from the field of advertising. Who really believes that eight out of ten people use brand X regularly? Who can seriously consider 'improved performance' graphs when they do not include scales? However, such examples are not really serious from our point of view – they are intended to be misleading! What is worrying is that many managers make policy decisions that have profound effects on society, and the decisions are made on the basis of faulty information! How do such serious blunders occur?

In many cases, decisions are based on incomplete information. If it is required to know, say, the number of defective items produced by a particular machine on a particular day, it is most unlikely that the entire output will be examined. What is far more likely is that the inspector will examine a portion of the output, and use the results from this test to predict the overall defective rate. It will be useful here to use the statistician's jargon: all the items which are under consideration go under the collective name *population*, and the items which are examined are called a *sample*. Our definition of 'population' is wider than the usually recognised meaning, as we can have a population of anything we wish, e.g. a population of rod diameters, a population of lengths of lives of lamps, a population of times of telephone calls. What you should notice is that although the members of a population can show variation, they do have a basic common denominator. It would be reasonable, for example to include an 8 second and an 8 minute phone call in the same population, but it would not be reasonable to include an 8 minute phone call in the same population as an 8 stone woman!

Why are samples examined and not the entire population? It may be too costly to examine the entire output: industry abounds with examples of machines being able to produce goods faster than inspectors can examine them. Again, 100% inspection may mean the destruction of the entire output — the only way to test artillery shells is to fire them! So there are good reasons for taking decisions on the basis of sample evidence. However, before we take such decisions we should know what happens when we take samples, and should be able to assess the degree of reliability of sample evidence.

The Normal Distribution

Firstly, we must extend our knowledge of probability. We examined the Binomial Distribution, and found that the probability of obtaining 0, 1, 2, ... 5 heads if 5 coins are cast together is given by the expansion of $(\frac{1}{2} + \frac{1}{2})^5$

i.e. $(\frac{1}{2})^5 + {}^5C_1(\frac{1}{2})^4(\frac{1}{2}) + {}^5C_2(\frac{1}{2})^3(\frac{1}{2})^2 + {}^5C_3(\frac{1}{2})^2(\frac{1}{2})^3 + {}^5C_4(\frac{1}{2})(\frac{1}{2})^4$
$+ (\frac{1}{2})^5 = \frac{1}{32} + \frac{5}{32} + \frac{10}{32} + \frac{10}{32} + \frac{5}{32} + \frac{1}{32}$

Now if we were to draw the histogram of the distribution, it would look like this:

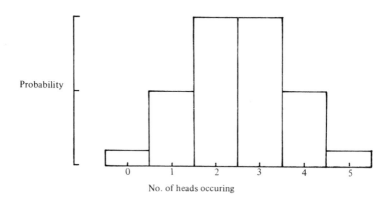

No. of heads occuring

The histogram has been drawn according to the rules outlined in the introductory section. Two important features of the histogram should be carefully noted: the area under the histogram is unity — the same as the total frequency. Also the area of each individual rectangle is the same as the probability that the event it encloses will occur. The area enclosing 'two heads occurring', for example, is $1 \times \frac{10}{32} = \frac{10}{32}$. This means that, given the histogram, we could find the probabilities from the areas. If we wanted to find the probability of 4 heads occurring, we would find the area under the histogram between $x = 3.5$ and $x = 4.5$. If we wanted to find the probability of less than 2 heads occurring, then we would find the area under the histogram where $x < 1.5$. You will soon realise how useful this method is.

Let us now consider the Binomial Distribution $(\frac{1}{2} + \frac{1}{2})^n$ given that n is very large. If we attempted to draw the histogram of this distribution, there would be a large number of rectangles. If we allowed n to be infinitely large, then the rectangles would be of negligible width and the distribution would look like this:

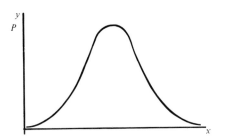

This is the so-called Normal Distribution. It differs from the Binomial and Poisson Distributions considered so far in that it is a continuous distribution. The variate x can have any value in the Normal Distribution, but in the Binomial and Poisson Distributions x must be an integer e.g. number of successes or number of defectives.

If we wished to find the height of any rectangle for the histogram of the Binomial Distribution, we use the appropriate term from the expansion. If, for example, we wished to find the height of the rectangle enclosing the third term of $(q + p)^n$, we would use $^nC_2 q^{n-2} p^2$. If we wished to find the height of the Normal Curve at a particular point x, we would have to use the rather fearsome expression:

$$y = \frac{1}{\sigma\sqrt{2\pi}} \, e^{-\frac{(x - \bar{x})^2}{2\sigma^2}}$$

where $\bar{x}$ is the arithmetic mean and σ the standard deviation of the distribution. Can we simplify this equation? Instead of considering the individual items x in the Normal Distribution, we could instead consider deviations from the arithmetic mean; that is, instead of considering x_1, $x_2, x_3, x_4, \ldots \ldots x_n$, we could consider $(x_1 - \bar{x}), (x_2 - \bar{x}), (x_3 - \bar{x})$ $\ldots \ldots (x_n - \bar{x})$ clearly, this will affect the mean of the distribution. We already know that $\Sigma(x - \bar{x}) = 0$. Hence we have transformed the distribution to have a zero mean.

Now suppose we divide the deviations from the mean by the standard deviation. The distribution now looks like this

$$\frac{x_1 - \bar{x}}{\sigma}, \frac{x_2 - \bar{x}}{\sigma}, \frac{x_3 - \bar{x}}{\sigma}, \ldots \frac{x_n - \bar{x}}{\sigma}$$

Let us calculate the variance of this new distribution. You will remember that the variance is given by

$$\frac{1}{n}\left[\Sigma x^2 - \frac{(\Sigma x)^2}{n}\right]$$

We will consider the bracketed part of the expression first: it tells us to find the sum of the squares of the items, i.e.

$$\left(\frac{x_1 - \bar{x}}{\sigma}\right)^2 + \left(\frac{x_2 - \bar{x}}{\sigma}\right)^2 + \left(\frac{x_3 - \bar{x}}{\sigma}\right)^2 + \ldots + \left(\frac{x_n - \bar{x}}{\sigma}\right)^2$$

$$= \frac{\Sigma(x - \bar{x})^2}{\sigma^2}.$$

From this we must subtract (the sum of the items)^2 divided by the number of items. But we already know that $\Sigma(x - \bar{x})$ i.e. the sum of the items is zero, so the variance of the new distribution is

$$\frac{\Sigma(x - \bar{x})^2}{n\sigma^2}$$

However, the expression $\dfrac{\Sigma(x - \bar{x})^2}{n}$ is σ^2, the variance of the original distribution, so the variance of the new distribution is

$$\frac{\sigma^2}{\sigma^2} = 1$$

and hence the standard deviation is also one. Thus we have transformed the Normal Distribution to have a zero mean and standard deviation of one. Writing $\bar{x} = 0$, and $\sigma = 1$ in the equation for the Normal Distribution we have:

$$y = \frac{1}{\sqrt{2\pi}} e^{-\frac{x^2}{2}}$$

Using logarithms, we can confirm that

$$\frac{1}{\sqrt{2\pi}} = 0.3989$$

and hence the equation becomes

$$y = 0.3989e^{-\frac{x^2}{2}}$$

Normal Distributions are usually described by their mean and standard deviation. The Normal Distribution that has a mean $\bar{x}$ and standard deviation σ is written for convenience like this: $N(\bar{x}, \sigma)$, and so the transformed Normal Distribution we have been considering would be written $N(0,1)$. Now the distribution $N(0,1)$ is called the *Standard Normal Distribution*, and it has special significance. We shall now graph this distribution. Using the equation obtained above, we can calculate the values of y when x has varying values.

x	$0.3989e^{-\frac{x^2}{2}}$
−3.5	0.0009
−2.5	0.0175
−1.5	0.1295
−0.5	0.3521
0	0.3989
0.5	0.3521
1.5	0.1295
2.5	0.0175
3.5	0.0009

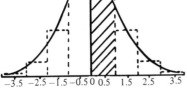

Of course had we included more values for x than this, then we would have obtained a more accurate sketch of the curve. However, the sketch is accurate enough to notice an important feature: that $N(0\ 1)$ is symmetrical about the y axis. The second thing to notice is that as the Normal Distribution is a limit of the Binomial Distribution, we would expect the area under this curve to be unity. Now the proof of this is beyond the scope of this book — as indeed is the derivation of the equation of this curve. However, we can use the areas of the rectangles superimposed on the curve as an approximation to the area under the curve. The rectangle drawn between $x = 0$ and $x = 1$ has a height the same as that of the curve at $x = 0.5$. The area of this rectangle is $1 \times 0.3521 = 0.3521$ sq. units. The other rectangles have been drawn in a similar fashion, and the area of the rectangles drawn between the points where $x = -4$ and $x = 4$ is

0.0009 + 0.0175 + 0.1295 + 0.3521 + 0.3521 + 0.1295 + 0.0175

+ 0.0009 = 1

Now although this method gives the expected result, we must remember that it is only an approximation. We have assumed that the area between $x = 0$ and $x = 1$ is the same as the rectangle drawn between these points, i.e. 0.3521 sq. units. In fact the true (shaded) area is 0.3413.

Earlier, we stated that areas can be used to solve probability problems. Suppose we chose an item V at random from the distribution $N(0,1)$ and we wanted to know the probability that V was less than 2. This probability would be the same as the area under the Normal curve where $x < 2$ i.e.

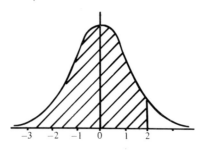

Mathematicians have calculated areas under the Normal curve, and the results are tabulated in the appendix, so all that we need do is to read-off the value we require from this table. The table gives the area to the left of the positive values of x for the Standard Normal Distribution. Thus, if an item V is chosen at random from $N(0,1)$, then reading from the table we obtain (say)

$P(V < 1) = 0.8413$
$P(V < 2) = 0.97725$
$P(V < 3) = 0.99865$
$P(V < 4) = 0.99997$

Can you see that

$$P(1 < V < 2) = 0.97725 - 0.8413 = 0.13595?$$

If an item V is chosen at random from $N(0,1)$ then the tables give

$$P(V < x) \text{ if } x \geqslant 0$$

and

$$P(V > x) = 1 - P(V < x), \text{ if } x \geqslant 0$$

also

$$P(x_1 < V < x_2) = P(V < x_2) - P(V < x_1) \text{ if } x_2 > x_1, \text{ and } x_1 \geqslant 0$$

The tables give the area to the left of the positive values of x but we might wish to know the area to the left of the negative values of x i.e.

$$P(V < -x)$$

Now as the curve is symmetrical, it follows that

$$P(V < -x) = P(V > x)$$

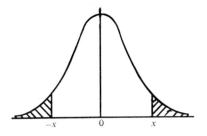

The diagram clearly shows this. It can also be seen that since

$$P(V > x) = 1 - P(V < x)$$

then

$$P(V < -x) = 1 - P(V < x)$$

Finally, we might wish to find $P(V > -x)$

$$\begin{aligned} P(V > -x) &= 1 - P(V < -x) \\ &= 1 - [1 - P(V < x)] \\ &= P(V < x) \end{aligned}$$

We can now summarise all the possibilities:

$P(V < x)$ is given in the tables
$P(V > x) = 1 - P(V < x)$
$P(V < -x) = 1 - P(V < x)$
$P(V > -x) = P(V < x)$

This may seem rather complicated, but in practice using the Normal Distribution Tables, is simplicity itself. The diagram above clearly shows that half the area under the Normal Curve lies to the left of $x = 0$. Hence the area under the Normal Curve to the left of a positive ordinate must be ≥ 0.5. If the probability we require is $< 0.5.$, (for example $P < -x$ or $P > x$ in the diagram above) then we must subtract from unity, the value obtained from the table. This gives us a simple operational rule for using the Normal Tables. Draw a sketch of the distribution, and decide whether the required area is greater than 0.5 or less than 0.5. If the area is greater than 0.5, then read the required probability directly from the table. If the area is less than 0.5, subtract from unity the probability obtained from the table. A few examples will clarify this.

EXAMPLE 1
An item V is drawn at random from $N(0,1)$. Find $P(V < 2)$

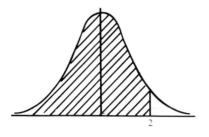

The sketch clearly shows that the required probability is greater than 0.5. Reading directly from the table

$P(V < 2) = 0.97725$

EXAMPLE 2
An item V is drawn at random from $N(0,1)$. Find $P(V < -1)$.

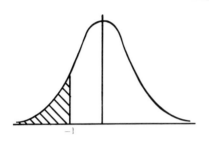

In this case, the sketch shows that the required probability is less than 0.5. From the table, we obtain

$P(V < 1) = 0.8413$

therefore

$P(V < -1) = 1 - 0.8413$
$\qquad\qquad = 0.1587$

TUTORIAL ONE
1. Consider the following distributions.
    (a) Heights of 5000 males chosen at random from the population of the U.K.
    (b) Distribution of incomes in the U.K.
    (c) Distribution of incomes in Soviet Russia.
    (d) Distribution of heads occurring if four coins are tossed, (1) 10, (2) 1000, (3) 1 million times.
    (e) Distribution of university mathematics graduates classed by IQ.
    (f) Distribution of Premium Bond numbers selected by ERNIE in a particular month.
    (g) Consumption of electricity between 9 a.m.–3 p.m. and 7 a.m.–7 p.m.
    (h) Distribution of the number of times a telephone number dialled at random is engaged.
    In each case draw a sketch of the distribution you think will occur, indicating those which are Normally Distributed.

2.   An item V is drawn at random from $N(0,1)$. Find

$P(V < 3)$
$P(V > -2)$
$P(V > 1.5)$
$P(V < -1.96)$
$P(-2.58 < V < 2.58)$
$P(-3.09 < V < 3.09)$
$P(-1.33 < V < 2.36)$

So far, we have only concerned ourselves with the Standard Normal Distribution $N(0,1)$, but of course most Normal Distributions are of the form $N(\bar{x}\ \sigma)$. However, the tables we have been considering can be used to solve probability problems for any Normal Distribution. We already know that if the items $x_1, x_2, x_3, \ldots\ldots x_n$ in the distribution $N(\bar{x}\ \sigma)$ are replaced by performing the operation.

$$\frac{x - \bar{x}}{\sigma}$$

then the distribution $N(\bar{x}\ \sigma)$ becomes $N(0,1)$. Now suppose we select an item t at random from $N(\bar{x}\ \sigma)$, and we require to know $p(t < x)$. Now x in $N(\bar{x}\ \sigma)$ is equivalent to $\dfrac{x - \bar{x}}{\sigma}$ in $N(0,1)$. Thus

$$p(t < x) \text{ in } N(\bar{x}\ \sigma) = p\left(V < \frac{x - \bar{x}}{\sigma}\right) \text{ in } N(0,1)$$

EXAMPLE 3
An item t is drawn at random from the distribution $N(10,2)$. Find the probability that t is greater than 12.

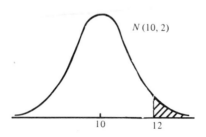

$$P(t > 12) \text{ in } N(10,2)$$
$$= P\left(V > \frac{12 - 10}{2}\right) \text{ in } N(0,1)$$
$$= P(V > 1) \text{ in } N(0,1)$$

We note from the diagram that the area required is less than 0.5.

From tables, $P(V < 1) = 0.8413$
and $P(V > 1) = 1 - 0.8413$
$= 0.1587$

We can conclude that there is a 15.87% chance of drawing an item greater than 12 if we select at random from $N(10,4)$.

Without doubt, the Normal Distribution is the most important of the three distributions we have considered — certainly it is very widely used. A considerable amount of data conforms quite well to this distribution (e.g. heights, weights, lengths), and later we will discuss methods of testing how well a Normal Distribution could describe a given distribution. The examples that follow will illustrate the power and versatility of this distribution.

EXAMPLE 4

A firm has a machine which turns metal cylindrical plugs. The machine is known to operate with a standard deviation of 0.001″. To meet an order, the machine is set to turn plugs with a 3 inch diameter, but the plugs must be within the *tolerance limits* of 2.998 to 3.002 inches. What proportion of the output produced to meet this order would fail to meet the tolerance limits, given that the diameters turned are Normally Distributed?

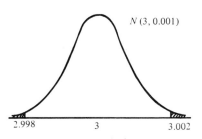

$N (3, 0.001)$

2.998   3   3.002

The shaded area represents the proportion of output rejected.

$P(t < 2.998)$ in $N(3, 0.001)$

$= P\left(V < \dfrac{2.998 - 3}{0.001}\right)$ in $N(0,1)$

$= P(V < -2)$ in $N(0,1)$

We note from the diagram that the area required is less than 0.5.

From tables, $P(V < 2)\ = 0.97725$

so, $P(V < -2) = 1 - 0.97725$

$= 0.02275$

and as the shaded areas, are equal (they are equally distributed about the mean)

$P(V > 2) = 0.02275$

total shaded area = 0.0455

We conclude that 4.55% of output fails to meet the tolerances.

EXAMPLE 5

A machine for filling bags of flour is known to operate with a standard deviation of 0.4 ozs. To satisfy the Weights and Measures Inspector, not more than 5% of 3lb. (48 ozs.) bags of flour may be underweight. To what weights should the machine be set?

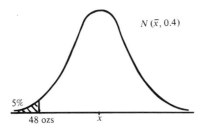

$N(\bar{x}, 0.4)$

5%

48 ozs x

To solve this problem we must find the mean of the distribution, and set the machine to this level. Consulting the diagram, we see that to meet the regulations

$P(t < 48 \text{ ozs.}) = 5\% = 0.05$ in $N(\bar{x}, 0.4)$

or $P\left(V < \dfrac{48 - \bar{x}}{0.4}\right) = 0.05$ in $N(0,1)$

Let us now examine the appropriate area in $N(0,1)$

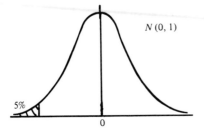

Now if $P(V < x)$ in $N(0,1) = 0.05$, then it surely follows that x is negative. The diagram clearly shows this. We can see the table to find the value of x for which $P(V < x) = 0.05$. Now we cannot look up 0.05 in the table, but we can look up $1 - 0.05 = 0.95$, and we find that for this value $x = 1.645$, so

$$P(V < 1.645) = 0.95 \text{ in } N(0,1)$$
$$\text{hence } P(V < -1.645) = 0.05 \text{ in } N(0,1)$$

But to solve this problem, we require

$$P\left(V < \frac{48 - \bar{x}}{0.4}\right) = 0.05 \text{ in } N(0,1)$$

so $\dfrac{48 - \bar{x}}{0.4} = -1.645$

and $\bar{x}$ $\quad = 48.658$ ozs.

If the machine is set at 48.658 ozs., then not more than 5% of packets will be less than 3 lb. in weight.

EXAMPLE 6

You will have realised by now that many problems involving the Binomial Distribution also involve tiresome calculations. Suppose we wished to find the probability of obtaining more than 220 defectives when sampling in 1000's from a 20% defective population. The appropriate Binomial Distribution is

$$(0.8 + 0.2)^{1000}$$

and we could solve this problem by adding 221st + 222nd + 223rd + .. + 1000th term of the expansion — an exercise that is not recommended! However, provided n is large, (as in this case) we can use the Normal

Distribution to give a very accurate approximation to the solution. The appropriate Normal Distribution is the one with the same mean and standard deviation as the Binomial Distribution. You should remember that the mean of a Binomial Distribution is given by np, and the variance by npq. In this case we have

$\bar{x} = 1000 \times 0.2 = 200$
$var = 1000 \times 0.2 \times 0.8 = 160$

We will use $N(200,\sqrt{160})$ as an approximation to $(0.8 + 0.2)^{1000}$. The diagram shows part of the histogram of $(0.8 + 0.2)^{1000}$. It is *not* drawn

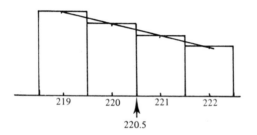

to scale. If we want to find the probability of obtaining more than 220 defectives, we must add the areas of the rectangles enclosing 221, 222, 223 . . . . . 1000. The line joining the mid-points of the top of each rectangle represents the Normal Distribution $N(200,\sqrt{160})$. If we wish to use this distribution to find the probability of more than 220 defectives, the diagram clearly shows that we require $P(V > 220.5)$. This 0.5 adjustment is always used when we use the Normal Distribution as an approximation to a discrete distribution.

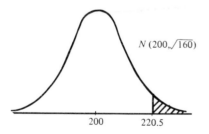

$P(t > 220.5)$ in $N(200, \sqrt{160})$

$= P\left(V > \dfrac{220.5 - 200}{\sqrt{160}}\right)$ in $N(0,1)$

$= P(V > 1.621)$ in $N(0,1)$

from tables.

$P(V < 1.621) = 0.9475$

and as the area we require is less than 0.5

$P(V > 1.621) = 0.0525.$

We conclude that the probability of more than 220 defectives is approximately 5.25%.

TUTORIAL TWO

1. An item t is selected at random from $N(10,4)$. Find

$P(t < 13)$
$P(t > 10.5)$
$P(t < 9)$
$P(8 < t < 11)$

2. Find the proportion rejected in example 4 if the tolerance limits were 2.999 to 3.001, the mean and standard deviation remaining the same as before.

3. Between which limits would you expect 95% of the distribution $N(10,4)$ to lie?

4. Suppose that for the machine quoted in example 5, the limit for underweight bags was (a) 10%, (b) 1%. Find the appropriate machine setting.

5. In a certain Normal Distribution it has been noticed that 93.32% of frequencies are less than 13 inches, and 15.87% are less than 8 inches. Find the distribution i.e. give its mean and standard deviation.

6. A firm requires a machine to cut metal plugs to within a tolerance of $3'' \pm 0.001''$. The firm would consider it acceptable if not more than 5% of plugs failed to meet the tolerances. What is the greatest possible standard deviation for the machine if the condition is to be satisfied?

Sampling Distributions: Sums and Differences

We can now examine precisely what happens when we take samples and evaluate the validity of sample evidence. What we will require initially is the *sampling distribution* of samples i.e. to find the form of the distribution (whether it is Binomial, Poisson, Normal or none of these) and the mean and standard deviation of the distribution. We will consider firstly the case of forming samples from sums and differences. How can we form a sample by taking a sum? Suppose that in the production of a certain component it is necessary to take two rods of different metals and weld them end-to-end. We must imagine that the welder has two boxes, one containing (say) steel rods, and the other containing copper rods. The welder selects a rod at random from each box and joins them end-to-end. The assembled rod forms the *sample sum.* What about a sample difference? Suppose that in the production of the same component the next stage is to fit a metal collar to the assembled rod, then if the rods and collars are selected at random the gap between the rod and collar forms a sample difference. Now if we know the mean and standard deviation of the parent populations, we can predict the mean and standard deviation of the distribution of sample sums or sample differences. Firstly let us consider the sample sum in general form.

We shall let the first population be $x_1 = a_1, b_1, c_1, \ldots$ containing n_1 items in all. The mean of this population is $\bar{x}_1$.

The second population is $x_2 = a_2, b_2, c_2, \ldots$ containing n_2 items, mean $\bar{x}_2$. The items $a \; b \; c \ldots$ can be considered as lengths. An item is selected at random from x_1, then an item from x_2, and they are joined. If this is repeated many times then a new distribution of lengths X will be formed, i.e.

$$X = x_1 + x_2$$

How many sample sums are possible? There are n_1 selections from population x_1 and n_2 selections from population x_2. You should realise, then, that there are n_1, n_2, sample sums in all.

Then $x = (a_1 + a_2), (a_1 + b_2), \ldots (b_1 + a_2)(b_1 + b_2) \ldots$ making n_1, n_2 items in all.

The mean of the sample sums is

$$\bar{x} = \frac{(a_1 + a_2) + (a_1 + b_2) + \ldots (b_1 + a_2) + (b_1 + b_2) \ldots}{n_1 \, n_2}$$

Now if you think about it, the items from the population x_1 will occur in the above expression n_2 times. Likewise, the items in x_2 will occur n_1 times. Hence

$$\bar{x} = \frac{n_2 \, (a_1 + b_1 + c_1 + \ldots) + n_1 \, (a_2 + b_2 + c_2 + \ldots)}{n_1 \, n_2}$$

$$\bar{x} = \frac{n_2 \Sigma x_1 + n_1 \Sigma x_2}{n_1 \, n_2}$$

$$\bar{x} = \frac{\Sigma x_1}{n_1} + \frac{\Sigma x_2}{n_2}$$

$$X = x_1 + x_2$$

i.e. the mean of the sample sums is the sum of the means of the population from which they were formed.

We also wish to know the standard deviation of the sample sums. Now we can considerably simplify this task by subtracting $\bar{x}_1$ from each item in x_1, $\bar{x}_2$ from each item in x_2 and $\bar{X}$ from each item in X. This will not affect the standard deviations, but this will make $\Sigma X = 0$, $\Sigma x_1 = 0$, and $\Sigma x_2 = 0$, also

$$\text{Var}(X) = \frac{\Sigma X^2}{n_1 n_2}, \ \text{Var}(x_1) = \frac{\Sigma x_1^2}{n_1}, \ \text{Var}(x_2) = \frac{\Sigma x_2^2}{n_2}$$

$$\Sigma X^2 = (a_1 + a_2)^2 + (a_1 + b_2)^2 + \ldots (b_1 + a_2)^2 + (b_1 + b_2)^2 \ldots$$

Using the fact that:

$$(a_1 + a_2)^2 = a_1^2 + 2a_1 a_2 + a_2^2$$

$$\Sigma X^2 = n_2(a_1^2 + b_1^2 + c_1^2 + \ldots) + n_1(a_2^2 + b_2^2 + c_2^2 + \ldots)$$

$$+ 2(a_1 + b_1 + c_1 + \ldots)(a_2 + b_2 + c_2 + \ldots)$$

$$\Sigma X^2 = n_2 \Sigma_1^2 x_1^2 + n_1 \Sigma x_2^2 + 2\Sigma x_1 \Sigma x_2$$

But we know that

$$\Sigma x_1 = \Sigma x_2 = 0$$

Hence

$$\Sigma X^2 = n_2 \Sigma x_1^2 + n_1 \Sigma x_2^2$$

$$\frac{\Sigma X^2}{n_1 n_2} = \frac{\not{n}_2 \Sigma x_1^2}{n_1 \not{n}_2} + \frac{\not{n}_1 \Sigma x_2^2}{\not{n}_1 n_2}$$

$$\text{Var}(X) = \text{Var}(x_1) + \text{Var}(x_2)$$

The variance of the sample sums is the sum of the variances of the populations from which they were formed. Moreover, if the parent populations are Normally Distributed, so will be the sample sums.

Now let us suppose that we use the same populations x_1 and x_2 to form sample differences X, then

$$\bar{X} = \bar{x}_1 - \bar{x}_2$$

and

$$\text{Var}(X) = \text{Var}(x_1) + \text{Var}(x_2)$$

The mean of the sample differences is the difference between the means of the populations from which they were drawn. The variance of the sample difference is the sum of the variance of the populations from which they were formed.

Notice that even when sample differences are taken, the variances are still added. Can you see why? If you cannot, try expanding $(a_1 - a_2)^2$ and notice the signs.

EXAMPLE 7

Metal rods type A have lengths distributed as $N(3.5'',0.03'')$ and type B as $N(4.5'',0.04'')$. A rod of each type is selected and placed in a slot 8.1'' long. If this operation is performed repeatedly, what proportion of rods will not fit?

Firstly, we require the distribution of lengths of two rods X. As both the separate distributions are Normally Distributed, we have

$$X = N(3.5'' + 4.5'', \sqrt{0.03 + 0.04})$$
$$N(8,0.05)$$

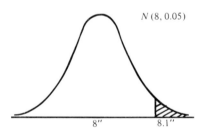

$N(8, 0.05)$

8''   8.1''

The probability we require is the same as

$$P(t > 8.1) \text{ in } N(8,0.05)$$
$$= P\left(V > \frac{8.1-8}{0.05}\right) \text{ in } N(0,1)$$
$$= P(V > 2) \text{ in } N(0,1)$$

From tables

$$P(V < 2) = 0.97725$$

And as the area required is less than 0.5

$$P(V > 2) = 0.02275$$

We conclude that 2.275% of assemblies will not fit the slot.

TUTORIAL THREE

1. Suppose we have two populations A = 6, 2 and 4 and B = 10 and 3. What is the mean and variance of both populations? Now suppose an item is selected at random from each population and they are added. In how many ways can this be done? Write out all the totals and find the mean and variance. Verify that the mean of the totals is the same as the sum of the means of the population, and that the variance of the totals is the same as the sum of the variance of the populations.

2. Let $x_1 = a_1 b_1 c_1 \ldots$ containing n_1 items, and $x_2 = a_2 b_2 c_2 \ldots$ containing n_2 items. Also let $x = (a_1 - a_2), (a_1 - b_1) + \ldots (b_1 - a_2), (b_1 - b_2) \ldots$ cnotaining $n_1 n_2$ items. Show that $\bar{x} = \bar{x}_1 - \bar{x}_2$ and Var (x) = Var (x_1) + Var (x_2).

3. Now you will have to think about this one! Part of the assembly of a machine involves placing a metal rod into a slot. The lengths of metal rods have the distribution $N(16, 0.03)$, and the lengths of slots have the distribution $N(16.1, 0.04)$. Find the distribution of the clearance between the bar and the slot. If a clearance of under 0.1″ is satisfactory find the proportion of assemblies that fail to meet this requirement.

Sampling Distribution of Sample Means

The second sampling distribution we must examine is called the distribution of sample means, and we shall apply this distribution to important techniques such as statistical estimation, significance testing and quality control. In essence, this distribution is extremely simple: we must imagine drawing from a certain population a large number of samples all containing the same number of items. If we calculate the mean of each sample, we have a distribution of sample means. Now if we know the mean and standard deviation of the population, we can predict the mean and standard deviation of the distribution of sample means.

Suppose we have a very large population, mean μ and standard deviation σ. From this population we draw samples of n items. Now the mean of the sample means will be μ (the mean of the population) and the standard deviation of the sample means (more usually called the

Standard Error of the sample means) will be $\sigma/\sqrt{n}$. Hence, as long as we know the mean and standard deviation of the population, we can state the sampling distribution of the means of samples drawn from it. Statisticians use the term Standard Error(s) when considering sampling distributions, and standard deviation when considering populations.

Now if you think carefully, you will realise that the mean of the sample means is quite likely to be the same as the mean of the population *if a large number of samples is drawn.* One would expect that the items in an individual sample would be representative of the items in the population, so the most likely value for any sample mean is the population mean. In practice, of course, some sample means will be greater than, and some less than the population. But the larger the number of samples drawn, the closer would we expect the average values of sample means to approach the population mean. In the next tutorial, you will be asked to undertake an experiment to verify this point.

It is not difficult to realise that the mean of the sampling distribution is the mean of the population, but its Standard Error is not so obvious. Certainly, we would expect the Standard Error to be *smaller* than the standard deviation of the population. Taking sample means would average out any extremities drawn from the population, so we could expect sample means to have a smaller *spread* than individual items. To illustrate this, let the population be the digits

0   1   2   3   4   5   6   7   8   9

then the spread of items is 0 to 9. If we were to draw samples of two items, then the smallest sample mean we could obtain is

$$\frac{0+1}{2} = 0.5$$

and the largest is

$$\frac{8+9}{2} = 8.5$$

a spread of 0.5 to 8.5.

Common sense then, leads us to believe that $s < \sigma$, but does not explain why $s = \sigma/\sqrt{n}$. To do this, we must consider some elementary algebra. Again let the population mean be μ and the population standard deviation be σ. From this population we draw a sample of n items. Let us call the items

$$x_1, x_2, x_3 \ldots \ldots \ldots \ldots x_n$$

so

$$n\bar{x} = x_1 + x_2 + x_3 + \ldots + x_n$$

Notice how the statistic $n\bar{x}$ is derived: we obtain it by *adding n samples of one item.* Using what we already know of the sampling distribution of sums, we can conclude that

$$\text{Var}(n\bar{x}) = \text{Var}(x_1) + \text{Var}(x_2) + \text{Var}(x_3) + \ldots + \text{Var}(x_n)$$

Now each of the single items selected could be any item in the population, so the variance of each item must be σ^2, the population variance. Hence

$$\text{Var}(n\bar{x}) = n\sigma^2$$

so   S.D. of $(n\bar{x}) = \sqrt{n}\sigma$

and S.D. of $\bar{x}$ $= \sigma/\sqrt{n}$

If the population is Normally Distributed, then we would expect the means of samples drawn from it also to be Normally Distributed. This probably does not surprise you. However, if the population is not Normally Distributed, the means of samples drawn from it becomes more like the Normal as the sample size increases – which is not such an obvious conclusion. Now the proof of both these statements is beyond the scope of this book, but we can demonstrate the truth of the latter statement by sampling from a non-normal population. Let the population be the digits

0   1   2   3   4   5   6   7   8   9

The Histogram of this distribution looks like this:

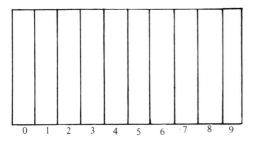

The shape of the histogram explains why we say the population has a rectangular distribution. There is an equal chance of drawing any item at random from this population.

Now let us suppose that samples of two digits are drawn, but the first is replaced before the second is drawn. You should realise that 100 selections are possible. For each sample, we could add together the two digits drawn. The smallest sample total would be 0 + 0 = 0, and the largest 9 + 9 = 18. Hence, all sample totals must fall within the range 0 to 18, which means that all sample means must be within the range 0 to 9. Now we must find the number of ways of obtaining each sample total. A sample total of zero can be obtained in one way only — if both the digits drawn are zero. A sample total of one can be obtained in two ways (0+1 or 1+0). A sample total of two could be obtained in three ways (0+2, 2+0 or 1+1). Continuing in this way we would obtain the following distribution.

Sample Total	Sample Mean	No. of Ways	Probability
0	0	1	0.01
1	0.5	2	0.02
2	1.0	3	0.03
3	1.5	4	0.04
4	2.0	5	0.05
5	2.5	6	0.06
6	3.0	7	0.07
7	3.5	8	0.08
8	4.0	9	0.09
9	4.5	10	0.10
10	5.0	9	0.09
11	5.5	8	0.08
12	6.0	7	0.07
13	6.5	6	0.06
14	7.0	5	0.05
15	7.5	4	0.04
16	8.0	3	0.03
17	8.5	2	0.02
18	9.0	1	0.01
		100	1.00

The sampling distribution of the means has been plotted in diagram 16. We can conclude that if samples of two items are drawn from a rectangular distribution, the distribution of sample means will be triangular.

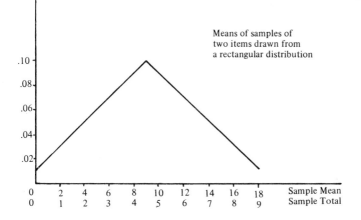

Means of samples of
two items drawn from
a rectangular distribution

| Sample Mean | 0 | 2 | 4 | 6 | 8 | 10 | 12 | 14 | 16 | 18 |
| Sample Total | 0 | 1 | 2 | 3 | 4 | 5 | 6 | 7 | 8 | 9 |

The number of ways of obtaining each sample total could have been obtained like this:

Sample Total											No. of Ways
0	1										1
1	1	1									2
2	1	1	1								3
3	1	1	1	1							4
4	1	1	1	1	1						5
5	1	1	1	1	1	1					6
6	1	1	1	1	1	1	1				7
7	1	1	1	1	1	1	1	1			8
8	1	1	1	1	1	1	1	1	1		9
9	1	1	1	1	1	1	1	1	1	1	10
10		1	1	1	1	1	1	1	1	1	9
11			1	1	1	1	1	1	1	1	8
12				1	1	1	1	1	1	1	7
13					1	1	1	1	1	1	6
14						1	1	1	1	1	5
15							1	1	1	1	4
16								1	1	1	3
17									1	1	2
18										1	1

The first column represents the sample totals that would be obtained if we selected the digits 00–09 inclusive (i.e. the range of sample totals would be from 0 to 9). The second column represents the sample totals that would have been obtained if we selected the digits 10–19 inclusive.

The range of sample totals in this group is 1 to 10. In other words, we add one to the highest and lowest of the previous range. In effect, each column moves down one line. Adding sideways we can obtain the number of ways each total can occur.

We can use a pattern like this to find the probability distribution of means of samples of three items drawn from a rectangular distribution. There would be 1000 samples, the smallest sample total being 0 + 0 + 0 = 0 and the largest 9 + 9 + 9 = 27. Firstly, let us consider the sample totals we would obtain if we selected the digits 000—009 inclusive. Surely, the distribution of sample totals would be the same as for samples of two the range of sample totals (0 to 18) is identical! Hence we can use the distribution of samples of two as our first column to find the distribution of samples of three. Now if we added one to all the sample totals in the range 0—18, we would obtain the range of totals if we selected from the digits 100—199. Moreover, the distribution would be the same, so we move the first column down one line. Repeating this for the remaining eight columns, we have:

Sample Total											No. of Ways
1	1										1
2	2	1									3
3	3	2	1								6
4	4	3	2	1							10
5	5	4	3	2	1						15
6	6	5	4	3	2	1					21
7	7	6	5	4	3	2	1				28
8	8	7	6	5	4	3	2	1			36
9	9	8	7	6	5	4	3	2	1		45
10	10	9	8	7	6	5	4	3	2	1	55
11	9	10	9	8	7	6	5	4	3	2	63
12	8	9	10	9	8	7	6	5	4	3	69
13	7	8	9	10	9	8	7	6	5	4	73
14	6	7	8	9	10	9	8	7	6	5	75
15	5	6	7	8	9	10	9	8	7	6	75
16	4	5	6	7	8	9	10	9	8	7	73
17	3	4	5	6	7	8	9	10	9	8	69
18	2	3	4	5	6	7	8	9	10	9	63
19	1	2	3	4	5	6	7	8	9	10	55
20		1	2	3	4	5	6	7	8	9	45
.							.	.	.		.
.							.	.	.		.
.							.	.	.		.
25								1	2	3	6
26									1	2	3
27										1	1

If we divide the number of ways each sample total can occur by 1000, we obtain the probability of each sample total. If we divide each sample total by three, we will obtain the sample means. However, it is probably more convenient to plot the probability distribution of sample totals. This has been done in diagram 17. We can conclude that the means of samples of three items drawn from a rectangular distribution form a bell-shaped distribution.

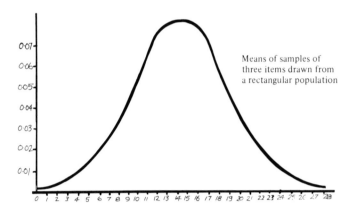

Means of samples of three items drawn from a rectangular population

This process has been repeated for samples of four items and samples of five items. The results are shown in diagrams 18 and 19. Notice the marked tendency towards Normality as the sample size increases. We can conclude that even when the population is not Normal, the means of samples approach Normality as the sample size increases. We can use the Normal Distribution to predict the likelihood that a sample mean is of a particular size.

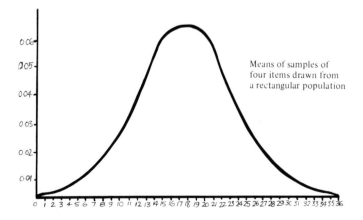

Means of samples of four items drawn from a rectangular population

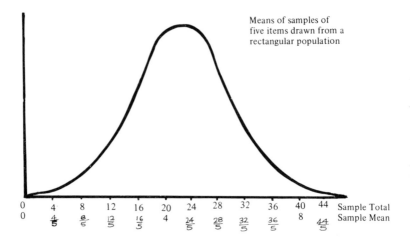

Means of samples of
five items drawn from a
rectangular population

| | | | | | | | | | | |
|0|4|8|12|16|20|24|28|32|36|40|44| Sample Total
|0|$\frac{4}{5}$|$\frac{8}{5}$|$\frac{12}{5}$|$\frac{16}{5}$|4|$\frac{24}{5}$|$\frac{28}{5}$|$\frac{32}{5}$|$\frac{36}{5}$|8|$\frac{44}{5}$| Sample Mean

EXAMPLE

Samples of 25 items are drawn from a population with a mean 20 and standard deviation 5. What proportion of samples will have a mean greater than 22?

Standard Error of sample means is $5/\sqrt{25} = 1$
Hence, sample means have a distribution $N(20,1)$

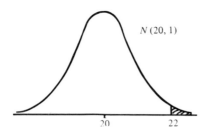

$N(20, 1)$

20   22

$P(x > 22)$ in $N(20,1)$

$= P\left(V > \dfrac{22-20}{1}\right)$ in $N(0,1)$

i.e. $P(V > 2)$ in $N(0,1) = 2.28\%$

EXAMPLE

It has been claimed by the inhabitants of Ruritania that their male population is stronger, healthier and taller than the males of its neighbour Industralasia. Now Industralasia practices conscription, and each potential conscript's height is measured, so Industralasia can reliably claim that the mean height of its young males is 168 cms. with a standard deviation of 7.5 cms. There are no comparable statistics for Ruritania, but Industralasian statisticians measured the heights of 100 randomly chosen Ruritanian young men, and found that their mean height was 173 cms. What can we conclude about the Ruritanian's claim?

Let us begin by supposing there is no difference in height between Industralasians and Ruritanians. If this is the case, then the sample mean 173 cms. could well have been obtained from Industralasian young men. How likely is this? In other words, if a sample of 100 items is drawn from a population with a mean 168 and S.D. of 7.5, how likely is it that the sample mean exceeds 173?

The standard error of sample means is $7.5/\sqrt{100} = 0.75$
The distribution of sample means is $N(168,0.75)$

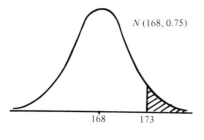

$N(168, 0.75)$

$P(x > 173)$ in $N(168,0.75)$

$= P\left(V > \dfrac{173-168}{0.75}\right)$ in $N(0,1)$

$= P(V > 6.66)$ in $N(0,1)$

The probability is so very small it can be discounted. Thus, it is extremely unlikely that a sample mean of greater than 173 cms. could be obtained from Industralasia, and it is most likely that the Ruritanian claim is justified.

TUTORIAL FOUR

1. If a population is believed to have non-normal characteristics what is the most important single step you can take to make sample results as accurate as possible?

2. If the average consumption of beef in Oxford is 31.4 lbs. a year, and in Manchester 33.7 lbs. a year, does this mean that people in

Manchester eat more beef than people in Oxford? Justify your answer.

3. A confectioner produces cakes which he packs in cartons of 100 for the catering trade. He prints on the carton: 'average weight not less than 95.5 gms.' If cakes have a mean weight of 100 gms. and S.D. 20 gms., what proportion of batches contravene the Trade Description Act?

4. A machine is cutting metal rods to a specified length of 2 inches, and it is known that the machine operates with a standard deviation of 0.02 inches. The quality control department test the output of the machine to ensure that the setting does not 'drift'. A sample of 100 rods gave a mean length of 2.005 inches. Does this suggest that the setting of the machine has drifted upwards?

5. A manufacturer produces metal rods, and the distribution of breaking strengths are $N(60,5)$ lb. He wishes to pack the rods in bundles and guarantee that the average breaking strength of rods exceeds 59 lbs., with 95% probability. How many rods should be in each bundle?

Sampling Distribution of Proportions

Let us consider once more the Binomial Distribution $(p + q)^n$. We stated earlier that by expanding this distribution, we could predict the probability of $0, 1, 2, \ldots \ldots n$, defectives if a sample of n items was drawn from a population containing $P\%$ defectives. Now suppose that instead of considering the actual number of defectives, we considered instead the proportions defective $0, \frac{1}{n}, \frac{2}{n}, \ldots . 1$. The probabilities of these proportions would be the same as the probabilities of their respective number of defectives.

Admittedly, the probabilities are the same in the two distributions, but the distributions themselves are essentially different. For the distribution of defectives, the range of values is 0 to n, but for the proportions it is 0 to 1. To obtain the distribution of the proportion defective, we must reduce the distribution of defectives by the factor $\frac{1}{n}$. Thus, the mean and standard deviation must also be reduced by the factor $\frac{1}{n}$.

Again, we use the term standard error rather than standard deviation when refering to sampling distributions of proportions.

You should remember that if a sample size n is drawn from a population that is $P\%$ defective, the mean number of defectives is np. Hence,

the mean proportion of defectives is $\dfrac{1}{n} \times np = p$

the same as the proportion defective in the population,

the standard error of the proportion defective is $\dfrac{1}{n} \times \sqrt{npq} = \sqrt{\dfrac{pq}{n}}$

Note the relationship between p and $P\%$: p is $P\%$ expressed as a proportion. So if $P = 5\%$, then $p = 0.05$.

EXAMPLE
Samples, size 1000, are drawn from a population that is 5% defective and the proportion defective in each sample is found. What is the expected sampling distribution of the proportion defective?

In this case $n = 1000$ $p = 0.05$

mean $= 0.05$

Standard error $= \sqrt{\dfrac{0.05 \times 0.95}{1000}} = 0.0069$

The distribution is $(0.05, 0.0069)$

Earlier in this chapter, we stated that the Normal Distribution could be derived from $(p + q)^n$ if $p = 0$ and n approaches infinity. Also, we noticed that the sampling distribution of sample means tends to Normality as the sample size increases. Now a similar feature also exists for the sampling distribution of proportions as the sample size increases. As n is large in the above example, we can state that the sampling distribution of porportions is $N(0.05, 0.0069)$.

Estimating the Population Mean and Variance

Let us now apply our knowledge of sampling distributions to estimate the population mean and variance from sample evidence. If our estimates are to be efficient, then they must be unbiased i.e. the estimate must be just as likely to be too low as too high. It follows, then, that if a large number of such estimates is made, the average of such estimates would tend to the correct value. We begin the investigation with two statements.

The best estimate of the mean of the population (μ) is the mean of the sample ($\bar{x}$). The best estimate of the variance of the population is the variance of the sample multiplied by $\dfrac{n}{n-1}$, where n is the sample size.

The first statement seems reasonable enough, but the second is rather surprising. Again, we will use some simple algebra to investigate the second statement.

Let the sample mean be $\bar{x}$,

the population mean be μ,

then items in the sample will be x_i,

the population standard deviation be σ,

the sample standard deviation be S

Consider any item in the sample x, now it follows that

$$(x - \mu) = (x - \bar{x}) + (\bar{x} - \mu)$$

and $(x - \mu)^2 = (x - \bar{x})^2 + (\bar{x} - \mu)^2 + 2(x - \bar{x})(\bar{x} - \mu)$

If we repeat this operation for all the items x_i in the sample, we have

$$\Sigma(x - \mu)^2 = \Sigma(x - x)^2 + n(\bar{x} - \mu)^2 + 2(x - \mu)\Sigma(x - \bar{x})$$

Now we already know that $\Sigma(x - \bar{x}) = 0$, so

$$\Sigma(x - \mu)^2 = \Sigma(x - x)^2 + n(\bar{x} - \mu)^2$$

If we want an unbiased estimate, we would need really to draw a large number of samples, and we could calculate the value $\Sigma(x - \mu)^2$ for each of the samples. Now the population variance σ^2 is

$$\frac{\Sigma(x - \mu)^2}{n}$$

So we would expect the average value of $\Sigma(x - \mu)^2$ obtained from all of the samples to be $n\sigma^2$. Likewise, if we calculate the value $\Sigma(x - \bar{x})^2$ for each sample, we would expect the average value of this quantity to be ns^2. Finally, we would expect the average value of $(\bar{x} - \mu)^2$ to be the variance of the sample means i.e. $\dfrac{\sigma^2}{n}$.

Now we can state that, on average,

$$n\sigma^2 = ns^2 + \sigma^2$$

$$s^2 = \frac{n\sigma^2 - \sigma^2}{n}$$

$$s^2 = \frac{\sigma^2(n - 1)}{n}$$

$$\sigma^2 = \frac{s^2 n}{n - 1}$$

The use of $\dfrac{n}{n-1}$ when estimating a population variance is extremely important and we will examine its implications in more detail later. For the moment, consider this point: if n is large then

$$\frac{n}{n-1} \simeq 1$$

and there is no need to adjust the sample variance. For example, if the sample size is 2, then we must multiply the population variance by

$$\frac{2}{2-1} = 2$$

which will profoundly affect its size. However, if n is 100, then the adjustment

$$\frac{100}{100-1} = 1.01$$

is negligible. We can use this fact as an operational definition for a large sample: a sample is large when

$$\frac{n}{n-1} \simeq 1$$

Confidence Limits for the Population Mean

Suppose we draw a random sample of 100 items from the population with a mean 5 and variance 0.01. The sampling distribution of the sample means would be

$$N\left(5, \frac{\sqrt{0.01}}{\sqrt{100}}\right) = N(5, 0.01)$$

Now let us suppose that many such samples were drawn, and ask between what limits we would expect 95% of the sample means to be. This is equivalent to finding the central (95% zone of the distribution $N(5, 0.01)$.

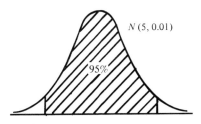

$N(5, 0.01)$

95%

From the diagram we require

$P(x_2 < x < x_1) = 95\%$ in $N(5,0.01)$

i.e.

$$P\left(\frac{x-5}{0.01} < V < \frac{x_1-5}{0.01}\right) = 95\% \text{ in } N(0,1)$$

The tables give

$P(-1.96 < V < 1.96) = 95\%$ in $N(0,1)$

so

$$\frac{x_2 - 5}{0.01} = -1.96 \text{ and, } \frac{x_1 - 5}{0.01} = 1.96$$

Giving

$x_1 = 5.0196$ and $x_2 = 4.9804$

We can expect 95% of sample means to lie in the range 4.99804 to 5.00196. Such limits are called the 95% *confidence limits* for sample means.

More usually, we will not know the population mean and variance and we will have to estimate them from sample evidence. Suppose a sample of 100 items had a mean 5 and variance 0.01. As n is large, the statistics can be used as an unbiased estimate of the population mean and variance. Now if our estimates are accurate, we could regard this sample as being one of many that have a 95% chance of having a mean between 4.99804 to 5.00196. In other words, we can be 95% sure that the true value of the population mean (μ) is between 4.99804 and 5.00196, even if our estimate is not perfectly accurate. Generalising, if a large sample of n items yielded a mean $\bar{x}$ and a standard deviation s, then we estimate that the population mean lies within the limits

$$\bar{x} \pm \frac{1.96s}{\sqrt{n}} \text{ with 95\% confidence}$$

Of course, we need not have chosen 95% confidence limits — the limits chosen depend upon the degree of accuracy you require. In the next tutorial you will be asked to work examples illustrating this.

The Sample Size

It is quite likely that if we estimate a population mean, then the limits we state to our estimate may be too wide. In the above example we estimated the population within ±0.0196, but it might have been necessary to estimate it within (say) ±0.01. Two options now face us: Either we can be satisfied with a confidence limit of less than 95% (which is not very satisfactory) or we can draw a second, larger sample. How large should this second sample be? This depends on the tolerable limits to our estimate, and the confidence limit we select.

We are instructed to estimate the mean of a certain population. Our estimate must be within an amount ±Z of the true value of the mean, and we must be 99% confident of our estimate. We draw a sample of (say) 100 items, set up the 99% confidence limits for the population mean, and find that our estimate is outside the limit ±Z. Clearly, we will have to draw a second, larger, sample. To find the size of this sample we use the standard deviation (S) of our first sample as an estimate of the population standard deviation. We could represent the problem like this:

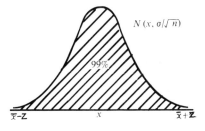

$N(x, \sigma/\sqrt{n})$

99%

$\bar{x} - Z$ x $\bar{x} + Z$

The quantity $\bar{x}$, which is the mean of our second sample, will be used as our new, improved estimate of the population mean. From the diagram, we can conclude that

$$P(x < \bar{x} + Z) \text{ in } N\left(\bar{x}, \frac{\sigma}{\sqrt{n}}\right) = 99.5\%$$

$$\text{or } P\left(V < \frac{(\bar{x} + Z) - \bar{x}}{\frac{\sigma}{\sqrt{n}}}\right) \text{ in } N(0,1) = 99.5\%$$

$$\text{i.e. } P\left(V < \frac{Z\sqrt{n}}{\sigma}\right) \text{ in } N(0,1) = 99.5\%$$

and from the tables

$$P(V < 2.576) \text{ in } N(0,1) = 99.5\%$$

so it must follow that

$$\frac{Z\sqrt{n}}{\sigma} = 2.576$$

and

$$n = \left(\frac{2.576\sigma}{Z}\right)^2$$

EXAMPLE

A manufacturer wishes to estimate a mean dimension of a certain component. He would be satisfied if he obtained an estimate within 0.01 cms. of the true mean, and was 99% confident of his result. An initial (large) sample has a standard deviation of 0.2 cms. What is the size of the sample he should examine?

Here we have $Z = 0.01$ and $\sigma = 0.2$

$$n = \left(\frac{2.576 \times 0.2}{0.01}\right)^2 = 2658$$

It is most important to use large samples when estimating population mean. If small samples are used then the distribution of sample means will not be Normal, and the analysis we have used would be inappropriate.

Estimating the Proportion Defective

Now let us see if we can estimate the proportion of items in a population with a certain characteristic (perhaps they are defective). We will follow the usual method of dealing in percentages rather than proportions, and so we must restate the sampling distribution we obtained earlier in this chapter. If we draw a sample of n items from a population that is $P\%$ defective,

The mean percentage of defectives in the sample is $P\%$

and the standard error of the percentage of defectives is $\sqrt{\dfrac{P(100-P)}{n}}\%$

Moreover, we stated that if n is large, the proportion of defectives in the sample will be Normally Distributed, so we can, as before, set up confidence limits. Conventionally, some rounding-off takes place of the values obtained from the Normal Tables.

$$P\% \pm 2\sqrt{\frac{P(100 - P)}{n}} \quad \text{marks the 95\% confidence limit and}$$

$$P\% \pm 3\sqrt{\frac{P(100 - P)}{n}} \quad \text{marks the 99.8\% confidence limit}$$

EXAMPLE

A firm is considering a modification to one of its products. The modification will reduce price, but adversely affect quality. In an attempt to assess consumer reaction, 100 people were chosen at random and tested the product. Eleven percent were not in favour of the change. Estimate the overall percentage of customers not in favour of the change.

$$11 \pm 2 \times \sqrt{\frac{11(100 - 11)}{100}}$$

$11 \pm 6.278\%$ with 95% probability

Clearly, the range of our estimate is too wide, and we can improve on it by increasing the sample size. Suppose we wish to estimate the proportion defective within $\pm r\%$ of the true value $P\%$ and be 95% certain of our result, then

$$r = 2\sqrt{\frac{P(100 - P)}{n}}$$

and

$$n = \frac{4P(100 - P)}{r^2}$$

We can now find the sample size to fix the overall percentage not in favour of modification within (say) 1%.

$$n = \frac{4 \times 11 \times (100 - 11)}{1} = 3916$$

TUTORIAL FIVE

1. How would you interpret the statistical expressions 'confidence limits' and 'certainty'.
2. Explain precisely, the relation between 'probability of an event' and the 'level of certainty that the event will occur'.
3. Let us suppose that a company makes a product to order and wishes to state with some degree of certainty the time between the placing of an order and the delivery of the product. A sample check of 81 orders shows that the average delivery time was 25 days with a

standard deviation of 6 days. What should it quote to customers about delivery dates? What assumption have you made in deriving your answer?

4. A random sample of 1000 people shows that at the next election 550 will support A. Is it certain that A will be elected, and if not, what proportion of the voters would you expect him to obtain?

5. A manufacturer wishes to estimate the mean weight of sacks of carbon black. He would be satisfied if his estimate was within ±5 lbs. of the true mean weight and be 99% sure of his estimate. An initial sample gives a S.D. of 15 lbs. What size sample yields the required estimate.

6. Let us examine a little more closely the assertion that only 30% of those of us who are drivers regularily wear seat belts. Such an assertion (unless it is mere guesswork) can be made only as a result of asking a number of people chosen at random. Suppose 100 people were chosen: estimate the proportion wearing seat belts with 95% confidence. How many people would have to be interviewed if we wished to fix the proportion within ±1% with 95% confidence.

EXERCISES TO CHAPTER SIX

1. An item X is drawn at random from the distribution $N(28,2)$. Find

    (a) $P(x > 30)$
    (b) $P(x < 27)$
    (c) $P(25.5 < x < 30.5)$
    (d) $P(29.8 < x < 31.2)$

2. A sample of 25 items is drawn from the distribution $N(100,25)$. If $\bar{x}$ is the sample mean, find

    (a) $P(\bar{x} < 106.5)$
    (b) $P(\bar{x} > 104.2)$
    (c) $P(\bar{x} < 97.6)$
    (d) $P(95.4 < \bar{x} < 105.8)$

3. A T.V. tube manufacturer wishes to guarantee a minimum life for his tubes, but he does not want more than 5% of his output to violate the guarantee. From past experience, he knows that the

life of tubes are Normally Distributed with a mean 8,200 hrs. and a S.D. 340 hrs. What guaranteed minimum life should he quote?

4. A firm wishes to manufacture components whose lengths must be within the tolerance limits 3" ± 0.01". It can purchase cutting machines which operate with a standard deviation of 0.002". Show that the machine is easily capable of meeting the tolerances. Cheaper (but less accurate) machines are available for purchase: what is the greatest standard deviation for such a machine if the firm wishes to be 99% sure that the tolerances are being met?

5. A machine automatically adds and mixes chloramphenicol to an ophthalmic ointment. The ointment must contain 1% ± 0.01% chloramphenicol, and any batch outside the tolerances must be destroyed at a cost of £15 per batch. The machine operates to a mean of 1% and S.D. of 0.004% and produces 10,000 batches per year. What is the total annual cost of destroyed batches? The firm is considering replacing its machinery with a new machine which costs an extra £500 per year to rent and operate. If the new machine operates with a standard deviation of 0.003%, is the machine worth purchasing?

6. A telephone switchboard handles an average 25 calls per hour. Assuming a Poisson distribution, what is the standard deviation of the number of calls? Suppose the switchboard is capable of handling 35 calls in an hour, what is the probability that the switchboard is overloaded? (Use a Normal Approximation.)

7. An automatic machine fills bags of sugar. The mean weight of the filled bags is 32 ozs., with a standard deviation of 0.2 ozs. The quality control department checks the weight of bags periodically. If a sample of 100 bags has a mean weight of 31.9 ozs., what would you conclude?

8. There are three distinct stages in the servicing of a particular machine. The times spent on each stage are

Stage	Mean	S.D.
Cleaning	18 mins.	1 min.
Greasing	12 mins.	0.5 mins.
Setting	10 mins.	0.5 mins.

Assuming that the times spent on each stage are independent of each other, and are Normally Distributed, estimate the probability that a machine is serviced in less than 37.5 mins.

9. A manufacturer wishes to estimate the mean diameter of a very large batch of ball bearings. In order to assess their suitability for

a particular component, a sample of 1000 was drawn at random and the diameters were carefully measured.

Diameter (cm)	f.
0.0030–0.0031	9
0.0031–0.0032	66
0.0032–0.0033	285
0.0033–0.0034	400
0.0034–0.0035	192
0.0035–0.0036	39
0.0036–0.0037	9

Give a 99% confidence limit for the population mean in future.

10. A bulk supplier of steel beams is required to quote the average breaking strength of each batch he produces and delivers. His estimate must be within ±10 lbs. weight of the true mean with 99% probability. From past experience, he knows that breaking strengths have a standard deviation of 30 lb. What size sample would yield the required estimate?

11. Your sales manager excitedly informs you that he has conducted a survey of competitors customers, and has discovered that 5% intend to switch to your product on the next purchase. He advises you to invest now to take advantage of this increase. The sales manager interviewed 100 people. What would you tell him?

12. To be sure that it is worthwhile to expand capacity, you must be convinced within ±½% of your sales increase. Using the information given by your sales manager in exercise 11, what size sample should be examined? What would you conclude?

Chapter Seven

Tests of Significance

Everyone who studies statistics for any length of time rapidly becomes aware that, apart from such rare exceptions as the Census of Population, statistical investigations are examinations of samples, some large, some small. Almost every day some opinion poll tells us what the population is thinking or how it is going to behave, or how people have reacted to a particular event; and all this by obtaining the opinions of a few thousand people or even less. You must have wondered, especially in circumstances when you know of no-one who agrees with the pollster, just how much reliance you can place on sample surveys. And, of course, when a survey is proved to be wrong by the course of events, the doubting Thomases have a field day.

One of the more important functions of statistical tests of significance is to give us a concrete quantitative measure of the reliability of sample results. Of course, in the very nature of things, we can never be absolutely sure that we have drawn the correct conclusion; but most people would be content if we were able to say that the probability of our being correct is .95 or .99 or some such figure. This is precisely what we did in the last chapter when we estimated population means and population proportions using a limited amount of information.

A secondary problem now arises however which is of crucial importance to those who use sample results. We live in a dynamic society in which people differ, and both over time and in different geographical areas sample surveys give us results which indicate that the pattern of behaviour is changing or that people in different areas behave differently and think differently. What we must know is whether such results are different from previous ones because behaviour patterns are changing or whether the differences are merely the results of errors which are inevitable in sampling. It is no use changing a successful sales policy unless you can be reasonably certain that the information you have really does indicate the need for a change.

Tests of Hypothesis

All the tests we have so far looked at have been concerned with an attempt to discover 'truth'; the question we have been trying to answer is 'what is the true average?' or 'what is the true proportion?' Now we must look at rather a different problem, in which we have not merely one sample result to consider, but two, or possibly more, results, each of which seems to lead us to a different conclusion.

The manufacturer of a new dry cleaning fluid may make the claim that his product is better than any other on the market, on the grounds that it will remove at least 90% of all spots on garments cleaned by his product. A test involving his fluid and another well known brand shows that the new fluid removes 92.1% of all spots while the older brand removed 87.3%.

A problem immediately arises. Is the difference between the two results sufficient evidence for us to say that the new fluid is better than the old? Without conducting any statistical tests four possible answers may arise:

a. The fluids are of equal efficacy and
   (i) we accept this hypothesis, or
   (ii) we reject this hypothesis.
b. Fluid A is better than fluid B or vice versa and
   (i) we accept this hypothesis, or
   (ii) we reject this hypothesis

Now, if you think about these two alternatives, it should be obvious that two main types of error may arise. The hypothesis we are considering may be true and yet we still reject it. Alternatively the hypothesis may be false yet we accept it as true. What we need is some means which will indicate to us whether a hypothesis is true or false. Once again it is important to stress that the tests we will consider will not provide a definitive answer to our problem. They can do no more than provide indicators, and only our own ability will enable us to use these indicators with effectiveness.

In this chapter we will always start by setting up a standard hypothesis:

$$H_o:$$
(The hypothesis to be tested)

$$\mu_1 = \mu_2$$
(The arithmetic means of the two populations from which the samples are drawn are the same)

$H_o: \mu_1 = \mu_2$. In other words we assume that the two samples are drawn from the same population, that the difference between the results is not significant, and that it arises as a result of sampling errors. But before we look at methods of testing this hypothesis let us consider more fully precisely what we mean by this expression 'significant'.

In normal everyday speech we tend to use this word with very little precision, as a synonym for important or meaningful. This is, however, too vague for the statistician.

Most of our work so far has been based on the laws of chance, or probability, and it will come as no surprise to you to learn that these concepts are fundamental to an understanding of significance. Faced with disparate results from samples, there is always a suspicion that the difference arises by random chance and does not indicate a genuine difference between the samples. If the difference could have arisen as a result of sampling errors it cannot really be accepted as being meaningful or as a basis for management action. Thus to the statistician, differences are only significant if they are unlikely to have occurred as a result of chance.

As before, however, we need to take the argument one step further. We can never be absolutely certain that chance has played no part without an examination of the whole population. As you are aware, there are degrees of certainty, and we must ask 'What is the probability that an event has occurred by chance?' If we decide that this probability is .05, the probability that it has not occurred by chance will be .95; i.e. we can be 95% certain that it has not occurred by chance. This level of certainty may satisfy you, in which case you could say that the event is significant at the 5% level.

Of course, only you or management can say what level of certainty will be satisfactory. Some will regard an event as significant if they are 95% certain that it could not have occurred by chance; others will demand a level of 99% certainty or even more, before they will accept that it can be regarded as significant. This decision should not be the task of the statistician. His job is to present the most accurate information possible, leaving others to act on it or not as they think fit.

One Tail and Two Tail Tests

The confidence limits we have looked at in the previous chapter have all been concerned with determining a range within which the true average will lie. In so doing we have accepted that there is a small percentage of sample results which may be both below and above the range we determine. The 95% certainty range, for example, implies that in any large group of sample averages $2\frac{1}{2}\%$ will lie above the range and $2\frac{1}{2}\%$ will lie below it. This is what we accept when we take 1.96 standard errors –

since if the distribution is normal the probability of this value being exceeded is .025 at each end of the range.

You will readily appreciate that many of the problems we meet are not concerned with values between limits, but rather with values above *or* below a given limit. A manufacturer of patent medicines for example, probably wishes to be sure that his products do not contain more than a certain quantity of a particular drug; the manufacturer of bread is concerned that each loaf is not below a certain weight, and is not so much interested in whether it exceeds the stated minimum weight by one or two ounces.

The type of test in which we are concerned with a range of values taking into account extremes at both ends of the range is known as a two tail test. Conversely, if we are interested only in extreme items at one end of the distribution it is a one tailed test. You must learn to distinguish very clearly between these two situations. The reason should be apparent but it is worth careful study.

Consider a manufacturer faced by the problem of producing within strictly defined limits. He may have tolerance limits of only plus or minus one ten thousandth of an inch. He is obviously interested to know what proportion of his output is likely to fall within these limits. In other words he desires to know the proportion of his output falling within the two tails of his distribution. He may be satisfied if 99% of his output is up to standard and would be willing to accept that 0.5% will be too large and 0.5% too small. He will then consider the range (arithmetic mean ± 2.58 standard errors).

If however he is interested only in that proportion of his output which falls below the required standard as he may be if he produces electric light bulbs with a guaranteed minimum length of life, he would consider only that proportion of output which fell below the level (arithmetic mean $- 2.58$ standard errors). He knows that there is a further 0.5% of output exceeding the level (arithmetic mean $+2.58$ standard errors) but is not interested in these since they must satisfy the objective criterion of being better than the required minimum standard.

EXAMPLE 1

Suppose that a manufacturer of a drug aims at a dose of 5 milligrams of the drug in each tablet. His equipment pressing the tablets can achieve this average but are known to have a variance of .09 milligrams when his chemist analyses a sample of 25 tablets. A dose of 6 milligrams can have serious results. What is the probability that his tablets will be harmful to a consumer?

Here we are concerned only with the possibility that a tablet contains more than 6 milligrams of the drug. If it contains less, it may be less effective but we are not worried over this possibility. We are, however, worried about the possibility that a tablet will exceed the average dose by one milligram or more; or in our notation that it will exceed (the arithmetic mean plus a standard errors) i.e. $\bar{x} + a$ s.e.'s.

In this case Standard error, $\dfrac{\sigma}{\sqrt{n}} = \dfrac{\sqrt{.09}}{\sqrt{25}} = .06$.

The upper limit of $\bar{x} + 1$ milligram is then $\bar{x} + \dfrac{1}{.06}$ standard errors

i.e. $\bar{x} + 16.67$ standard errors. Our tables do not, unfortunately show the probability of exceeding a value of $\bar{x} + 16.67$ s.e.'s. You can see however that the probability of exceeding $\bar{x} + 4$ s.e.'s is $1 - 0.99997$ or .00003, or three in a hundred thousand, and it is quite obvious that the probability of exceeding $\bar{x} + 16.67$ s.e.'s (or of administering a harmful dose) must only be one in many millions — a negligable risk.

Whether this risk, small as it is, should be accepted, is of course a matter for executive decision rather than for the statistician.

Differences Between Means

If you have understood what has gone before the procedure we are about to use should be perfectly comprehensible to you. We will first set up a hypothesis and determine at what level we will accept our figures as being significant. We will then test it as before by asking what is the maximum difference from $\bar{x}$ that we would expect to occur at this level of certainty by pure chance.

EXAMPLE 2

Suppose that a group of 30 students not using this book obtain an average examination mark of 56 with a variance of 64. A second group of 40 students using this book obtain an average mark of 63 with a variance of 49. Can we say that the second group are better statisticians than the first? Is the difference in examination performance significant?

Firstly we must set up the hypothesis that we wish to test — in this case that there is no difference between the two groups of students.

$$H_o: \bar{x}_1 = \bar{x}_2$$

Now we must examine and determine the expected difference in the two averages which could arise by chance, and to do this we will need to determine the *standard error of the difference between means*. As you are aware the combined variance of two samples is the sum of the individual variances. Since standard error is directly related to the variance and to the number in the sample you would expect the standard error of the difference between means (s.e. diff. bet means) to be the sum of the individual standard errors. In fact

$$\text{s.e. diff. bet. means} = \sqrt{\frac{s_1^2}{n_1} + \frac{s_2^2}{n_2}}$$

where s_1 and s_2 are the standard deviations of the two samples and n_1 and n_2 are the number of items in the two samples

$$\text{The s.e. diff. bet. means} = \sqrt{\frac{64}{30} + \frac{49}{40}} = \sqrt{2.133 + 1.225} = 1.832$$

Now we must decide on a level of certainty that we require. Since it is possible that we are considering changing the text book we use we would demand a high degree of certainty, say 99%. Since $\bar{x}_1 - \bar{x}_2$ can be either positive or negative, this is a two tailed test. The expected difference between the means could be in either a positive or a negative direction – hence the expected maximum difference arising by chance at the 99% certainty level is 2.58 s.e.'s. Hence we would expect a difference of up to (2.58)(1.83) to arise by chance i.e. up to 4.72. Now in fact the observed difference is $63 - 56 = 7.0$, and we are 99% certain that this could not occur by chance. We must reject the hypothesis that $\bar{x}_1 = \bar{x}_2$, and regard the students in the second group as being the better statisticians.

An alternative procedure would be to say that the observed difference of 7 marks is a difference of $\frac{7}{1.832}$ standard errors i.e. 3.8 s.e.'s. The probability of a difference as great as this occurring is shown by the tables to be only .00014 – it would occur only 14 times in 100,000. We are in fact 99.986% certain that the second group of students are better. Without further tests of course we can not be equally certain that their standard is a result of using this book!

Difference Between Proportions

EXAMPLE 3

Let us consider the timekeeping of young articled clerks. A sample group of 50 examined in Bristol over a period of time showed that 18 of them were late for work on at least one occasion. A similar group of 80 in London showed that 32 of them were late on one or more occasions.

The London clerks claim that these figures do not show them to be worse timekeepers than their colleagues in Bristol. What validity has this assertion?

Firstly consider the problem in percentage terms. 36% of the Bristol clerks were late and 40% of the London clerks.

Our hypothesis is that there is no difference in the timekeeping of the two groups:

$$H_o: \; p_1 = p_2$$

We will take our certainty level as 95% and assess 1.96 × s.e. diff. bet. proportions.

Now the s.e. diff. bet. props. $= \sqrt{\dfrac{p_1 q_1}{n_1} + \dfrac{p_2 q_2}{n_2}}$

where p_1 and p_2 are the proportions of the sample having the attribute we are considering in each sample, q_1 and q_2 are the proportions of each sample not having that attribute ($q = 1 - p$), and n_1 and n_2 are the numbers in the samples.

The s.e. diff. bet. props. $= \sqrt{\dfrac{(.36)(.64)}{50} + \dfrac{(.40)(.60)}{80}}$

$$= \sqrt{.00461 + .003} = .08724$$

The expected difference which could arise by chance is, at the 95% level

1.96 × .087 = .17052 or 17%

The observed difference is in fact only 4%, very much less than we would expect to arise by chance and we must accept the hypothesis. London articled clerks are on this evidence no worse than their brethren.

TUTORIAL ONE

1. A recent poll covering 200 people showed that only 30% of a given population supported entry into E.E.C. A pro-Common market organisation disbelieves the figures and conducts its own survey. A random sample of 400 voters drawn from another area showed that 160 favoured entry. What conclusions would you draw?

2. In question 1 it is known that the two samples exhibit dissimilar characteristics. What is the maximum difference between the two sample results that you would expect to find? (Assume a 90% certainty level.)

3. The mean score at a widely sat economics examination is 65 with a variance of 121. An economics lecturer at a certain Polytechnic wishes to obtain statistical evidence that the 33 students he entered for the examination were not average students. The only information he can obtain in addition to the national average mark is the marks of his own students. Is there sufficient evidence available to support or reject his opinion?

4. In testing a new drug on a control group of 60 patients in a large hospital it was found that 42 of them made a rapid and complete recovery. In a second test of a drug that was already in general use it was found that 268 out of 400 patients recovered. Can we say that the new drug is more effective than the old?

The Chi-Squared (χ^2) Test

χ = the Greek letter Chi, pronounced 'ky'.

We have already seen how to test the significance of the difference between two proportions. Suppose however we want to test a situation in which our data is divided not merely into two groups, but into four or five or more. This is the great advantage of the Chi Squared test — it is of wider applicability than the ones you have already learned.

EXAMPLE 4

Suppose we want to test the hypothesis that reading habits depend on the level of education. A sample of 1000 is taken and each is classified according to his educational background and the number of books read in the last month. The results are:

| Books Read | Education | | | |
	Comprehensive	Technical College	University	Total
Under 3	480	300	120	900
3 or more	20	30	50	100
Total	500	330	170	1000

Firstly we will set up the null hypothesis — that reading habits do not depend on educational background. This means that we are hypothesising that the same proportion of people in each category read three or more books as in the sample as a whole (i.e. 10%). Let us now rewrite the table inserting in brackets the figures we would expect to find if our hypothesis is correct.

Books Read	Education			
	Comprehensive	Technical College	University	Total
Under 3	480	300	120	900
	(450)	(297)	(153)	(900)
3 or more	20	30	50	100
	(50)	(33)	(17)	(100)
Total	500	330	170	1000

We are vitally concerned with the difference between these expected figures and the figures we actually obtained from our sample. Yet it should be obvious that we cannot merely look at total differences. If our calculations are correct they must total zero. (Why?) Nor can we merely ignore the sign of the difference. So as we did in calculating standard deviation we will obtain the sum of the differences squared. In other words if O = the observed frequency and E = the expected frequency we will use as a basis for analysis $\Sigma (O - E)^2$.

Now if this total is negligable we need go no further. You should be able to see that in that case the null hypothesis is proved. But what is negligable? You should have noted mentally by this time that we get the same value of $(O - E)^2$ from $(400 - 396)^2$ as we do from $(6 - 2)^2$, and it is difficult to deny that the former difference is of much less relative importance than the latter. We ought in other words to be concerned with relative differences. So what we consider is the relative magnitude of the difference of the observed from the expected frequency.

Thus we measure the value of χ^2 as:

$$\chi^2 = \Sigma \frac{(O - E)^2}{E}$$

In our example

$$\chi^2 = \frac{(480 - 450)^2}{450} + \frac{(300 - 297)^2}{297} + \frac{(120 - 153)^2}{153} + \frac{(20 - 50)^2}{50}$$
$$+ \frac{(30 - 33)^2}{33} + \frac{(50 - 17)^2}{17}$$

$$= 2.0 + .03 + 7.1 + 18.0 + .27 + 64.0 = 91.4$$

What does this figure mean? In the Appendix you will find a table showing values of χ^2 which you expect to arise by chance in certain circumstances. Examine this table carefully. You will see firstly in the left hand column something you have not met before — 'degrees of freedom'.

To explain this odd term consider the simple 2 × 2 table below.

	A	B	Total
a	x	20 − x	20
b	35 − x	5 + x	40
Total	35	25	60

We have freedom to insert any figure we wish (less than 20) in the top left hand box, or any appropriate figure in any of the other boxes. But once we have inserted this figure all the other figures in the table are predetermined by the totals which are given. We have here one degree of freedom.

Consider now the 2 × 3 table

	A	B	C	Total
a	x	p	40 − x − p	40
b	y	20 − p	x + p − 10	60
Total	50	20	30	100

As before we can insert any figure we wish below 40 in the top left hand box. But having decided the value of x the value of y is determined for us as $(50 - x)$. Similarly we may insert any figure we wish at p, provided only that p is less than 20 and that $(x + p)$ is less than 40. But having decided on a value for x and p all other figures are determined. Here then we have two degrees of freedom.

Now do the same thing for a 3 × 3 table and a 4 × 2 table. Can you see a pattern emerging? So long as the table is a 2 × 2 or larger table, the number of degrees of freedom is always equal to $(R - 1)(C - 1)$ where R is the number of rows and C is the number of columns in the table.

If you look now at the top row of the χ^2 table you will see headings

$$\chi^2_{.05} \quad \chi^2_{.025} \quad \chi^2_{.01} \quad \chi^2_{.005} \text{ etc.}$$

The subscripts .05, .025 etc. refer to probability and the columns give us the probability of a given value of χ^2 being exceeded by random chance.

The results of our example were $\chi^2 = 91.4$ with two degrees of freedom. Looking at our table we see that the probability of obtaining a value of greater than 10.597 with two degrees of freedom is only .005. Thus the probability of obtaining a value as high as 91.4 is infinitely small and we may say with some degree of certainty that it does not arise by chance. We must therefore reject our hypothesis and accept that there is a relationship between education and reading habits.

TUTORIAL TWO

1. How many degrees of freedom do the following tables give us:

    (a) 4 X 4   (b) 1 X 3   (c) 2 X 2   (d) 2 X 1   (e) 5 X 3?

2. Construct tables having the following degrees of freedom:

    3,   6,   2,   1,   4

3. A die is thrown 120 times and the face noted. The following results were obtained:

Face Value	1	2	3	4	5	6
No. of throws:	17	24	23	16	18	22

    Is the die biased?

4. A survey of smoking habits among hospital patients shows the following consumption of cigarettes and the incidence of lung cancer:

	Daily cigarette consumption			
	Under 5	5–14	15 and over	Total
Lung Cancer	7	19	15	41
Other Diseases	12	10	6	28

    Do these figures show a connection between smoking and lung cancer?

Goodness of Fit

A different but very important use of the χ^2 test is to establish whether observed data conform to a pattern which we could have predicted using the distributions we have already studied — the Binomial and Poisson distributions. If we can show that there is a significant relationship between the frequency of occurrence of observed data and the frequency as predicted by one of these distributions we will have at our disposal an important management aid in planning.

Suppose a factory was considering the establishment of an on the spot medical team to deal with accidents rather than relying on the local hospital. One of the factors it would be necessary to consider would be the balancing of cost against the use that would be made of the service. Now if it finds from experience that the pattern of accidents follows either a Binomial or a Poisson distribution it has the means at its disposal of assessing the probability of any given number of accidents occurring daily, and this will be of material help in determining the facilities that are needed.

Imagine that a firm worried about the number of accidents that occur is about to take such a step. Its records will show the accident pattern that had occurred in the past and our task is to determine whether this pattern approximates to one of our standard distributions. Firstly we must calculate the average number of accidents ($\bar{x}$) and the standard deviation (s). Given these we can obtain the probability of, and hence the expected frequency of, any given number of accidents from the Poisson distribution using the expression

$$P(x) = e^{-a} \cdot \frac{a^x}{x!}$$

To obtain the expected frequencies from the Binomial distribution is more awkward, but if you remember that for the Binomial distribution

$$\mu = np \quad \text{and} \quad \sigma = \sqrt{npq}$$

it is apparent that

$$q = \frac{npq}{np} = \frac{\sigma^2}{\mu}$$

$$p = 1 - q$$

and

$$n = \frac{\mu}{p}$$

and we can obtain the probability of any given number of accidents happening by expanding $(q + p)^n$.

Thus we can derive an expected frequency for each number of accidents and the calculation of χ^2 using $\chi^2 = \Sigma \dfrac{(E - O)^2}{E}$ is straightforward. The process will become clear if we consider an example.

EXAMPLE 5
The following figures relate to accidents occurring in a given factory over a period of 320 days. We are required to test whether the distribution is binomial at a level of significance of .05.

Daily no. of accidents x	Frequency of occurrence f	fx	$x - \bar{x}$	$(x - \bar{x})^2$	$f(x - \bar{x})^2$
0	13	0	−2.5	6.25	81.25
1	49	49	−1.5	2.25	110.25
2	87	174	−0.5	0.25	21.75
3	109	327	+0.5	0.25	27.25
4	56	224	+1.5	2.25	126.00
5	6	30	+2.5	6.25	37.50
	320	804			404.00

$$\bar{x} = \frac{804}{320} = 2.5 \qquad \sigma = \sqrt{\frac{404}{320}} = \sqrt{1.2625}$$

Hence

$$q = \frac{\sigma^2}{\bar{x}} = \frac{1.2625}{2.5} = .505 \qquad p = (1 - q) = .495$$

$$n = \frac{\bar{x}}{p} = \frac{2.5}{.495} = 5.05$$

The Binomial expression we are concerned with $(q + p)^n$ is so close to $(.5 + .5)^5$ that we are justified in using this as a basis for calculating the expected frequency of accidents. This expansion is:

$$(.5)^5 + 5.(.5)^4(.5) + \frac{5.4.(.5)^3(.5)^2}{2.1} + \ldots\ldots$$

and this is:

$$(.5)^5 + 5.(.5)^5 + 10.(.5)^5 + 10.(.5)^5 + 5.(.5)^5 + (.5)^5$$

$$= \frac{1}{32} + \frac{5}{32} + \frac{10}{32} + \frac{10}{32} + \frac{5}{32} + \frac{1}{32}$$

Since the total observed frequency is 320 the expected frequency of each class is merely the probability obtained from the binomial expression multiplied by 320 and we have:

No. of accidents	Expected frequency	Observed frequency
0	10	13
1	50	49
2	100	87
3	100	109
4	50	56
5	10	6

We can now apply the χ^2 test as we did before. But there is one problem — how many degrees of freedom are there in a table of this nature. The standard rule is this:

Degrees of freedom = the number of values of $\dfrac{(O-E)^2}{E}$ used to

obtain the value of χ^2, less the number of values calculated from the original data

In this case we have six values of $\dfrac{(O-E)^2}{E}$ and we have calculated from the original data Σf, $\bar{x}$, and σ. Thus we have $6 - 3 = 3$ degrees of freedom. Our calculations would proceed as follows:

O	E	$(O-E)$	$(O-E)^2$	$\dfrac{(O-E)^2}{E}$
13	10	3	9	.90
49	50	−1	1	.02
87	100	−13	169	1.69
109	100	9	81	.81
56	50	6	36	.72
6	10	−4	16	1.60
				5.74

So the value of χ^2 is 5.74. Reference to our tables shows that at the .05 level with 3 degrees of freedom the critical value of χ^2 is 7.81. Do you remember what this critical value is? It is the maximum value we would expect to arise by chance. Since our calculated value of χ^2 is 5.74 (< 7.81), we may assume that the differences have arisen by chance or perhaps it is better to say that we may not assume that they have not arisen by chance. It would seem that the pattern of accidents can be represented by the Binomial Distribution $(.5 + .5)^5$.

Let us now apply a similar test to the data to see if it conforms to the Poisson distribution:

$$P(x) = e^{-a} \cdot \frac{a^x}{x!}$$

We have already calculated a, the arithmetic mean and know that it is 2.5. With this information we calculate the probability of any given number of accidents occurring as:

$$e^{-2.5} + e^{-2.5} \cdot \frac{2.5}{1} + e^{-2.5} \cdot \frac{(2.5)^2}{2.1} + e^{-2.5} \cdot \frac{(2.5)^3}{3.2.1} + \ldots$$

$$= .0821 + .0821(2.5) + \frac{.0821(6.25)}{2} + \frac{.0821(15.625)}{6} + \frac{.0821(39.06)}{24}$$

$$+ \frac{.0821(97.65)}{120} + \ldots$$

$$= .0821 + .2052 + .2565 + .2138 + .1336 + .0668 + \ldots$$

This gives us the probability of any given number of accidents per day. Our total observed frequency of accidents was 320 so we can now obtain an estimate of the frequency of occurrence of each accident level by multiplying the probability by 320. Our expected and observed frequencies are now as in the table:

(O)	(E)	$(O - E)$	$(O - E)^2$	$\dfrac{(O - E)^2}{E}$
13	26	−13	169	6.5
49	66	−17	289	4.38
87	96	− 9	81	0.84
109	68	+41	1681	24.70
56	43	+13	169	3.90
6	21	−15	225	10.70
				51.02

Now it is apparent at a glance that the Poisson distribution does not fit the series anything like so well as did the Binomial, but we must still apply the χ^2 test since we cannot see by inspection whether the differences are significant or not. In this case we have had to obtain only the total frequency and the arithmetic mean from the original data, so we have $6 - 2$ degrees of freedom. At the .05 level with four degrees of freedom the critical value of χ^2 is 9.49. Even at the .001 level it reaches only 18.47, still very much below our calculated value. We can be certain that the differences that arise are not a result of chance. The Poisson distribution does not fit the data and should not be used.

TUTORIAL THREE

1. The values of $\bar{x}$, and σ for a binomial distribution are, $\bar{x} = 12$ and $\sigma = 3$. Find p, q, and n and hence write the second term of the expansion $(q + p)^n$.

2. You are required to undertake an investigation to discover if the rate of sickness in your factory is greater for men than for women. State the appropriate null hypothesis and one other hypothesis.

3. What exactly does the critical value of χ^2 obtained from your tables represent? What does it imply when the calculated value of χ^2 exceeds this figure?

4. Toss four coins in the air and note the number of heads occurring. Repeat the experiment 20 times. Assuming that the coins have an equal chance of coming down heads or tails, calculate the theoretical distribution from the expansion $20(\frac{1}{2} + \frac{1}{2})^4$. Test the results for goodness of fit.

5. Repeat the experiment 100 times and again test the results for goodness of fit on the same assumptions.

EXERCISES TO CHAPTER SEVEN

1. A manufacturer uses a machine to pack his products. A large random sample shows that the average weight of each pack is 12 ozs. with a standard deviation of 1.02 ozs. The pack bears the words 'minimum net weight 10 ozs.' but a complaint is received from a customer that a pack he bought weighs under 10 ozs. The manufacturer wishes to know what proportion of his output is likely to weigh less than 10 ozs.

2. In a market survey it is found that 125 people out of 625 do not like the taste of a new type of mousse. You are required to estimate the true proportion of the population who do not like the new product. If it is now necessary to estimate the true proportion to within 1%, what size of sample should be taken?

3. An investigation into the beef eating habits of northern England yields the following results:

	Yorkshire	Lancashire
No. in sample	480	320
Average annual consumption of beef	52.5 lbs.	49.7 lbs.
Standard Deviation	4.2	3.6

Do these figures show that beef consumption per head is higher in Yorkshire than in Lancashire?

4. Two manufacturers are producing identical products by different methods. The first claims that a sample of 900 chosen at random shows that only 2% of his output is defective. The second manufacturer conducts a similar test on a random sample of 600 items and finds that 3.0% are defective. Is there sufficient evidence to show that the process used by the first manufacturer is superior?

5. A recruiting officer for the Scots Guards believes that Scotsmen
    on the whole are taller than Englishmen. The records show that in
    the last year he has interviewed 1250 Scotsmen and 4125 English-
    men. His medical officer gives the average height of the former as
    5 ft. $8\frac{1}{2}$ ins. with a standard deviation of 2.5 inches; and of the
    latter 5 ft. $7\frac{1}{2}$ ins. with a standard deviation of 2.5 inches. Do
    these records justify his belief?

6. If in question 5, 24% of Scotsmen and 27% of Englishmen are
    rejected on health grounds, does this show that Scotsmen are
    more healthy than Englishmen?

7. A group of university graduates are classified according to their
    results in the Advanced Level examinations and their class of
    honours awarded in their Degree Examinations. The results are
    shown below. Test whether there is a connection between
    performance at Advanced Level and performance in the Degree.

	A Level Grading			
	2 'A's or more	2 'B's or more	2 'C's or more	Totals
1st class Honours	7	2	1	10
2nd class Honours	12	34	14	60
Pass	1	14	15	30
Totals	20	50	30	100

8. The following table shows the number of deaths from road
    accidents in eight towns during a period. Fit a Poisson distribution
    to the data and test for goodness of fit

No. of deaths per week	0	1	2	3	4
No. of weeks	15	64	85	58	18

9. Using the data of question 8 obtain the theoretical Binomial ex-
    pansion and test whether the distribution of accidents is binomial.

10. Statistical problems are worked by a group of 8 students. The
    results are as follows:

Student	A	B	C	D	E	F	G	H
Incorrect answers	20	26	8	16	22	32	16	20

    Is there sufficient evidence to show that *C* is a better student than
    the others?

11. An investigation into accidents in a group of factories gives the
     following results:

Accidents per year	No. of occasions
0	53
1	30
2	9
3	5
4	3

Determine the average number of deaths per annum, calculate the
theoretical Poisson frequencies and test the series for goodness of
fit.

Chapter Eight

Quality Control

If you conducted a 'popularity poll' of all the jobs in a factory, it is an odds on bet that the inspector would be low down on the list. The factory operatives dislike the inspector because they see him as the person who rejects part of their output, and so reduces bonus earnings. Management all too often (though incorrectly) see the inspector as a necessary evil — a cost they would dearly like to dispense with! Some managers even call inspectors 'unproductive labour'. Fortunately, rational people recognise that inspectors perform a most important function. After all, quality *does* matter. Mass production has meant that goods are available to us as consumers at prices we can, on the whole, manage to afford. Unfortunately, mass production can also mean that a proportion of production will inevitably be inferior in quality. We have all had experience of annoying faults developing in the goods we buy. The inspector can, and should be the guardian of quality.

Now we do not wish to give you the impression that inspectors should solely be seen as fulfilling an important social function — they also serve the firm well (despite the cynical attitude of some sectors of management). Quality also matters to the firm — for the firm that produces shoddy goods will eventually find its sales suffering, and will find an unacceptable part of its output returned as rejects. The efficient inspector is a positive asset to the firm.

Having recognised the need for inspection, we must now decide on the form that the inspection will take. Some firms inspect finished goods in special inspection shops. The fault with this method is that as faults are found after the goods have been made, it is possible that the firm has produced large batches of defective items that have not been discovered until it is too late to do much about it. The purpose of inspection is not merely to identify defective items, but also to stop further defectives occurring. Inspection should occur as near to the point of production as possible, and inspection should be frequent. A fairly common, and commendable, system of inspection is the 'patrolling inspector'. The 'patrolling inspector' visits each machine frequently and inspects items produced by that machine, so if lapses from quality are occurring the trouble can be rectified before too many defectives have been produced. Also, the inspector working to a sensible system will keep good records of how each machine is performing. This often enables him to spot the possibility of lapses from accepted standards *before* they occur.

In few cases will there be 100% inspection — not only would this be costly, but in many cases machines can produce faster than inspectors can inspect! Quality Control then, in the main, depends upon sampling theory. The actual inspection system used will depend upon the nature of the possible faults that can occur. Industry abounds with examples of goods and components that must be within certain measurable dimensions (or tolerance limits) and initially we will look at a quality control system for such goods. Later, we will look at systems for goods which are either classed as defective or non-defective (either the product works or it does not).

We shall consider first the case of a manufacturer producing metal rods. The metal rods must have diameters within the tolerance limits 1″ to 1.01″ i.e. 1.005 ± .005. When we dealt with the mean and standard deviation, we found it convenient to take an arbitary origin, and remove the decimal point by taking a different unit of measure. We shall resort to this method again. Taking 1 inch as an origin, and one ten thousandth of an inch as unit, the tolerance limit becomes 50 ± 50. How can we set up a suitable quality control system for this manufacturer.

Step 1

Firstly, we must obtain some idea of the mean and standard deviation of the diameters of the metal rods. We do this by sampling. Twenty samples, each of five items are drawn, and their diameters measured and recorded in 10,000 ths of an inch above 1 inch.

Sample No.	1	2	3	4	5	6	7	8	9	10
	59	24	58	58	54	67	55	22	68	53
	57	28	52	67	45	43	44	54	59	74
	44	66	32	46	79	33	45	62	49	39
	53	53	35	76	46	39	40	35	42	58
	82	43	11	46	69	53	74	69	44	67
Sample Total	295	214	188	293	293	235	258	242	252	291
Sample Mean	59	43	38	59	59	47	52	48	50	58
Sample Range	38	42	47	30	34	34	34	47	26	35

Sample No.	11	12	13	14	15	16	17	18	19	20
	52	66	42	46	63	71	43	34	47	53
	73	63	58	53	65	67	66	36	45	48
	48	30	44	40	70	55	52	41	78	64
	37	68	65	31	51	36	49	31	55	46
	62	49	50	33	58	52	40	47	77	53
Sample Total	272	276	259	203	307	281	250	189	302	264
Sample Mean	54	55	60	41	61	56	50	38	60	53
Sample Range	36	38	23	22	19	35	26	16	33	18

To estimate the mean diameter of all the rods, we calculate the mean of the sample means ($\bar{\bar{x}}$ — pronounced 'x double-bar').

$$\bar{\bar{x}} = \frac{59 + 43 + 38 + \ldots}{20} = \frac{1041}{20} = 52$$

You should realise that in the original units this represents 1.0052 correct to the nearest $\frac{1}{10,000}$ inch.

We also wish to use the standard deviation of the items sampled as an estimate of the standard deviation of the entire output. We could of course, calculate the S.D. of the 100 items drawn, but this would be quite a task! In practice, it is done like this: firstly we calculate the range within each sample (in case you have forgotten, the range is the difference between the greatest and the smallest item. So the range of the first sample is $82 - 44 = 38$). Then we calculate the mean range R

$$\bar{R} = \frac{38 + 42 + 47 + \ldots}{20} = \frac{633}{20} = 32$$

(or 0.0032 in original units).

We estimate the standard deviation by multiplying the mean range by an amount A_n. Now A_n is a statistical constant that varies according to the sample size. Unfortunately, the derivation of the distribution of the constant A_n is beyond the scope of this book, but tables are available, and an extract is printed below.

The distribution of A_n for samples of n items

Sample size n	A_n	Sample size n	A_n
2	0.8862	7	0.3698
3	0.5908	8	0.3512
4	0.4857	9	0.3367
5	0.4299	10	0.3249
6	0.3946		

We have drawn samples of 5 items, so the estimated S.D. is

0.4299 × 32 = 13.76

(or 0.001376 in original units).

Step 2

Check that the machine can meet the tolerances, because if it cannot, there is little point in continuing. To do this we will assume that the distribution of the diameters is Normally distributed, i.e. $N(52,13.76)$. Think back to when we examined the Normal Distribution: we learnt that we can be 99.8% sure that the range of items within the distribution is within the limits.

$\bar{x} \pm 3.09\sigma$

or, what amounts to the same thing, the range is

$\pm 3.09\sigma$

In this case, then, we can expect (with 99.8% probability) that the diameters will lie within the range

$\pm 3.09 \times 13.76 = \pm 43$

Now to meet the tolerances, the range of diameters must be within the limits ±50, so we can conclude that the machine is capable of meeting the tolerances.

Step 3 Calculate the Warning Limits for Sample Means

In the section on sampling theory, we saw that if samples size n were drawn from the population $N(\bar{x},\sigma)$, then 95% of sample means would lie within the range

$$\bar{x} \pm \frac{1.96\sigma}{\sqrt{n}}$$

In this case, we have the 95% confidence limits as

$$52 \pm \frac{1.96 \times 13.76}{\sqrt{5}}$$

52 ± 12

i.e. 40 to 64 (1.004 − 1.0064 in original units)

We can expect 19 samples out of 20 to have Sample means within the range 40 to 64. We call this range the WARNING LIMITS for sample means.

Step 4 Calculate the Action Limits for Sample Means
Similarly, we know that 99.8% of sample means lie within the range

$$\bar{x} \pm \frac{3.09\sigma}{\sqrt{n}}$$

in this case

$$52 \pm \frac{3.09 \times 13.76}{\sqrt{5}}$$

i.e. 33 to 71 (1.0033 to 1.0071 in original units)

We can expect 499 out of 500 sample means to lie within the range 33 to 71 and we call these limits the ACTION LIMITS.

We now have a basis for controlling the quality of output. As long as only 1 sample mean in 20 lies outside the warning limits, and one sample mean in 500 is outside the action limits, we are producing a uniform quality and the process is said to be *under control*. We will consider later what action should be taken if these limits are violated.

Step 5 Calculate the Allowable Width of the Control Lines
In step two we verified that the machine was *capable* of meeting the two tolerances, but we must also check that the tolerances are actually being met. Further, we want to build into our system a method of detecting any lapses from the tolerance limits which occurs during production. This is the sort of situation we could get if the machine setting was such that the mean diameter of our rods was rather high (say, 1.006). Our samples show that with a standard deviation of 13.76, $3.09\sigma = \pm43$. But with $\bar{x}$ at 1.006, $\bar{x} \pm 3.09\sigma = 1.0103$ which is outside tolerance limits. Look carefully at the diagram below.

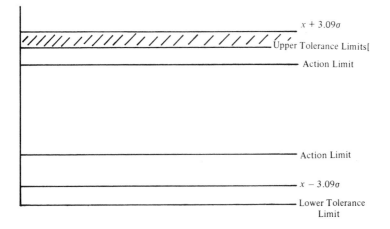

The first thing to notice is that the upper tolerance minus the lower tolerance exceeds the range of measurements allowed ($2 \times 3.09\sigma$). According to the test conducted in stage one, we can conclude that the machine is capable of meeting the tolerances. However, the upper tolerance is below the upper range for individual items, so it is certain that the upper tolerance will be exceeded (see shaded area). If we were now to use the Action Limits as our quality control system i.e. check that the means of samples do not fall outisde these limits, then the samples could indicate that production was under control and *lapses from the upper tolerance limits would go undetected.*

Can you see that if the tolerances are outside the limits $\bar{x} \pm 3.09\sigma$, then no undetected lapses from tolerance will occur? More formally, we can state that the tolerance limits must be at least

$$3.09\sigma - \frac{3.09\sigma}{\sqrt{n}}$$

outside the Action Limits if no undetected lapses from tolerance are to occur. Putting it another way, the upper action limit must be at least this amount below the upper tolerance, and the lower action limit at least this amount above the lower tolerance. This is called the *Allowable width of the control lines* (A.W.C.L.). Writing T_u for the upper tolerance and T_L for the lower tolerance, we have,

$$\text{AWCL} = T_L + \left[3.09\sigma - \frac{3.09\sigma}{\sqrt{n}} \right] \text{ to } T_u - \left[3.09\sigma - \frac{3.09\sigma}{\sqrt{n}} \right]$$

Substituting, we have

$$0 + \left[3.090 \times 13.76 - \frac{3.09 \times 13.76}{\sqrt{5}} \right]$$

$$\text{to } 100 - \left[3.090 \times 13.76 - \frac{3.09 \times 13.76}{\sqrt{5}} \right]$$

i.e. 24 to 76

As the Action Limits (33 to 71) lie well within this range, we conclude that no undetected lapses from tolerance will occur.

Step 6

We can now set up the *control chart for sample means*. In the diagram the Action and Warning Limits have been inserted, and the means of the twenty samples have been plotted. All of the samples are well within the Action Limits. Notice that two sample means fall outside the warning limits — but we would only expect one to do so. Is this significant? Probably not — it is quite likely that all of the means of the next

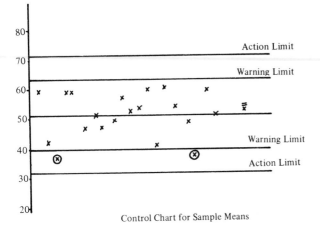

Control Chart for Sample Means

twenty samples will be within the Warning Limits.

Suppose we use the means chart as our basis for controlling quality. We draw samples of five items every (say) 15 mins. and plot the means on the chart. Suppose our results looked like this.

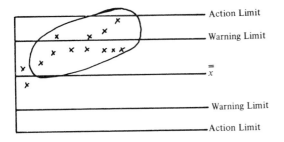

Carefully compare the two diagrams. In the first case, the sample means are clustered fairly evenly about the mean. In the second case, the means are clustering about the upper warning limit. What does this signify? It appears that the mean diameter has increased — the machine has *drifted* upwards. We would be justified in halting production and lowering the setting of the machine. Now breaks in production are expensive, and we must be sure that drifting has occurred before a resetting is justified. Commonsense tells us that in the first diagram there is insufficient evidence of a drift.

Now suppose we reset the machine and continue to draw samples. The first sample drawn has a mean less than the lower Action Limit.

What shall we do? We expect only one sample mean in 500 to be outside these limits — yet the first sample drawn is outside the lower Action Limit. It is possible that the first sample drawn is the 1 in 500 — the rogue sample is just as likely to be the first as any other. However, exceeding the Action Limits by definition demands that we take action. The appropriate action is to draw more samples immediately — we certainly would not be justified in waiting 15 minutes to draw the next sample. If their means are clustered about $\bar{\bar{x}}$, then it seems likely that we have indeed met the odd 1 in 500 case. However, if their means are clustered about the lower warning limit, then trouble is indicated.

Step 7 Set up a control chart for ranges

The means chart in itself is insufficient to control quality. It is quite possible for lapses from quality to occur without any indication from the means chart. This is because sample means cannot detect changes in the *range* of diameters. Consider the two samples below

Sample (1) 46, 48, 50, 52, 54 mean 50
Sample (2) 2, 16, 38, 50, 62, 84, mean 50

This would be plotted in the same position on the means chart, but whereas the range of the first sample is 8, the range of the second is 68! Clearly, a deterioration in quality can go undetected without a ranges chart.

The derivation of the ranges chart depend upon the factor A_n — whose derivation was stated earlier to be beyond the scope of this book. All we can do here is to state that only one sample range in 40 will exceed $\bar{R}R_w$ (the warning limit), and only one in 1000 will exceed $\bar{R}R_a$ (the action limit). The table below gives the values of R_w and R_a for varying sample sizes.

n	R_w	R_a
2	2.81	4.12
3	2.17	.299
4	1.93	2.58
5	1.81	2.36
6	1.72	2.22
7	1.66	2.12
8	1.62	2.04
9	1.58	1.99
10	1.56	1.94

We have in our example $n = 5$, $\bar{R} = 32$, so

Warning Limit = 32 × 1.81 = 58
Action Limit = 32 × 2.36 = 76

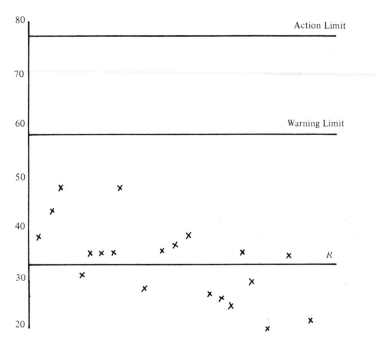

Control Chart for Sample Ranges

In the diagram, the control chart for sample ranges has been drawn. Notice that we insert only an upper Warning and Action Limit. (Why?) As only upper limits are used, the probabilities on which these limits are based are the same for both charts (i.e. if there is a 1 in 20 chance of a sample mean being outside the Warning Limits, there is a 1 in 40 chance it exceeds the upper Warning Limit). The probabilities we have used to position the limits for both charts follow *British Standards Specifications*.

Notice that the range chart indicates that production is under control. Suppose that the ranges chart indicated trouble, what remedial action could we take? The variability of output has increased and no amount of resetting would put this right! Probably the machine is wearing out (it contains some *slack*) and we may have to replace parts, renew bearings, sharpen cutting edges etc.

There are a number of constants that can be used to help you set up a quality control system. The tutorial that follows is designed to help you derive the constants for yourself.

TUTORIAL ONE

1. In step 1 we stated that if the machine was to be capable of meeting the tolerances, then

$$6.18\sigma < T_u - T_L$$

Now we know that $\sigma = \bar{R}A_n$, so

$$6.18\,\bar{R}A_n < T_u - T_L$$

i.e. $\bar{R} < \dfrac{T_u - T_L}{6.18\,A_n}$

The quantity $\dfrac{T_u - T_L}{6.18\,A_n}$ can be regarded as the maximum permitted value for the mean range. Calculate this value for the example worked in the text, and compare with the actual value of $\bar{R}$.

2. Now we know that

$$\text{Max } \bar{R} = \frac{T_u - T_L}{6.18\,A_n}$$

$$= T_u - T_L \times \frac{1}{6.18\,A_n}$$

Writing R_m for $\dfrac{1}{6.18\,A_n}$

$$\text{Max } \bar{R} = R_m(T_u - T_L)$$

Find the value of R_m when $n = 5$ (remember that A_n depends solely on the sample size). Find Max $\bar{R}$ for the example in the text, and compare with our answer to question one.

3. The Warning Limits for sample means are:

$$\bar{\bar{x}} \pm \frac{1.96\sigma}{\sqrt{n}}$$

or $\bar{\bar{x}} \pm \dfrac{1.96\,\bar{R}A_n}{\sqrt{n}}$

if $\dfrac{1.96\,A_n}{\sqrt{n}} = M_w$, then

Warning Limits $= \bar{\bar{x}} \pm M_w\bar{R}$

Evaluate M_w when $n = 5$. Use the expression above to agree the Warning Limits of the example quoted in the text.

4. If the Action Limits for sample means are

   $$\bar{\bar{x}} \pm M_a \bar{R}$$

   Write an expression for M_a and evaluate when $n = 5$. Agree the action limits for the example quoted in the text using the constant M_a.

5. If the AWCL is

   $$T_u + A\bar{R} \text{ to } T_u - A\bar{R},$$

   Write an expression for A and evaluate when $n = 5$. Agree the AWCL in the text using the constant A. You can now complete the following table for $n = 5$ and use it for the other examples you will be asked to work.

R_m	M_w	M_a	A.	R_w	R_a
				1.81	2.36

6. Some manufacturers put lower warning and action limits on their ranges chart. What could their motive be for doing this?

7. Sometimes it is useful to quote the minimum tolerance to which a machine is capable of working. What is the minimum tolerance for the machine quoted in the text. (Hint: consider its Max $\bar{R}$ and AWCL).

8. A particular component is produced to a design dimension of 20 ± 12 (in some convenient units). A large number of samples of 5 items are drawn, giving $\bar{x} = 22.5$ and $\bar{R} = 8$. Calculate the Action Limits and the AWCL. What do you notice? What action would you recommend?

Fractional Defective Quality Control – (F.D.Q.C.)

We shall now examine a quality control system for goods that are classed as either defective or non-defective. Much of what we have said for the first system also applies here – the Warning Limits and the Action Limits will be fixed at the same probability levels, and records will be kept in a similar fashion. The main difference is the lack of tolerance limits, and hence there will be neither a range chart nor an A.W.C.L.

Electric light bulbs are packed in consignments of 500 and periodically a batch is exhaustively examined for defectives. So far, 50 batches have been examined, and have yielded the following number of defectives:

```
10  8  7  6  6  6  8  8  6  6
 8  6  6  7  5  8  8  7  8  8
 6  7  9  8  8  6  6  9  6  7
 6  7  3  7  5  7  6  7  9  8
 6  8  7  6  6  6  8  8  6  6
```

We will use this information to set up a Quality Control system. Firstly, we put the data into the form of a frequency distribution, and calculate the mean and the standard deviation:

No. of defectives per sample x	No. of Samples f	fx	fx^2
3	1	3	9
4	0	0	0
5	2	10	50
6	19	114	684
7	10	70	490
8	14	112	896
9	3	27	243
10	1	10	100
	50	346	2472

$$\bar{x} = \frac{346}{50} = 6.92$$

$$\sigma = \sqrt{\frac{2472}{50} - (6.92)^2} = 1.245$$

Now, in fact, a distribution of defectives is typically Binomial, but it is more usual to assume that a Normal Distribution is a good approximation. So, we have the Warning Limit at:

$6.92 + (1.96 \times 1.245) = 9.36$ defectives

Only one sample in forty can be expected to have more than this number of defectives.

The Action Limit is:

$6.92 + (3.09 \times 1.245) = 10.77$ defectives

Only one sample in a thousand will have more than this number of defectives.

On the chart the results of the first 50 samples have been recorded. As you can see, the process is well under control.

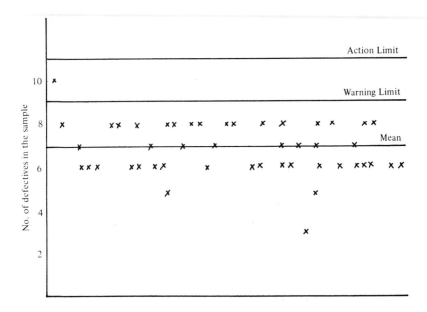

EXERCISES TO CHAPTER EIGHT

1. A manufacturer knows that a supplier of components produces one defective item in 100. He instructs his quality control department to open each batch delivered and randomly select 100 items. If more than one defective item is found, the batch is returned to the supplier. The manager of the quality control department criticises this instruction – why?

2. A manufacturer makes metal washers, and a large number of samples of 5 items have been kept. The mean range of washers is 0.0002 cms. The firm receives an order for washers within the tolerance limits of 1 ± 0.0003 cms. Should the order be accepted?

3. From a particular productive process twenty samples, each of twenty items were selected. The following number of defectives were recorded: 3, 1, 0, 2, 1, 0, 0, 1, 1, 2, 0, 0, 4, 1, 0, 3, 0, 0, 1, 0. Set up a control chart for the number of defectives in a sample. *Note.* The mean number of defectives per sample is too small to

be able to use a Normal Approximation, so the Poisson Distribution must be used. Find the appropriate Poisson Distribution, and find the number of defectives x such that $P(x$ defectives) is less than 5% (the inner limit) and find the number of defectives x_1 such that $P(x_1$ defectives) is less than 1% (the outer limit).

4. In order to set up a quality control system, 50 samples were exhaustively examined for the number of defectives. The results were:

No. of defectives	5	6	7	8	9	10	11	12	13
No. of samples	2	0	8	12	16	8	0	3	1

Set up a quality control system for the number of defectives per sample.

5. A machine is filling bags of fertiliser automatically. Samples of five were drawn, the grand mean weight being 25.02 lbs., and the mean range 0.06 lbs. Set up a quality control system using sample means and ranges. The firm receives an order on the understanding that bags will not contain less than 25 lbs. What conclusions would you draw?

6. A machine produces and gaps spark plugs to tolerance limits of 25 thousandths of an inch ±2 thousandths. Plugs outside this range would give erratic running of the engine. It is decided to set up control charts for the machine. Records reveal that the average range for samples of 5 items is 1, and the mean of the sample means is 25.1. Set up the mean and range charts, conducting a check that the tolerances are being met.

7. For the machine quoted in the last question, find the minimum tolerance to which it can operate.

8. Again for the machine in question 6, find the maximum shift that could occur in the mean before there was danger of exceeding the tolerances.

9. Samples of 5 items are drawn at periodic intervals from the output of spark plugs. On a particular day, the first ten samples were:

Sample No.	1	2	3	4	5	6	7	8	9	10
	24.8	26.0	26.1	25.6	25.7	25.2	25.3	25.5	25.1	25.9
	25.6	25.4	25.0	25.5	25.8	26.1	25.9	25.6	25.8	25.7
	25.5	24.9	25.0	25.0	25.9	25.6	25.7	26.2	26.3	25.1
	25.8	24.8	25.4	26.0	25.5	25.4	26.0	25.1	25.2	25.8
	24.8	24.9	25.5	24.9	25.1	25.2	25.1	26.1	25.1	25.5

Plot these samples on a mean and range chart. What action would you recommend?

10. Suppose that the first 10 samples had been

Sample No.	1	2	3	4	5	6	7	8	9	10
	25.6	25.2	25.5	25.1	25.0	23.2	24.3	24.8	23.7	24.9
	25.7	25.8	24.9	24.6	25.3	26.2	25.3	25.5	26.3	25.5
	24.9	25.0	24.7	24.3	26.2	25.6	24.3	26.3	26.7	26.5
	25.0	24.3	24.9	26.0	24.7	25.3	26.3	23.1	25.0	25.4
	24.8	24.7	25.5	24.5	24.8	24.7	23.8	25.8	24.8	22.7

Again plot the samples on a mean and range chart. What action would you recommend?

Chapter Nine

Inventories and Stock Control

We have discussed at length problems involving an optimum solution. Linear Programming enables us to achieve an optimum objective, given certain restrictions. The method rested on an important qualification: that the problem can be translated into a series of linear relationships. Now while it it true that many problems are linear and can be solved in this fashion, many are not. Can we solve the objective if the relationships are non-linear?

In recent years, mathematicians have given much thought to such problems. They have derived a method called DYNAMIC PROGRAMMING which promises to make a great impact in this decade. Although Dynamic Programming uses methods beyond the scope of this book, we can still examine such problems, concentrating on simple non-linear relationships.

Inventories

Let us again draw an example from the business world. Irrespective of the nature of his firm, a businessman will need to hold stock. Retailers need stocks to satisfy their customers. Manufacturers need stocks of components, raw materials and spare parts for machinery if production is to be continuous. A manufacturer may also need stocks of finished goods to satisfy his customers. Even an accountant must hold stocks of stationery!

Now the problem of holding stock (or inventories as they are more usually called) involves the firm in costs. It costs money to place an order for stock (clerical costs, telephone charges etc.), and such costs are called *ordering costs*. If production for stock is taking place, then ordering costs are replaced by the cost of preparing the machinery. Such costs are called *setup costs*. The actual holding of stock also incurs costs in the form of warehousing, stock depreciation and obsolesence. We call such costs *holding costs*. Again, costs are incurred when stocks run out. Such cost (called *stockout costs*) result from the loss of orders or customer goodwill. If the firm is producing, then stockout costs also arise from the cost of idle machinery and labour. Thus, a policy to minimise inventory costs is a rational objective.

An Inventory Model

Let us first consider a rather artificial example. A firm is selling a particular product, and the demand for the product is 60,000 units a year. The demand is at a constant rate throughout the year. The firm replenishes its stock by ordering a batch size q, and there is immediate delivery. It costs £10 to place an order and £4 to hold a unit of stock for a year. The firm wishes to know the batch size q that will minimise inventory costs.

The steady demand implies that the firm can order stock at uniform intervals, and the immediate delivery implies that re-ordering will occur when stock runs out. We could represent the situation in a diagram like this:

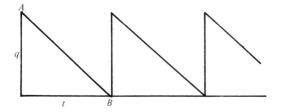

The firm orders a batch of q items and the line AB shows that stocks are depleted at a steady rate. After a time t, stock is zero and a new batch is ordered. The time interval between orders t is called the *inventory cycle*.

How many orders will the firm have to place each year? If the demand is 60,000 per annum, and the batch size ordered is q, then there will be $\dfrac{60,000}{q}$ orders per year.

As it costs £10 to place an order, the annual ordering cost will be

$$\frac{60,000}{q} \times 10$$

i.e. $\dfrac{600,000}{q}$

What are the stock holding costs? If a batch q is ordered and stock is depleted at a steady rate, then the average stock held is

$$\frac{q}{2}$$

and as it costs £4 per annum to hold a unit of stock, the annual stock holding cost will be

$$\frac{q}{2} \times 4$$

i.e. $2q$

Thus the total inventory cost per annum is

$$\frac{600,000}{q} + 2q$$

Graphical Representation of the Model

If we wish to minimise inventory costs, then we must find a value of q that makes

$$\frac{600,000}{q} + 2q$$

as small as possible. Let us select certain values of q and calculate what the total inventory cost per annum would be.

q	ordering costs (£)	holding costs (£)	total costs (£)
100	6000	200	6200
200	3000	400	3400
300	2000	600	2600
400	1500	800	2300
500	1200	1000	2200
600	1000	1200	2200
700	857	1400	2257
800	750	1600	2350

We can now graph total inventory costs (T.I.C.) against batch size q. This is done in Fig. 2. It is obvious that the optimum batch size is somewhere between 500 and 600, but it is very difficult to read its exact value from the graph. We will find later that the optimum batch size is 548. Could you have deduced this from the graph?

Figure 2

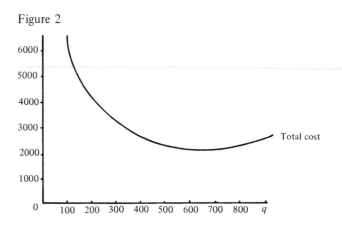

Marginal Cost

What shall we do? We could calculate T.I.C. for all values of q between 500 and 600, but the object of mathematics is to eliminate trial and error solutions. Let us raid the economists toolbox, and borrow the concept of marginal cost. When an economist considers marginal cost, he is considering the change in total cost that results from a change in output. To make marginal cost relevant to our problem, we must define marginal cost as the change in total cost that results from a change in the size of the batch ordered. Thus, if we increase batch size from 100 to 200 items, then total cost falls from 6200 to 3400. The marginal cost is

$$3400 - 6200 = -£2800$$

We can calculate the marginal cost of other changes in q.

q	Total Cost	Marginal Cost
100	6200	
		−2800
200	3400	
		− 800
300	2600	
		− 300
400	2300	
		− 100
500	2200	
		0
600	2200	
		+ 57
700	2257	
		+ 93
800	2350	

Notice that the first marginal cost has been placed between 100 and 200, and not opposite 200. This is to show that marginal cost refers to a change in batch size from 100 to 200 units. The other marginal costs have been positioned in a similar fashion.

Figure 3

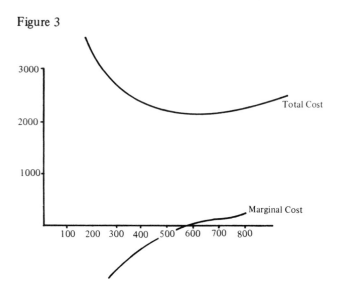

Figure 3 shows the relationship between total cost and marginal cost. Again, marginal cost has been positioned between the relevant batch sizes. Study the diagram carefully: does anything strike you, bearing in mind the problem we are trying to solve? It would seem that when total cost is at a minimum, marginal cost is zero. Now if you think about this statement, it is easy to prove logically. If total cost is falling then marginal cost must be negative (remember, marginal cost measures the change in total cost as batch size increases). Likewise, if total cost is rising, marginal cost must be positive. Now when total cost is at a minimum, it can neither be rising nor falling. Hence, marginal cost can neither be positive nor negative i.e. it must be zero.

Now there is an inconsistency here. If we agree that total cost is minimised when marginal cost is zero then, reading from the graph, the optimum batch size is 550. But if you have been reading carefully, you will remember that earlier the optimum batch size was stated to be 548. Which is correct? How has the difference arisen? The optimum batch size is 548, and 550 is an approximation. If we had considered the marginal cost of increasing the batch size by 10, or better still, by considering unit increases in batch size the approximation would have been better. The more accurate is our measurement of marginal cost, the better would be our estimate.

Has the concept of marginal cost really helped us to find minimum inventory costs? If we are to get an accurate result for the optimum batch size, this involves accurate measurement of marginal cost. Hence, we must obtain the total cost of every conceivable batch size between 500 and 600, and calculate the marginal cost of each unit change. What a waste of effort! If we had to calculate the total cost of each batch size between 500 and 600 we could see by inspection which was the optimum batch size. There would be no need to calculate marginal cost! If marginal cost is to help us, we must be able to calculate it without having to calculate total cost at all.

The Gradient of a Curve

Can we do this? Let us first consider what we are doing when we calculate marginal cost. The marginal cost of increasing the batch size from 200 to 300 units is $-$ £800. This gives a marginal cost of

$$-\frac{£800}{100} = -£8$$

Now refer back to figure 3. It is easy to see that the marginal cost per unit could have been obtained by dividing AB by BC. In other words, the marginal cost is the gradient of the line AC. Think back to the introductory chapter when it was stated that the gradient of a straight line was obtained by dividing dy by dx. In this case, dy is -800 and dx is 100. Hence

$$\frac{dy}{dx} = \frac{-800}{100} = -8$$

Now let us see whether we can calculate marginal cost without calculating total cost. Consider the diagram below. It shows how inventory costs (y) vary as batch size (q) varies.

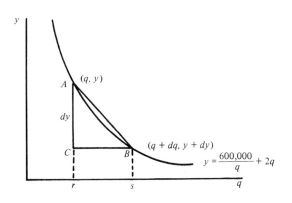

Suppose the batch size was increased from r to s, then the marginal cost would be given by the gradient of the line AB. Let the co-ordinates of the point A be (q,y). The change in total cost is AC (i.e. dy) and the change in batch size is CB (i.e. dq). Now it must follow that the co-ordinates of the point B are $(q + dq, y + dy)$, and the gradient of the line is dy/dq. Now we know that

$$y = \frac{600,000}{q} + 2q$$

Hence $y + dy = \dfrac{600,000}{q + dq} + 2(q + dq)$

Now if we subtract y from the left-hand side of this equation, we can subtract $\dfrac{600,000}{q} + 2q$ from the right hand side.

$$dy = \frac{600,000}{q + dq} + 2q + 2dq - \frac{600,000}{q} - 2q$$

$$dy = \frac{600,000q - 600,000(q + dq)}{q(q + dq)} + 2dq$$

$$dy = \frac{-600,000dq}{q(q + dq)} + 2dq$$

To obtain the gradient, we divide dy by dq

$$\frac{dy}{dq} = \frac{-600,000}{q(q + dq)} + 2$$

We have obtained a formula for calculating marginal cost that is expressed only in terms of q, the batch size.

EXAMPLE 1
Check that the marginal cost of increasing the batch size from 200 to 300 units is $-£8$.

We have

$$q = 200, \ dq = (300 - 200) = 100$$

$$\frac{dy}{dq} = \frac{-600,000}{200(200 + 100)} + 2$$

$$= -£8$$

It was stated earlier that to obtain the optimum batch size we must obtain the marginal cost more accurately. This is done by considering smaller changes in batch size, i.e. by pushing B further up the curve. How far up the curve could we push B? Suppose we pushed B to its furthest extent, then B and A would coincide. Thus, the change in batch size would be zero, and the marginal cost (dy/dq) becomes.

$$\frac{dy}{dq} = \frac{-600,000}{q(q+0)} + 2$$

$$\frac{dy}{dq} = \frac{-600,000}{q^2} + 2$$

This is the most accurate estimate we can obtain for marginal cost. Now we are able to solve the problem of the optimum batch size. We know that when inventory costs are at a minimum, marginal cost is zero. Hence,

$$\frac{-600,000}{q^2} + 2 = 0$$

$$q^2 = 300,000$$
$$q = 548 \text{ to the nearest integer}$$

Let us now summarise what we have done. We have shown that an algebraic expression can be obtained which describes the inventory model. We can find the batch size q which minimises inventory costs without calculating total cost if we find an expression for marginal cost and put it equal to zero. When we found marginal cost we found the gradient of the total cost curve at any point. We can use precisely the same method for finding the gradient of any curve.

The Gradient of $y = x^2 + 2x$

Let us use the same method to find the gradient of $y = x^2 + 2x$.

Again, let us start by finding the gradient of the line AB.

Now $y = x^2 + 2x$
so $y + dy = (x + dx)^2 + 2(x + dx)$
$y + dy = x^2 + 2xdx + (dx)^2 + 2x + 2dx$

If we subtract y from the left hand side, we can subtract $x^2 + 2x$ from the right hand side.

$$dy = 2xdx + (dx)^2 + 2dx$$

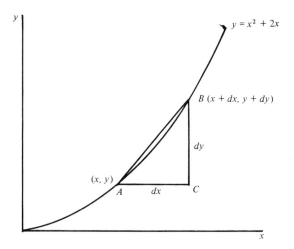

To find the gradient of the line AB, divide dy by dx.

$$\frac{dy}{dx} = \frac{2xdx + (dx)^2 + 2dx}{dx}$$

$$\frac{dy}{dx} = 2x + dx + 2$$

Now if we let B approach A, then dx approaches zero, and the gradient of the curve at any point A becomes

$$\frac{dy}{dx} = 2x + 2$$

EXAMPLE 2

Find the gradient of the curve $y = x^2 + 2x$ at the point where $x = 2$.

$$\frac{dy}{dx} = 2x + 2$$

When $x = 2$

$$\frac{dy}{dx} = 2 \times 2 + 2$$

The gradient is 6.

This method of finding the gradient of a curve is called DIFFEREN-TIATING and the gradient is called the DERIVATIVE. To use the mathematicians' jargon, we find the gradient of the curve in which y is some function of x by differentiating y with respect to x. This is precisely what the expression $\dfrac{dy}{dx}$ means.

TUTORIAL ONE

1. Use the method of the text to differentiate $y = x^2$. What is the gradient of the curve when $x = 1$?
2. Consider the sketch of the curve $y = x^2$ shown below

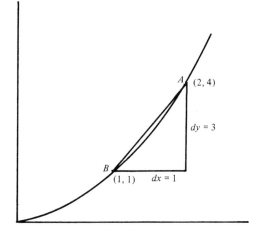

It is easy to see that the line AB has a gradient 3. We could allow A to approach B, and recalculate the gradient of the line AB (can you see that the gradient would become progressively smaller)? The results are summarised in the table below, where x and y refer to the point A, dx is x minus one and dy is y minus one.

x	2	1.5	1.2	1.1	1.01	1.001	1.0001
y	4	2.25	1.44				
dx	1	0.5	0.2				
dy	3	1.25	0.44				
dy/dx	3	2.5	2.2				

Complete the table.
What do you notice about the gradient? Does this confirm your answer to question 1.

3.   Differentiate from first principles

    (1) $y = 3x^2$
    (2) $y = 3x^2 + 3x + 2$
    (3) $y = x^3 + x^2$
    (4) $y = 2x^3 + 4x^2$
    (5) $y = \dfrac{1}{x}$

    (6) $y = \dfrac{1}{x^2} + \dfrac{1}{2x}$

Can you deduce a rule for finding dy/dx?

General Differentiation

In the last tutorial you were asked to deduce a rule for finding dy/dx. The rule is this: multiply the index by the coefficient of x (which gives the coefficient of the derivative) and subtract one from the index (which gives the index of the derivative).

    Thus, if $y = mx^n$

$$\frac{dy}{dx} = mnx^{n-1}$$

EXAMPLE 3

    Differentiate $y = 4x^4 + 2x^3 - 5x^2 - 6x + 7$
                     $y = 4x^4 + 2x^3 - 5x^2 - 6x^1 + 7x^0$

(The inclusion of x^0 does not alter the equation as $x^0 = 1$. It has been included to show how to deal with terms that are not functions of x)

$$\frac{dy}{dx} = 16x^3 + 6x^2 - 10x - 6$$

Notice that the constant 7 does not affect the derivative – why?

EXAMPLE 4

    Differentiate $y = \dfrac{600,000}{q} + 2q$ with respect to q.

Firstly, we write the expression in negative index form.

$$y = 600{,}000q^{-1} + 2q$$

$$\frac{dy}{dq} = -600{,}000q^{-2} + 2$$

$$\frac{dy}{dq} = \frac{-600{,}000}{q^2} + 2$$

This is the inventory model, and you can now see how easy it is to obtain an expression for marginal costs.

The method is not merely applicable to inventory models; we can use the derivative to find the maximum and/or minimum value of any non-linear function. Mathematicians would express the problem this way: if we are given that y is a function of x and wish to find the value(s) of x that make y a maximum, minimum or both, then differentiate y with respect to x and put the derivative equal to zero.

Second Derivatives

Two questions present themselves from the last statement. Firstly, how can we tell if an expression has either a maximum or a minimum value? (You should be able to answer this for yourself). Secondly, how can we distinguish between maximum and minimum values? To investigate this problem, let us consider the expression

$$y = 4x^3 + 3x^2 - 36x + 24$$

$$\frac{dy}{dx} = 12x^2 + 6x - 36$$

Putting the derivative equal to zero

$$12x^2 + 6x - 36 = 0$$
$$2x^2 + x - 6 = 0$$
$$2x^2 + 4x - 3x - 6 = 0$$
$$2x(x + 2) - 3(x + 2) = 0$$
$$(2x - 3)(x + 2) = 0$$
$$\text{either } x = 1.5 \text{ or } x = -2$$

Now if we wish to know the maximum or minimum values of the expression, the derivative tells us to concentrate our attention where $x = 1.5$ and where $x = -2$, now look at diagram 7 which shows the graph of

$$y = 4x^3 + 3x^2 - 36x + 24$$

together with the graph of its derivative. Clearly the expression has a maximum value when $x = -2$, and a minimum value when $x = 1.5$.

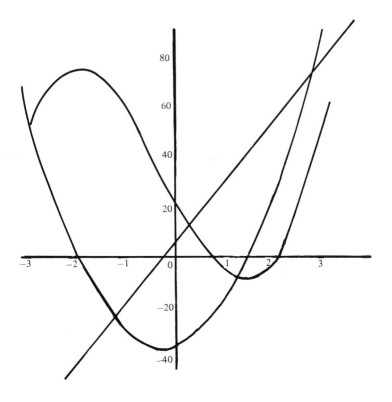

Notice that at these points the derivative has a zero value.

So far we have considered the derivative as a measure of the gradient of a curve, but if you think carefully you will realise that it also gives the rate of change of y with respect to x. Look again at the graph of the derivative. If x is less than -2, the derivative has a POSITIVE value, and the expression $y = 4x^3 + 3x^2 - 36x + 24$ is INCREASING. If x is between -2 and 1.5, the derivative has a NEGATIVE value and the expression is DECREASING. If the value of x exceeds 1.5, the derivative again becomes POSITIVE and the expression is again INCREASING.

What can we conclude from this? At the maximum value of an expression its derivative changes from positive to negative i.e. the derivative must be DECREASING. At the minimum value of an expression, its derivative changes from negative to positive i.e. the derivative must be INCREASING. How can we tell if the derivative is increasing or decreasing? Surely by differentiating the derivative. This is called taking the SECOND DERIVATIVE, $\left(\text{notation}\dfrac{d^2y}{dx^2}\right)$

If $y = 4x^3 + 3x^2 - 36x + 24$

$$\frac{dy}{dx} = 12x^2 + 6x - 36$$

$$\frac{d^2y}{dx^2} = 24x + 6$$

Using the same argument as before, if the derivative is decreasing then the second derivative is negative. Again this is shown in diagram 7. If the derivative is increasing, the second derivative is positive. We now have a method for determining whether a TURNING POINT is a maximum or minimum value. For a MAXIMUM VALUE, the derivative will be decreasing and the second derivative NEGATIVE. For a MINIMUM VALUE, the derivative will be increasing and the second derivative POSITIVE. In practice this is much simpler than it sounds. We know that the expression

$$y = 4x^3 + 3x^2 - 36x + 24$$

has turning points where $x = 1.5$ and where $x = -2$. Also we know that

$$\frac{d^2y}{dx^2} = 24x + 6$$

when $x = -2$

$$\frac{d^2y}{dx^2} = 24 \times -2 + 6 = -42$$

Thus the second derivative is negative, so a maximum value occurs when $x = -2$. When $x = 1.5$

$$\frac{d^2y}{dx^2} = 24 \times 1.5 + 6 = 42$$

The second derivative is positive, so a minimum value occurs when $x = 1.5$.

TUTORIAL TWO
1. Show that the total cost equation for the inventory model has a minimum value.

2. Determine which of the following have a maximum and which has a minimum value

   (a) $y = mx^2 - c$
   (b) $y = c - mx^2$

3. How many turning points do the following equations have?

   (a) $y = mx + c$
   (b) $y = ax^2 + bx + c$
   (c) $y = ax^3 + bx^2 + cx + d$

4. A motorist pays £25 road tax and £30 per year insurance. His car does 20 miles to the gallon, and petrol costs 35p per gallon. His car is serviced every 3000 miles at a cost of £10. Depreciation per mile increases as mileage increases, and can be calculated by multiplying the square of the mileage by 0.001p. If he does x miles per year, derive an expression for his motoring cost per mile. Find the mileage that would minimise cost per mile.

5. The fixed costs of producing a commodity is £50,000. This is made up from the cost of obtaining premises and plant. The variable costs (labour, raw materials etc.) have been calculated to be 50p per unit. How much should be produced to minimise average production costs? What can you conclude?

6. Economists tell us that under certain conditions (which they call Perfect Competition) a producer would maximise his profit by equating marginal cost with price. Their argument runs something like this: if at a particular level of output marginal cost is £1 and price is £2 and if an extra unit is produced, total cost would rise by £1 but total revenue would rise by £2. Profit would rise by £1. Thus, if price exceeds marginal cost it is worthwhile increasing output. Now suppose the magnitudes are reversed: marginal cost is £2 and price is £1. If the producer reduces output by 1 unit, the total cost falls by £2, but total revenue falls by £1. Thus profit would again rise by £1. If marginal cost exceeds price, then reducing output increases profit. Thus, so claim the economists, profit would be maximised at the output where price and marginal cost are equal. Suppose that a producer is operating under such condition, and has a total cost of $TC = 0.0025x^2 + 0.005x$ per week where x is the output. If he can sell his output at £1 per unit, how much should he produce to maximise profit?

7. A container manufacturer receives an order for 400 cu. ins. containers to be made in cylindrical form. The manufacturer wishes to

construct the containers from the smallest area of sheet steel. Find the internal dimensions of the can that would achieve the objective. You may leave your answer in terms of π.

Now let us see if we can solve more difficult inventory problems using the differential calculus. The previous example was concerned with a firm buying goods for stock. Now let us consider a case where production for stock takes place. It costs a firm £50 to set up its machinery for production, and the machinery, once started, can produce 4000 units per week. Demand for the product is a steady 2000 units per week and it costs £0.001 to hold a unit of stock per week. We wish to know the production run size q that would minimise inventory costs, and the inventory cycle i.e. the time between production runs.

We could represent the inventory cycle like this:

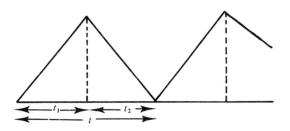

You should note that the triangle is isosceles because stock accumulation is at the same rate as stock depletion. Generally, this is not the case and later you will be asked to work examples illustrating this.

Production will take place for t_1 weeks during which inventories will be accumulating at a rate of $(4000 - 2000)$ i.e. 2000 per week. After t_1 weeks, production stops and inventories are at a peak level. Inventories will be depleted for t_2 weeks at a rate of 2000 per week, after which time the stock held will be zero and a new production run will be started. Thus, the inventory cycle is $t_1 + t_2 = t$ weeks.

Can we deduce anything about the time of the production run? If the total amount produced in t_1 weeks is q, and the rate of production is 4000 per week, it follows that

$$t_1 = \frac{q}{4000}$$

Likewise the amount produced q will be used up at a rate of 2000 per week, so all the output will be used in $\dfrac{q}{2000}$ weeks.

Now we know that stock will be zero in t weeks, hence

$$t = \frac{q}{2000}$$

If you think carefully, you will realise that the area of one of these triangles represents the average stock held during the period of t weeks, i.e. during the cycle. Now to find the area of the triangle, we must multiply half the base by the height. Half the base is $\dfrac{t}{2}$, but what is the height? It is the peak inventory held during the cycle. The peak inventory will be reached when the production run ends (i.e. after t_1 weeks), and the inventory held at this time is the amount produced minus the amount demanded. The amount produced is q, and the amount demanded is 2000 per week for t_1 weeks i.e. $2000 t_1$. Thus, the peak inventory is

$$q - 2000 t_1$$

We already know that t_1 is the same as $\dfrac{q}{4000}$, and so the peak inventory is

$$q - 2000 \times \frac{q}{4000}$$
$$= q - \frac{q}{2}$$

Thus, the area of the triangle is $\dfrac{t}{2}\left(q - \dfrac{q}{2}\right)$, and as this also represents the average inventory held during the cycle, we can now obtain an expression for stock holding costs during the cycle. As it costs £0.001 (or $£10^{-3}$ using index form) to hold a unit of stock for 1 week, stock holding costs per cycle are

$$\frac{10^{-3} \times t}{2}\left(q - \frac{q}{2}\right)$$

If we add £50 to this, i.e. the setup cost for a production run we will have the total inventory cost per cycle i.e.

$$50 + \frac{10^{-3} \times t}{2}\left(q - \frac{q}{2}\right)$$

Dividing this expression by t will give the weekly inventory costs.

$$C = \frac{50}{t} + \frac{10^{-3}}{2} \left(q - \frac{q}{2} \right)$$

or $\dfrac{50}{t} + \dfrac{10^{-3}q}{4}$

But we already know that $t = \dfrac{q}{2000}$. Thus, the inventory costs per week are,

$$C = \frac{50 \times 2000}{q} + \frac{10^{-3}q}{4}$$

Or, using index form again,

$$C = \frac{10^5}{q} + \frac{10^{-3}q}{4}$$

To find the run size q that minimises inventory costs, we must differentiate C with respect to q and put the derivative equal to zero.

$$\frac{dc}{dq} = \frac{-10^5}{q^2} + \frac{10^{-3}}{4}$$

$$q = \sqrt{\frac{4 \times 10^5}{10^{-3}}}$$

$$= \sqrt{4 \times 10^8}$$

$$= 20,000$$

Thus, the machinery will operate until 20,000 units are produced. The units will be sold in $\dfrac{20,000}{2000} = 10$ weeks, which is the length of the inventory cycle.

So far our analysis of inventories has ignored the possibility of stock-out costs, and the time has now come to rectify this. We shall consider an example similar to the initial problem. A firm uses components at a steady rate of 60,000 per year. The firm replenishes its stock by ordering a batch size q and delivery is immediate. It costs £10 to place an order, and 40p to hold a unit of stock for 1 year. If the firm runs out of stock, then stocks of semi-finished goods pile up. It has been estimated that it would cost 40p per year for every unit that is out of stock for 1 year. This cost varies because it is more expensive to fit the component at a later date: the finished good itself must be stored until completion. Again we wish to know the batch size q to minimise inventory costs.

Now it may be cheaper for the firm to deliberately run out of stock. If this is the case, the situation could be represented like this:

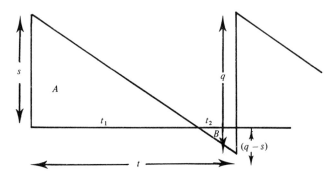

The inventory cycle i.e. the time interval between orders will last for t years. An initial batch size s will be ordered, but this is not sufficient to keep production going for t years. After t_1 years, stocks are zero and production of semi-finished goods will continue for a further t_2 years. The cycle is then complete and a new batch q is ordered. This batch must be larger than s: it must contain not only the s units to start the next cycle, but a further $(q - s)$ units to make up the stock shortage in the last cycle. It is assumed that the $(q - s)$ units are used immediately to clear the backlog of semi-finished goods. The cyclical pattern then repeats itself.

If q is the batch size ordered, then the number of orders placed per year will be

$$\frac{60,000}{q} = \frac{6 \times 10^4}{q}$$

And as it costs £10 to place an order, annual ordering costs will be:

$$\frac{6 \times 10^4 \times 10}{q} = \frac{6 \times 10^5}{q}$$

The average stock held during the period t_1 years will be given by the area of triangle A, i.e.

$$\frac{St_1}{2}$$

Now as the cycle lasts t years, the average stock held per year is

$$\frac{St_1}{2t}$$

We know that it costs £0.4 a unit for one year, so stock holding costs per year will be:

$$\frac{0.4St_1}{2t}$$

Likewise, the area of triangle B gives the components short during the time period t_2 i.e.

$$\frac{t_2(q-S)}{2}$$

and the average stockout per year is

$$\frac{t_2(q-S)}{2t}$$

Thus the stockout cost per year is

$$\frac{0.4t_2(q-S)}{2t}$$

If we call C the annual inventory cost, then

$$C = \frac{6 \times 10^5}{q} + \frac{0.4St_1}{2t} + \frac{0.4t_2(q-S)}{2t}$$

As the usage of stock is 6×10^4 per year, in t years $t(6 \times 10^4)$ would be the stock requirement. Now we know that the stock requirement for t years is q, hence

$$q = t(6 \times 10^4)$$

$$\text{and } t = \frac{q}{6 \times 10^4}$$

$$\text{likewise } t_1 = \frac{S}{6 \times 10^4}$$

$$\text{and } t_2 = \frac{(q-S)}{6 \times 10^4}$$

We can substitute these values in the cost equation, and this will leave only two unknown quantities, q and s.

$$C = \frac{6 \times 10^5}{q} + \frac{0.2S^2}{q} + \frac{0.2(q-s)^2}{q}$$

$$C = \frac{6 \times 10^5 + 0.4S^2}{q} + 0.2q - 0.4S$$

We wish to find values of q and S that would minimise C.

$$\frac{dc}{ds} = \frac{0.8S}{q} - 0.4 \text{ and } \frac{dc}{dS} = 0$$

$$\frac{0.8S}{q} = 0.4$$

$$S = \frac{q}{2}$$

$$\frac{dc}{dq} = \frac{-6 \times 10^5 - 0.4S^2}{q^2} + 0.2, \text{ and } \frac{dc}{ds} = 0$$

$$0.2q^2 = 6 \times 10^5 + 0.4S^2$$

$$q^2 = 3 \times 10^6 + 2S^2$$

But $S = \dfrac{q}{2}$

$$q^2 = 3 \times 10^6 + \frac{q^2}{2}$$

$$\tfrac{1}{2}q^2 = 3 \times 10^6$$

$$q = \sqrt{6 \times 10^6}$$

$$q = 2450$$

$$\text{and } S = \frac{2450}{2}$$

$$= 1225$$

To obtain the complete picture, we wish to know the inventory cycle. As $t = \dfrac{q}{6 \times 10^4}$, $t = 0.04$ years, and assuming a 365 day year, $t = 0.04 \times 365 = 2$ weeks approximately.

We can now summarise the solution to the problem. The firm will make an initial order of 1225 items (which will be used in a week). Two weeks later 2450 items will be ordered, of which half will be used up immediately to clear the backlog, and the remainder used during the week. Orders will then take place every two weeks.

It must be stressed that most inventory problems are unique, and each will require a unique method of solution. The examples worked in this chapter, however, are well-known and in the tutorial you will be asked to derive general expressions for each model. The weakness with each model is the assumption of constant demand. This is seldom met in practice, and in later chapters we shall see how to handle inventory models when demand varies.

TUTORIAL THREE

1. Consider the following notation,

   D = demand per time period
   R = output per time period
   C_s = ordering costs (or setup costs)
   C_1 = cost of holding a unit of inventory per time period

   Obtain expressions for q from the first two inventory models. Check your expressions by substituting and comparing with the solutions given in the text.

2. Let the inventory at the beginning of the cycle be S, and let C_2 be the stockout costs per time period. Find general expressions for q and S in the third model, again checking your answers by substitution.

3. The most convenient expression for S is model III is

$$S = \frac{C_2 q}{C_1 + C_2}$$

   What can you deduce about C_1 and C_2 if stockouts are to occur?

4. A firm must supply 50 components per day to a customer. Output is 500 per day, and because of engineering factors, the machinery must be kept running for exactly 3 days. It cost £3 to hold a unit of stock for one month, and each failure to deliver a unit costs £0.5 per day in penalties. Design a rational inventories policy (assume a 30 day month).

EXERCISES TO CHAPTER NINE

1. Differentiate:

   (a) $y = 3x^3 + 4x^2 + 5x + 3$

   (b) $y = \dfrac{2x^3 + 3x^2 + 6x + 36}{12}$

(c) $y = \dfrac{4}{x^3} + \dfrac{3}{x^2} + \dfrac{2}{x}$

(d) $y = {}^4\sqrt{x} + {}^3\sqrt{x} + \sqrt{x}$

2.   What do we obtain if we differentiate:

     (a) total cost with respect to output

     (b) distance moved with respect to time.

     (c) velocity with respect to time.

and these two are for economists.

     (d) consumption with respect to income.

     (e) national income with respect to investment.

3.   The range of a spacecraft is the distance it has travelled since lift-off. The range (in miles) after t mins. can be calculated from the expression:

$$r = \frac{60t^2 - t^3}{3}, \quad t \leqslant 8$$

Find its range after 8 minutes (to the nearest mile) and its average velocity in miles per minute.
After 8 minutes the spacecraft enters earth orbit. Find its orbital velocity and hence deduce an expression for r when $t > 8$. If the first orbit is completed after 150 mins., find the length of the orbit.

4.   After completing one orbit, the spacecraft accelerates again at the same rate as when it entered orbit. Find an expression for the velocity of the spacecraft when $t > 150$. If the spacecraft is to leave Earth's orbit, it must reach an escape velocity of 300 miles per minute — find the time when this velocity is reached.

5.   A manufacturer has found that if he wants to increase his output he must lower his price. His total revenue £R from an output x, is given by the expression: $R = x(148 - x)$. Find:

     (1) The output that would maximise total revenue.

     (2) The maximum total revenue.

     (3) The price he would have to charge to maximise total revenue.

6.   The manufacturer in the last question calculates his production costs to be £1000 in fixed costs and £36 in average variable costs per unit, so if C is total costs, $C = 1000 + 36x$. If we find $R-C$, we have an expression for Total Profit:

(a) Find $R - C$ and put this equal to P.

(b) Find the output x_1 that would maximise profit.

(c) What would be the profit earned by producing x_1 units and what price would be charged per unit?

7. Economists call the change in total revenue due to a unit change in output – 'marginal revenue'. You will probably remember that marginal cost is the change in total cost resulting from a unit change in output. Find expressions for marginal cost and marginal revenue for the product in the last question.

Moreover, economists state that profit is maximised by producing that output where marginal cost and marginal revenue are equal. Verify this using the expressions you have just derived.

8. Suppose the demand for a commodity is 2000 per year, at a steady rate. It costs £10 to place an order, and 10p to hold a unit for a year. Find the batch size q to minimise inventory costs, the number of orders placed per year, and the length of the inventory cycle.

9. It costs £100 to set up the machine for a production run, and once started, 5000 units per week can be produced. Demand is 1000 units a week, at a steady rate. It costs 0.1p per week to hold a unit of inventory, find the run size which minimises inventory costs, the run length, and the inventory cycle.

10. A manufacturer uses 25,000 components from stock per year at a steady rate. It costs him £15 to place an order and 40p to hold a unit of stock for 1 year. Stockout costs are 80p per unit per year.

    (a) Will it pay the manufacturer to run out of stock?

    (b) Find the batch size ordered to minimise inventory costs.

    (c) What is the opening inventory level s?

    (d) What is the length of the inventory cycle?

Chapter Ten

Simulation Techniques

Have you noticed how someone latches on to a particular word or phrase, and then it is picked up by the mass media and hammered at us? War 'escalates', people 'commute' to work, enforcement officers avoid 'no go' areas. We no longer have spares; we have back-up systems; we no longer spoil the countryside; we pollute the environment. It would be easy to write a long list of such words and phrases. Nowadays, we are all familiar with the word 'simulation' – how did its popularity arise?

During World War II many raw materials were in short supply or not obtainable, and were produced artificially. Such goods were stamped 'imitation', and this proved to be a marketing disaster! 'Imitation' is a highly emotive word. It means 'made in the likeness of', but it did not mean this to potential purchasers! It meant a rather inferior substitute, to be avoided at all costs. If people were to buy such goods then the 'imitation' label would have to be dropped. Thus, 'imitation' silk became 'synthetic' silk, and sales were much healthier. Although the word synthetic was an improvement, it still did not produce the required product image. Now we have 'simulated' fur and 'simulated' suede. Why did the word 'simulated' succeed when 'imitation' and 'synthetic' failed? Probably because 'simulated' is more pleasant sounding, and almost certainly because few people knew what the word meant!

The exploration of space has certainly extended the familiarity of the word simulation. To most people, it means a cartoon of an orbital docking or the re-entry of the capsule into the Earth's atmosphere. To N.A.S.A., simulation means a vital part of mission planning and mission control. At the Space Centre in Houston there is a mock-up of the space vehicles in which the astronauts familiarise themselves with layout and control systems. Clearly, it is better to learn in the simulator than to learn in the vehicle. The simulators are not merely used for astronaut training – if something goes wrong during the mission the failure can be reproduced on the simulators and experiments performed to rectify the fault. Again, it is better to experiment on the ground than in the vehicle.

The motives of simulation are clearly indicated by N.A.S.A. A problem is solved by simulation if it is too costly, too dangerous or just impractical to solve directly. Let us consider a simple example. In the previous chapter you learned that if a coin was spun three times, then the probability distribution of the number of heads occurring could be given by the Binomial Distribution $(\frac{1}{2} + \frac{1}{2})^3$. If we knew nothing about the Binomial Distribution, could we obtain the probabilities required? We could, of course spin three coins a large number of times and record the results. But this is not simulation. Is any other method available?

Monte Carlo Methods

We could use the so-called 'Monte-Carlo' method to solve this problem. The method substitutes for the problem to be solved, a simulated process with the same probabilities. The simulated probabilities are obtained by some random device (for example a roulette wheel – hence its name). The most commonly used random device is the Random Digit Table, of which an extract has been reproduced in the Appendix. The digits have been grouped for convenience, and for no other reason.

In what sense is the table random? Any digit chosen from an arbitary place in the table is equally likely to have any value from zero to nine. Further, if we were to read off a series of digits from such an arbitrary place, we could not predict the sequence of digits obtained. When deriving a simulated process it is necessary to read-off a chain of such digits, and it is most important not to use the same part of the table over and over again. Having chosen an arbitrary place to start, move up or down the table, or to the right or left; but be consistant. In other words, devise your rule for selecting random digits in advance of entering the table.

We require a simulated process with the same probabilities as the Binomial Distribution $(\frac{1}{2} + \frac{1}{2})^3$. Now the probability of a head is $\frac{1}{2}$, but so is the probability of drawing an even random digit. Hence the occurrence of an even random digit can simulate the occurrence of a head. The rules for simulating $(\frac{1}{2} + \frac{1}{2})^3$ could be:

(1) Select random digits in groups of 3.
(2) Record the following results:

3 even digits	= 3 heads
2 even and 1 odd	= 2 heads
1 even and 2 odd	= 1 head
3 odd	= no heads

Suppose, for example, that the first 10 random groups were

419	581	499	221	933	956	408	421	086	031
HTT	THT	HTT	HHT	TTT	TTH	HHH	HTH	HHH	HTT

Tabulating these results,

No. of heads	Frequency	
3	11	2
2	11	2
1	̶H̶H̶t̶	5
0	1	1

If a large number of groups was chosen, then a distribution very similar to $(\frac{1}{2} + \frac{1}{2})^3$ would be obtained. Of course, it would be much simpler to expand $(\frac{1}{2} + \frac{1}{2})^3$, but this simple example does show how simulation works.

An Inventory Model

Let us use simulation to solve an inventory problem, which in some respects is similar to the inventory model considered in chapter 9. A firm uses 200 components per day at a steady rate. Stock is replaced by purchasing from a supplier. It costs £10 to place an order. Stock holding costs are 0.1p per day. We can find the batch size q that would minimise inventory costs. You should remember that inventory costs per day will be given by the expression

$$C = \frac{200 \times 10^3}{q} + \frac{0.1q}{2}$$

and the marginal cost $\frac{dc}{dq}$ is

$$\frac{dc}{dq} = \frac{-2 \times 10^5}{q^2} + \frac{0.1}{2}$$

To find the batch size q which minimises inventory costs, marginal cost is put equal to zero.

$$\frac{-2 \times 10^5}{q^2} + \frac{0.1}{2} = 0$$

$$q = \sqrt{(4 \times 10^6)} = 2000$$

The time interval between orders (the inventory cycle) is $2000 \div 200$ i.e. 10 days.

In previous examples, we have assumed the delivery is immediate, but this is unlikely. Usually, there is a time lag between placing an order and the delivery of the goods. This time lag is called the LEAD TIME. Now if the lead time is constant, it will not affect the way we have analysed inventories. This will not be the case however, if the lead time is subject to random variations. Suppose in the example above, the lead time was never less than 2 days nor more than 4 days. The lead time distribution might look like this:

Lead Time	Frequency	Lead Time Demand
2 days	29%	400 components
3 days	48%	600 components
4 days	23%	800 components

It will no longer be sufficient to re-order when stock falls to zero. The firm has the choice of selecting three re-order levels: 400, 600 or 800. If a re-order level of 800 components is chosen (i.e. an order is placed when stock falls to 800) then no stockouts will occur. However, stock holding costs will be higher than with re-order levels of 400 or 600. If re-order levels of 400 or 600 are used, then stockouts are inevitable. We shall simulate re-order policy such that an order is placed when the stock level is 600.

Before considering the simulated system, it will be useful to consider some preliminaries. Let us call the stock at the beginning of the cycle the opening stock, and suppose it is 2000 components. This means that with a re-order level of 600 components, an order will be placed at the end of the seventh day. Now the length of the cycle will depend on the lead time. If the lead time is 2 days, the cycle length would be 9 days. Lead times of 3 or 4 days would give cycle lengths of 10 and 11 days respectively. Thus an opening stock of 2000 components could give three different cycle lengths. Must the opening stock be 2000 components? Suppose the opening stock is 2000 and the cycle length is nine days. When the new order arrives there will still be 200 components in stock, so the opening stock for the next cycle will be 2200 components. With an opening stock of 2200 components, stock will be re-ordered at the end of the eighth day and the cycle length would be 10, 11, or 12 days depending on the lead time. Thus there are six possible different cycles: a 2000 or a 2200 component opening stock each with cycle lengths of 10, 11, or 12 days.

The stock holding cost will be different for each of the six cycles. Suppose we consider the cycle which has an opening stock of 2000, and a cycle length of 11 days. The average stock held on each of the 11 days would be:

Day	Average Stock
1	1900
2	1700
3	1500
4	1300
5	1100
6	900 ——————— average
7	700
8	500 ——————— re-order
9	300
10	100
11	0

It can be easily seen that the average stock per day over the cycle is 900 components. The holding cost over the cycle would be

$$900 \times 11 \times 0.1 = £9.90$$

Stock holding costs for other cycles could be calculated in a similar fashion

Opening Stock	Cycle Length	Cost of Cycle
2000	9 days	£9.90
2000	10 days	£10.00
2000	11 days	£9.90
2200	10 days	£12.00
2200	11 days	£12.10
2200	12 days	£12.00

How could a stockout occur? This will happen if the lead time demand exceeds the re-order level. With a re-order level of 600, the lead time demand would have to be 800 for a stockout to occur i.e. if the lead time is four days. Now suppose the lead time was two days. With a re-order level of 600, lead time demand would be 400 and there would be 200 units of inventory held at the end of the cycle (or the cycle has a closing stock of 200 components.)

In order to simulate the re-order policy, we must obtain a sequence of lead times. What we require is a simulated process with the same probabilities as the lead times. Suppose we selected random digits in pairs, then there would be 100 different selections we could make (00 to 99 inclusive). The probability of selecting any one pair would be 1%.

The probability of selecting any pair in the range 00–28 inclusive would be 29% (note that the range includes 29 pairs, not 28). This is the same probability as a two day lead time. Thus if a number in this range was selected, it could simulate a two day lead time. Similarly, a selection in the range 29–76 (probability 48%) could simulate a 3 day lead time, and in the range 77–99 (probability 23%) could simulate a 4 day lead time.

We can now simulate a re-order level of 600 components. The rules for the simulated system would be as follows:

1. If opening stock is 2000, re-order at the end of the seventh day, and at the end of the 8th day if the opening stock is 2200.
2. Select a pair of random digits. The lead time is determined as follows:

Random digits	Probability	Record lead time of
00–28	29%	2 days
29–76	48%	3 days
77–99	23%	4 days

3. Add the lead time to the re-order day to obtain cycle length.
4. Enter the cost of the cycle from the table obtained earlier.
5. If lead time is 3 days, record a zero closing stock and an opening stock of 2000 for the next cycle.
6. If the lead time is 4 days record a stockout. Also record a zero closing stock and a 2000 opening stock for the next cycle.
7. If the lead time is 2 days, record a closing stock of 200 and an opening stock of 2200 for the next cycle.

If the opening stock for the first cycle is assumed to be 2000 components, the simulation could be like this:

Cycle No.	Opening Stock	Random Number	Lead Time	Cycle Length	Holding Cost	Closing Stock	Stockout
1	2000	43	3	10	10	0	NO
2	2000	40	3	10	10	0	NO
3	2000	45	3	10	10	0	NO
4	2000	86	4	11	9.9	0	YES
5	2000	98	4	11	9.9	0	YES
6	2000	03	2	9	9.9	200	NO
7	2200	92	4	12	12	0	YES
8	2000	18	2	9	9.9	200	NO
9	2200	27	2	10	12	200	NO
10	2200	46	3	11	12.1	0	NO
11	2000	57	3	10	10	0	NO
12	2000	99	4	11	9.9	0	YES
13	2000	16	2	9	9.9	200	NO

Cycle No.	Opening Stock	Random Number	Lead Time	Cycle Length	Holding Cost	Closing Stock	Stockout
14	2200	96	4	12	12	0	YES
15	2000	58	3	10	10	0	NO
16	2000	30	3	10	10	0	NO
17	2000	33	3	10	10	0	NO
18	2000	72	3	10	10	0	NO
19	2000	85	4	11	9.9	0	YES
20	2000	22	2	9	9.9	200	NO
21	2200	84	4	12	12	0	YES
22	2000	64	3	10	10	0	NO
23	2000	38	3	10	10	0	NO
24	2000	56	3	10	10	0	NO
25	2000	90	4	11	9.9	0	YES
				258	259.2		

Now 25 orders have been placed at a cost of £10 each. The ordering costs per day are:

$$\frac{250}{258} = £0.968 \text{ per day}$$

The daily holding costs are:

$$\frac{259.2}{258} = £1.005$$

Thus the simulated process gives a daily inventory cost of £1.973, with eight stockouts.

Now let us compare this result with an 800 component re-order level, when stockouts could not occur. You should satisfy yourself that the simulation would be governed by the following rules.

1. The opening stock could be 2000, 2200 or 2400 giving re-orders at the end of the 6th, 7th and 8th day respectively.
2. There are nine possible cycles, the cost of which are:

Opening Stock	Cycle Lengths	Cost
2000	8	£9.6
2000	9	£9.9
2000	10	£10
2200	9	£11.7
2200	10	£12
2200	11	£12.1
2400	10	£14
2400	11	£14.3
2400	12	£14.4

3. A lead time of two days gives a closing stock of 400 and an opening stock of 2400 for the next cycle. Lead times of 3 days and 4 days give closing stocks of 200 and zero and opening stocks for the next cycle of 2200 and 2000 respectively.
4. The cycle lengths are obtained in the same way as the previous simulation.

Using the 800 component re-order level, an opening stock of 2000 and the same lead times the simulation would look like this:

Cycle No.	Opening Stock	Lead Time	Cycle Length	Closing Stock	Holding Cost
1	2000	3	9	200	9.9
2	2200	3	10	200	12.0
3	2200	3	10	200	12.0
4	2200	4	11	0	12.1
5	2000	4	10	0	10.0
6	2000	2	8	400	9.6
7	2400	4	12	0	14.4
8	2000	2	8	400	9.6
9	2400	2	10	400	14.0
10	2400	3	11	200	14.3
11	2200	3	10	200	12.0
12	2200	4	11	0	12.1
13	2000	2	8	400	9.6
14	2400	4	12	0	14.4
15	2000	3	9	200	9.9
16	2200	3	10	200	12.0
17	2200	3	10	200	12.0
18	2200	3	10	200	12.0
19	2200	4	11	0	12.1
20	2000	2	8	400	9.6
21	2400	4	12	0	14.4
22	2000	3	9	200	9.9
23	2200	3	10	200	12.0
24	2200	3	10	200	12.0
25	2200	4	11	0	12.1
			250		294.0

Inventory costs are $\frac{250}{250} + \frac{294.0}{250} = £2.176$ per day, an increase of £0.203 over the previous model. However, the previous model contained eight stockouts which must also be costed. Suppose that if a stockout occurs, the firm can send its own van to collect the 200 components required for that day. Collection will be on a cash-and-carry basis, and the firm will obtain a discount of £6. However, it costs £10 to send the van so the net cost of cash-and-carry would be £4. Hence, the stockout cost would be £4 plus the cost of holding 1 days inventory, i.e. £4 + 100 × £0.001 = £4.10. Thus the cost of the eight stockouts would be £32.80, which gives a daily stockout cost of

$$\frac{32.80}{258} = £0.127$$

We can summarise the results like this:

Policy 1. Re-order level 800 components, inventory costs £2.176 daily.

Policy 2. Re-order level 600 components, inventory costs £2.100 daily.

On the basis of this simulation, policy 2 would be chosen.

TUTORIAL ONE

1. Using random numbers, derive simulated systems with the same probabilities as the following distributions.

(a) Lead Time	Frequency
5 days	27%
6 days	28%
7 days	35%
8 days	10%

(b) Daily Demand	Frequency
17 units	7%
18 units	8%
19 units	13%
20 units	43%
21 units	18%
22 units	11%

(c) No. of breakdowns	Frequency
1	22.3%
2	49.4%
3	28.3%

(d) No. of defectives per sample	Frequency
0	5
1	15
2	22
3	22
4	26
5	10

2. Using distributions (a) and (b) in the first question, simulate the lead time demand for 10 days.
3. Consider again the inventory model quoted in the text. Suppose policy 3 is to re-order when the stock falls to 400 components. Find the rules for the system. Using the same lead times as in the text, simulate 25 cycles and find the daily inventory cost. Compare the result with policy 2.

A Queueing Model

The final stage in the production of a certain product is inspection and rectification of faults. Previously, this was done by the distributor, but the firm making the product has decided that in future it will perform this operation itself. The production is scheduled such that one product leaves the production line every 20 minutes, but the actual time varies randomly like this:

Minutes late	3	4%
	2	6%
	1	18%
On time		36%
Minutes early	1	21%
	2	9%
	3	6%

Attempting to match production times, the inspector is supplied with sufficient mechanical aids to enable him to complete an inspection and to rectify any faults in 20 mins., on the average. Some products will be relatively free from defects and will pass through inspection in less than 20 mins., while others will take longer. The following distribution of inspection times was obtained

Time	Frequency
18	9%
19	26%
20	32%
21	23%
22	10%

The factory operates for 8 hours each day, so 24 inspections are possible. The management wishes to know whether the 24 inspections will be completed on time, and for how long during any day would the inspectors be idle. Again, we can solve this type of problem by simulation.

Firstly, let us find the distribution of random numbers with the same frequencies as production times.

Minutes late	3	4%	00—03
	2	6%	04—09
	1	18%	10—27
On time		36%	28—63
Minutes early	1	21%	64—84
	2	9%	85—93
	3	6%	94—99

The production times give the times when the products arrive at the inspection shop. It will be logical to call them 'arrival times' and mark them with a minus when late, a plus when early, and zero when on time. Let us suppose that a product is waiting for inspection at the beginning of the day (it will be the last product produced on the previous day). This means we will require 23 arrival times. The following digits were selected from a table of random numbers: 72, 94, 52, 78, 12, 21, 25, 20, 34, 93, 27, 89, 99, 97, 03, 95, 31, 50, 91, 91, 34, 46, 73.

The digits simulate the following arrival times

+1, +3, 0, +1, −1, −1, −1, −1, 0, +2, −1, +2, +3, +3, −3, +3, 0, 0, +2, +2, 0, 0, +1

If we note that the first digit is the second arrival time (the first product is waiting inspection), and arrivals are scheduled at 20 minute intervals, actual arrival times are:

0, 19, 37, 60, 79, 101, 121, 141, 161, 180, 198, 221, 238, 257, 277, 303, 317, 340, 360, 378, 398, 420, 440, 459

Now we require a sequence of inspection times. The distribution of random pairs with the same frequency as the inspection times is:

Time	Frequency	Random pairs
18	9%	00—08
19	26%	09—34
20	32%	35—66
21	23%	67—89
22	10%	90—99

Selecting twenty four random pairs will give inspection times.

27, 49, 14, 34, 05, 99, 17, 69, 11, 84, 20, 65, 05, 23, 31, 21, 79, 72, 67, 65, 53, 75, 45, 69

So the inspection times will be (in minutes)

19, 20, 19, 19, 18, 22, 19, 21, 19, 21, 19, 20, 18, 19, 19, 19, 21, 21, 21, 20, 20, 21, 20, 21

There are only three rules for this simulated system. The starting time for any inspection will be determined by either the finishing time of the previous inspection or the arrival time of the product; whichever is the later. The finishing time for any inspection will be the starting time plus the inspection time. If the arrival time of any product exceeds the time the previous inspection was finished, then the difference represents the inspector's idle time.

The simulation shows that on this particular day, the inspector was idel for 11 minutes, and the twenty-four inspections took 7 minutes longer than scheduled

Product Number	Arrival Time	Start	Inspection Time	End	Idle Time
1	0	0	19	19	0
2	19	19	20	39	0
3	37	39	19	58	0
4	60	60	19	79	2
5	79	79	18	97	0
6	101	101	22	123	4
7	121	123	19	142	0
8	141	142	21	163	0
9	161	163	19	182	0
10	180	182	21	203	0
11	198	203	19	222	0
12	221	222	20	242	0
13	238	242	18	260	0
14	257	260	19	279	0
15	277	279	19	298	0
16	303	303	19	322	5
17	317	322	21	343	0
18	340	343	21	364	0
19	360	364	21	385	0
20	378	385	20	405	0
21	398	405	20	425	0
22	420	425	21	446	0
23	440	446	20	466	0
24	459	466	21	487	0

$$\overline{11}$$

TUTORIAL TWO

1. In the last example, the inspector's idle time was examined, but no account was taken of the time products had to wait for inspection (WAITING TIME). Using the same system, find the total waiting

time. Can you deduce any relationship between waiting time and
idle time? Which, in this case, is the more important to the firm?

2. A National Health Service doctor decides to introduce an appointments system for daily consultations. A colleague supplies him
with the following information as to patient punctuality.

Minutes early	3	6%
	2	29%
	1	41%
On time		12%
Minutes late	1	7%
	2	5%

The doctor times his consultations over a period, and derives the
following frequency distribution:

12 minutes	10%
13 minutes	15%
14 minutes	28%
15 minutes	34%
16 minutes	13%

For convenience, he would like to issue appointments at 15 minute
intervals. He wishes to have an idea of his idle time, the patients
waiting time, and whether he can complete his appointments on
schedule. Simulate sixteen consultations and derive the required
information.

3. The distributions of arrivals and services at a supermarket checkout
per time period are given below

Arrivals	Frequency	Services	Frequency
8	8%	7	9%
9	22%	8	19%
10	38%	9	42%
11	26%	10	25%
12	6%	11	5%

Simulate 40 time periods and find the queue length.

4. In the previous example, the queue will become very long. If the
queue exceeds 10, it spills over into the selling area. Under such
conditions, it is necessary to open extra checkouts. Find the appropriate time periods when it is necessary to do this.

EXERCISES TO CHAPTER TEN

1. The mean number of breakdowns in a factory has a Poisson Distribution with a mean of 2 per week. Find the probabilities of 0, 1, ... 8 breakdowns per week (answers as percentages, correct to 1 decimal place). Suppose we wish to simulate a sequence of breakdowns. Using random numbers – what size group of random numbers must we choose? Find the distribution of random numbers that fits the Poisson Distribution.

2. Using the information in question one, obtain a sequence of 100 breakdowns, and form a frequency distribution. Test this distribution against the Poisson frequencies for goodness of fit.

3. The repair shop is capable of handling three breakdowns per week. On how many occasions in a hundred weeks is the repair shop overtaxed using (a) the Poisson Distribution, (b) the simulated sequence of breakdowns.

4. Items arriving per time period for servicing at a servicing shop have the following distribution:

Arrivals	0	1	2	3	4	5	6	7	8	9
%		2	7	15	20	20	16	10	6	3	1

Servicing times vary with the complexity of the fault, and the distribution of the number of services completed in each time period is

Services	1	2	3	4	5	6	7	8	9	10
%	4	8	14	18	18	15	10	7	4	2

Simulate 25 time periods and assess the maximum queue length.

5. This question refers to the production model analysed in the text. Suppose that instead of giving a distribution of deviations from scheduled production times, the problem had given *actual* production times. If the actual production times were:

Times (mins.)	17	18	19	20	21	22	23
Frequency %	4	6	18	36	21	9	6

then very different results would have been obtained. Use the same random numbers as used in the text, and rework the simulation using actual production times.

6. The time it takes a cashier to clear a customer at a supermarket checkout varies according to the number of items purchased. The

manager times the cashier, and derives the following frequency distribution.

Time taken to clear customer (secs.)	40	45	50	55	60	65	70
Frequency %		5	8	20	31	22	10	4

He has only one cashier.

The manager notices that during the busiest times, a long queue is forming and overspilling into the selling area. He fears that this is causing people to shop elsewhere. The distribution of 'inter-arrival times' of customers is

Inter-arrival time (secs.)	30	35	40	45	50
Frequency (%)	12	21	29	28	10

Simulate 25 arrivals using random numbers and find the queue length after 25 arrivals.

The manager decides that the maximum queue length allowable before customers go elsewhere is 5. Would a second cashier achieve this?

7. A manufacturer uses a component at a rate of 50 per day. It costs £6 to place an order, and £0.015 to hold a unit of inventory for a day. Find the batch size q that would minimise inventory costs, on the assumption that delivery is immediate.

   Suppose that on 52% of occasions delivery is on the day following the placing of the order, and on 48% of occasions delivery is within two days. Simulate, and cost, 25 cycles for 50 and 100 re-order levels on the assumption that it cost the firm 50p per unit of stock short per day.

8. If the machine breaks down, it costs a firm £50 in lost profit and wages for unemployed operatives if the machine is not repaired immediately. A machine usually breaks down because of the failure of a particular component, so the firm carries a stock of components to effect repairs as quickly as possible. The frequency of weekly failures has been as follows:

No. of failures	Frequency
0	26%
1	42%
2	19%
3	13%

The policy of the firm has been to make the stock of components up to 3 each week. However, the component is very large, and takes up a lot of space. If the firm would be willing to hold one component less it could install another machine which would earn the firm an extra £20 profit per week. Simulate 100 weeks and determine whether it would be worth holding 2 components per week.

Chapter Eleven

Discounted Cash Flow
and
Investment Appraisal

Suppose that when you started work as a very junior office boy you had been given the choice as to the salary scale you were to receive for the rest of your life. You might have been offered a starting salary of £1000 per year, with an annual increase of £150 at the end of each year of service. Alternatively you could have chosen the same starting salary with an annual increase of 10% of your previous years' salary. Which would you choose? It is not an easy problem to solve off the cuff, because it involves two elements. Firstly, it is obvious that the young man who has the increment of £150 a year will soon have a far higher salary than the one who takes 10% increase each year. Eventually, however the latter will catch up and pass him. The question is how long will it take? Secondly, of course, the total salary drawn by our first young man over the initial years will be the greater of the two, and we would have to determine whether there is adequate time in a working life for the second young man not only to catch up in terms of annual salary, but to earn more in terms of total salary.

We could, of course, calculate the results the hard way as follows:

| | Jones | | Smith | |
Year	Salary	Total	Salary	Total
1	1000	1000	1000	1000
2	1150	2150	1100	2100
3	1300	3450	1210	3310
4	1450	4900	1331	4641
5	1600	6500	1464	6105
6	1750	8250	1610	7715
7	1900	10,150	1710	9425
8	2050	12,200	1948	11,373
9	2200	14,400	2143	13,516
10	2350	16,750	2357	15,873

After nine increases, Smith has caught up in terms of annual salary, although his total salary is still well behind Jones's. But enough of this!

Finding the break-even annual and total salary is obviously going to be a long and tedious calculation.

Let us approach the problem the way a mathematician would, firstly writing the salaries down this way:

Jones: $1000 + 1150 + 1300 + 1450 + \ldots\ldots\ldots\ldots$ and so on.
Smith: $1000 + 1100 + 1210 + 1331 + \ldots\ldots\ldots\ldots$ and so on.

When we look at the figures we see that in both cases the salary figure is changing according to a regular pattern, and we could forecast the next figures merely by examining those we have and discovering this pattern. Figures such as this, behaving according to a pattern are known as *series*.

Further, if we think carefully these two series are of different types. In Jones's case the figures are rising by a given *amount* each time (such that any figure is the previous figure + 150). In Smiths' case, however, the figures are increasing by a given *ratio* or percentage each year (such that any figure is the previous figure + 10% of that figure, or, the previous figure multiplied by $\frac{11}{10}$).

To understand the nature of the 'simple' decision our office boy had to take, we must examine both of these series carefully.

Arithmetic Progressions

If the values of a series can be obtanied by adding a common quantity to each value, the series is called an Arithmetic Progression. The annual salary figures of Jones form the Arithmetic Progression with a first term 1000 and a common difference 150. Let us consider again this series.

The salary he earns in the second year is $1000 + 150 = 1150$
The salary he earns in the third year is $1150 + 150$
$$\text{or } 1000 + (2 \times 150) = 1300$$
The salary he earns in the fourth year is $1300 + 150$
$$\text{or } 1000 + (3 \times 150) = 1450$$

Thus, instead of writing down the salaries like this:

$1000 + 1150 + 1300 + 1450 + \ldots\ldots\ldots$

We can write them down like this:

$1000 + 1000 + (1 \times 150) + 1000 + (2 \times 150) + 1000 (3 \times 150) + \ldots$

Can you see the pattern that has emerged? Can you see that in the eighth year his salary would be

$1000 + (7 \times 150)$?

Now let us consider the general form of the Arithmetic Progression. It has a first term 'a' and a common difference 'd'.

the first term is 'a'
the second term is $a + d$
the third term is $(a + d) + d = a + 2d$
the fourth term is $(a + 2d) + d = a + 3d$
the nth term is $a + (n - 1)d$

EXAMPLE 1

Brown has a similar job to Jones. His salary scale is the same except for a long-service increment which brings his salary up to £3500 after 15 years' service. What is the long service increment?

After 15 years' service, Jones's salary will be

$$1000 + (15 - 1) \times 150 = £3100$$

The long-service increment must be £400.

Jones retires after 40 years' employment: how much has he earned during his working life? We can solve this problem by summing the series

$$1000 + 1150 + 1300 + 1450 + \ldots\ldots$$

but again this would be very tedious. Can we derive a general expression that will enable us to solve the problem without writing out all the terms in the series? Again, let 'a' be the first term and 'd' the common difference of an Arithmetic Progression. If 'S' is the sum, and there are 'n' terms, then

$$S = a + (a + d) + (a + 2d) + \ldots\ldots\ldots + a + (n - 1)d$$

If the expression is re-written with the right hand side reversed then

$$S = [a + (n - 1)d] + [a + (n - 1)d - d] + [a + (n - 1)d - 2d] + \ldots$$

Adding the two equations gives:

$$2S = [2a + (n - 1)d] + [a + d + a + (n - 1)d - d] + [a + 2d + a$$
$$+ (n - 1)d - 2d] + \ldots\ldots\ldots$$
$$2S = [2a + (n - 1)d] + [2a + (n - 1)d] + [2a + (n - 1)d] + \ldots$$

and as there are 'n' terms:

$$2S = n[2a + (n - 1)d]$$

$$S = \frac{n}{2}[2a + (n - 1)d]$$

Thus, the total amount earned by Jones during his working life is

$$\frac{40}{2} [2000 + (40 - 1) 150] = £157,000$$

Geometric Progressions

If the values of a series can be obtained by multiplying each successive term by a common quantity, the series is called a Geometric Progression. Smiths' salary forms the Geometric Progression with a first term 1000 and a common ratio 1·1.

Smith's salary in the second year is $1000 \times 1 \cdot 1 = 1100$
Smith's salary in the third year is $1100 \times 1 \cdot 1 = 1000 \times (1 \cdot 1)^2 = 1210$
Smith's salary in the fourth year is $1210 \times 1 \cdot 1 = 1000 \times (1 \cdot 1)^3$
$$= 1331$$

It should be obvious to you that Smith's salary in the 'n'th year is $1000 \times (1 \cdot 1)^{n-1}$. Likewise, the value of the 'n'th term of the geometric progression with a first term 'a' and a common ratio 'r' is

$$ar^{n-1}$$

EXAMPLE 2

Previously, we saw that Smith's salary in the 10th year was £2357. We can check this figure by substituting in ar^{n-1}, i.e.

$1000 \times (1 \cdot 1)^9 = £2358$
(the difference is due to rounding-off)

Now suppose that we wish to find the total salary earned by Smith during his 40 years working life. Again we require a general formula to solve the problem without having to write out all the terms in the series. If 'S' is the sum of 'n' terms of the Geometric Progression with a first term 'a' and a common ratio 'r' then,

$$S = a + ar + ar^2 + ar^3 + \ldots\ldots\ldots + ar^{n-1}$$

Multiplying both sides by r gives . . .

$$rS = ar + ar^2 + ar^3 + ar^4 + \ldots\ldots\ldots + ar^n$$

and subtracting the second equation from the first . . .

$$S - rS = a - ar^n$$
$$S(1 - r) = a(1 - r^n)$$

$$S = \frac{a(1 - r^n)}{1 - r}$$

The total amount earned by Smith during his working life is

$$S = \frac{1000(1 - 1 \cdot 1^{40})}{1 - 1 \cdot 1} = 442,900$$

We are now in a position to assess which of the two salary schemes is the better. If we assume that both Smith and Jones have a 40 year working life, then Smith earns a total salary of £442,900 and Jones earns £157,000. Obviously, we would prefer a 10% increase to a flat £150 per year increase.

Let us consider the G.P.

$$1 + 2 + 4 + 8 \ldots \ldots$$
$$a = 1, \quad r = 2$$

The sum of the first 4 terms is

$$\frac{1(1 - 2^4)}{1 - 2} = 15$$

The sum of the first 10 terms is

$$\frac{1(1 - 2^{10})}{1 - 2} = 1023$$

Clearly the larger the number of terms we take the greater will be the sum of the G.P.

Now consider the G.P.

$$1 + \tfrac{1}{2} + \tfrac{1}{4} + \ldots \ldots$$
$$a = 1, \quad r = \tfrac{1}{2}$$

The sum of the first two terms is $1 + \tfrac{1}{2} = 1\tfrac{1}{2}$
The sum of the first three terms is $1\tfrac{1}{2} + \tfrac{1}{4} = 1\tfrac{3}{4}$
The sum of the first four terms is $1\tfrac{3}{4} + \tfrac{1}{8} = 1\tfrac{7}{8}$
The sum of the first five terms is $1\tfrac{7}{8} + \tfrac{1}{16} = 1\tfrac{15}{16}$
The sum of the first six terms is $1\tfrac{15}{16} + \tfrac{1}{32} = 1\tfrac{31}{32}$

Can you draw any conclusion about the sum of this G.P.?

Let us try to generalise: if $r > 1$, then the sum of a G.P. will increase without limit; if $0 < r < 1$ then there will be a definite limit on the sum of the G.P. In the example above, it appears that the total can never exceed two. This can be proved as follows. Suppose we take an infinite number of terms, then the sum of the G.P. is

$$\frac{1(1 - 0)}{1 - \tfrac{1}{2}}$$

because as n approaches infinity, $(\frac{1}{2})^n$ approaches zero. Thus, the sum can never exceed

$$\frac{1(1-0)}{\frac{1}{2}} = 2$$

The process of finding the maximum sum of a G.P. is called 'summing to infinity'. If it is possible to sum to infinity (i.e. if $0 < r < 1$) then the G.P. is called convergent. If $r > 1$, then it will not be possible to sum to infinity, and the series is called divergent. Now consider the convergent G.P. with first term 'a' and common ratio 'r'.

$$s_\infty = \frac{a(1-r^n)}{1-r}$$

If $r < 1$ and $n = \infty$

$$r^n = 0$$

and $a(1 - r^n) = a$

thus $s = \dfrac{a}{1-r}$

EXAMPLE 3

Suppose that people spend $\frac{9}{10}$ ths and save $\frac{1}{10}$ th of any money received. If £1000 is received, then £900 is spent and £100 saved. The people receiving the £900 will spend £810 and save £90. This process will continue. Can you see that to find the total amount of spending generated by an initial receipt of £1000 involves summing to infinity the Geometric Progression with a first term 1000 and common ratio 'r' = 0.9

$$s = \frac{1000}{1 - 0.9} = £10,000$$

TUTORIAL ONE

1. Show that the sum of the first n odd numbers always forms a perfect square.
2. In a later chapter it will be useful to sum the series $1 + 2 + 3 + \ldots . n$. Find an expression for the first n integers.
3. In 1960 I received £500 dividend from securities I own. The dividend increased by a constant 10% per annum. What was the total dividend I earned between 1960 and 1969 inclusive?
4. You decide to increase your saving by £200 per annum. If you save £500 in the first year, how long will it take to save £5300?

5. A well known paradox is the race between the hare and the tortoise. The hare can run 100 yards in the same time that it takes the tortoise to run 1 yard. The hare gives the tortoise 100 yards start. When the hare runs the first 100 yards, the tortoise is 1 yard in front. When the hare runs the next yard, the tortoise is $\frac{1}{100}$ yard in front. When the hare runs the next $\frac{1}{100}$ yard, the tortoise is $\frac{1}{1000}$ yard in front. Thus, the hare never catches the tortoise. Can you resolve the paradox?

Compound Interest

Closely linked to geometric progressions are compound interest problems. Suppose £100 is invested at 5% per annum, then the interest earned after the first year is

$$£100 \times \frac{5}{100} = £5$$

and the value of the investment at the end of the first year would be

£100 + £5 = £105

If the rate of interest is expressed as a proportion rather than a percentage, the value of the investment can be calculated thus:

£100 + 100 × 0.05
£100 (1 + 0.05) = £105

Expressing the rate of interest as a proportion is more convenient than expressing it as a percentage. Instead of a rate of interest of $r\%$ per annum (the usual form in textbooks) we shall let the rate of interest be $100r\%$ per annum. This presents little difficulty: if the rate of interest is 7% per annum, then

$100r = 7$
$r = 0.07$

Suppose £P is invested at $100r\%$ per annum compound. At the end of the first year, the value of the investment is

$P + Pr$
$P(1 + r)$

At the end of the second year, the investemnt is

$P(1 + r) + rP(1 + r)$
$(1 + r)(P + rP)$

$(1 + r)(1 + r)P$

$P(1 + r)^2$

At the end of the third year, the value of the investment is

$P(1 + r)^2 + rP(1 + r)^2$

$(1 + r)^2(P + rP)$

$(1 + r)^2 P(1 + r)$

$P(1 + r)^3$.

It should be obvious to you from the pattern emerging that the value of the investment at the end of the nth year is

$P(1 + r)^n$ .... (1)

This formula has a disadvantage: it involves calculating $(1 + r)^n$, which is not an easy calculation. Fortunately, there is a discount table to help you. It gives values of $(1 + r)^{-n}$, (it will be realised later that this is a more convenient form than $(1 + r)^n$). The discount table in the appendix gives sufficient values for all the work you will be asked to do in this book. Rates of interest are measured horizontally and time is measured vertically. Most books of logarithms include a reciprocal table. The reciprocal table, together with the discount table will give values of $(1 + r)^n$.

The reciprocal of a number is the number divided into unity:

The reciprocal of $x = \dfrac{1}{x} = x^{-1}$

The reciprocal of $x^{-1} = \dfrac{1}{x^{-1}}$

$= \dfrac{1}{\frac{1}{x}}$

$= x$

(if this presents difficulty you should re-read the section on indices). hence, reciprocal of $(1 + r)^{-n} = (1 + r)^n$.

EXAMPLE 4

Suppose £1000 is invested at 7% per annum compound. The value of the investment at the end of the tenth year is

$1000(1 + 0.07)^{10}$

From the discount tables $(1 + 0.07)^{-10} = 0.5083$.

$(1 + 0.07)^{10} = \dfrac{1}{0.5083}$

From the reciprocal tables, $\dfrac{1}{0.5083} = 1.968$.

Hence, the value of the investment is

$1000 \times 1.968 = £1968$

Increasing the sum invested

Suppose that you decided to invest £1000 on the first of January in a certain year, and to invest a further £100 at the end of each year. If interest is compounded at 10% per annum we can deduce the following:

The amount invested at the end of the first year is
$1000(1 + 0.1) + 100$
The amount invested at the end of the second year is
$1000(1 + 0.1)^2 + 100(1 + 0.1) + 100$
The amount invested at the end of the nth year is
$1000(1 + 0.1)^n + 100(1 + 0.1)^{n-1} + 100(1 + 0.1)^{n-2} + \ldots$
$\ldots + 100$

Again it is useful to generalise. If an amount P is invested at the beginning of a year, and a further amount a is invested at the end of each year, and if s is the sum invested after n years then

$$s = P(1 + r)^n + a(1 + r)^{n-1} + a(1 + r)^{n-2} + \ldots + a$$

If the first term of the right-hand side is ignored, then the remainder forms a geometric progression with a first term $a(1 + r)^{n-1}$ and a common ratio $\dfrac{1}{1 + r}$. Substituting in the G.P. formula, the series can be summed for n years.

$$S = P(1 + r)^n + \frac{a(1 + r)^{n-1}\left[1 - \left(\dfrac{1}{1+r}\right)^n\right]}{1 - \dfrac{1}{1 + r}}$$

It now remains to tidy up this expression. Writing the denominator as a single fraction gives

$$S = P(1 + r)^n + \frac{a(1 + r)^{n-1}\left[1 - \left(\dfrac{1}{1+r}\right)^n\right]}{\dfrac{r}{1 + r}}$$

or $S = P(1 + r)^n + \dfrac{a(1 + r)^{n-1}[1 - (1 + r)^{-n}](1 + r)}{r}$

Using the rules of indices,

$$S = P(1 + r)^n + \frac{a(1 + r)^n [1 - (1 + r)^{-n}]}{r}$$

$$S = P(1 + r)^n + \frac{a(1 + r)^n - a}{r}$$

Removing the common factor $(1 + r)^n$

$$S = \left(P + \frac{a}{r}\right)(1 + r)^n - \frac{a}{r} \qquad \dots \dots (2)$$

We can now use the formula to solve the original problem. Suppose we wish to know the sum invested after 4 years, then,

$$S = \left(1000 + \frac{100}{0.1}\right)(1 + 0.1)^4 - \frac{100}{0.1}$$

$$S = 2000 (1.1)^4 - 1000$$

From the discount tables, $(1.1)^{-4} = 0.6830$.

And using reciprocal tables $(1.1)^4 = \frac{1}{0.6830} = 1.464$

$$S = 2000 \times 1.464 - 1000 = £1928$$

EXAMPLE 5
The formula can also be used when the investment is reduced by a constant amount each year. Suppose £20,000 is invested at the beginning of a year at 5% per annum compound, and £2000 is withdrawn at the end of each year for four years. In this case, 'a' is a withdrawal and will be negative. The sum remaining is

$$S = \left(20,000 - \frac{2000}{0.05}\right)(1 + 0.05)^4 - \frac{-2000}{0.05}$$

$$S = -20,000(1.05)^4 + 40,000$$

$$S = -20,000 \times 1.216 + 40,000$$

$$S = £15,680$$

The basic formula can be manipulated by changing the subject, but it is advisable to first ask what new statement would be most useful. Often, we wish to know how much must be invested now to give a specified income for a specified period, given the rate of interest. Clearly, the sum invested at the end of the period will be zero, i.e.

$$0 = \left(P - \frac{a}{r}\right)(1 + r)^n + \frac{a}{r}$$

(remember that as money is withdrawn, 'a' will be negative).

$$0 = P(1 + r)^n - \frac{a}{r}(1 + r)^n + \frac{a}{r}$$

$$P(1 + r)^n = \frac{a}{r}(1 + r)^n - \frac{a}{r}$$

$$P(1 + r)^n = \frac{a}{r}[(1 + r)^n - 1]$$

$$P = \frac{a[(1 + r)^n - 1]}{r(1 + r)^n}$$

$$P = \frac{a[1 - (1 + r)^{-n}]}{r} \qquad \ldots \ldots (3)$$

EXAMPLE 6

Suppose an investment is required to yield £1500 at the end of each year for five years, and money can be invested at 5% per annum compound, then P, the sum that must be invested at the beginning of the first year is:

$$P = \frac{1500[1 - (1 + 0.05)^{-5}]}{0.05}$$

$$P = \frac{1500 \times 0.2165}{0.05}$$

$$P = £6495$$

TUTORIAL TWO

1. Your rich uncle dies and leaves you £1500 invested in unit trusts. Your accountant advises you that the value of your investment can be expected to grow by 7% per annum (income and capital growth). Estimate the value of the investment after 10 years.
    You now decide to increase the value of the investment, and instruct your bank manager to invest £100 annually in unit trusts. If the first purchase is made twelve months after the death of your

uncle, estimate the value of your investment after ten years.

2. The managing director is due to retire at the end of the year, and the board vote than an income of £2000 per annum be paid to him or his family for 10 years. The accountant is instructed to set aside a sum of money now from which the income will be paid. If the fund can be invested at 8% per annum, how much should the accountant set aside?

3. Make 'a' (the amount withdrawn per annum) the subject of basic formula (3).
    A building society grants a £3000 mortgage at 7% per annum. The borrower is to repay the loan in 10 annual installments. How much must he pay each year? (You should satisfy yourself that the modification to formula (3) is suitable to answer this question.)

4. Basic formula (3) is more usually called the ANNUITY formula (of which more later) and you can derive this from first principles in the following fashion:
    Suppose £P is invested at the beginning of a year at $100r\%$ per annum, and at the end of that year an amount £a is withdrawn. What is the sum invested at the end of the first year?
    If a further £a is withdrawn at the end of the second year, what is the sum invested at the end of the second year?
    If the process is repeated for n years, what is the sum invested at the end of the nth year? You should notice that if the first term is ignored, you have obtained a G.P. which should now be summed to n terms.
    At the end of the nth year, the investment will have a zero value. If you put the expression for the value of the investment after n years equal to zero, and make P the subject, you will obtain an equation identical to formula (3).

5. At the end of each year a sum £a is invested at $100r\%$ per annum. If this process is repeated for n years, for how many years will the first investment have been earning interest? Write an expression for the value of the investment after n years, and put this equal to T. If the resulting G.P. is simplified, then the SINKING FUND FORMULA is derived.

$$T = a \left[\frac{(1 + r)^n - 1}{r} \right]$$

Make a the subject of this formula.
A machine has an expected life of seven years, and the buying manager estimates that its replacement price will be £5000. The

accountant is instructed to set aside a fixed sum at the end of each year to cover replacement cost. How much will be need to set aside each year if the money can be invested at 7% per annum?

The Concept of Present Value

Were some philanthropist to offer you the choice of receiving £1000 cash now, or alternatively, £1000 cash in twelve months time, it is highly likely that you would prefer to take the money now. Most people would. It seems that we have a strong preference for holding cash now as against the prospect of receiving cash in the future. This seems to be true no matter how certain we are of obtaining the money in the future. Economists call this preference, 'liquidity preference'. To the mathematician, who in many aspects of his work is trying to quantify human behaviour, the concept implies that people place a higher value on money received now than they do on the same amount of money received in the future. We may say that the £1000 we receive today has a greater value to us than the £1000 we are going to receive next year. It is still, of course, 1000 pound notes, no more, no less. But the mere fact that you can express a preference is sufficient evidence that these one pound notes have a different value according to when they are received. Preference alone, however, is of little use in mathematics. If this aspect of human behaviour is to be analysed, and if the analysis is to help us, we must find some means of quantifying the preference; that is of saying by how much we prefer £1000 now to £1000 next year, or the year after.

Fortunately the use of the interest formula developed earlier gives us a simple method of assessing the difference in value between money now and money in the future. We know that if we were to invest £1000 today at say 10% interest, at the end of twelve months it will have grown to £1100 and at the end of two years to £1210. It is thus possible for us to say that £1000 is the present value both of £1100 received one year hence, and also of £1210 received two years hence, when the rate of interest is 10%. Generalising, we may define present value as follows:

The present value of £a receivable n years hence is that sum of money which, invested at the current rate of interest r%, will amount to £a after the expiry of n years.

This fearsome looking concept becomes more manageable if you consider the formula already developed. You know already that if £P is invested for n years at r%

$$a = P(1 + r)^n \ldots \ldots$$

From this it is easy to deduce that:

$$P = \frac{a}{(1 + r)^n} = a(1 + r)^{-n} \ldots \ldots$$

These two equations are the basis of all investment appraisal. You will have realised that they are in reality merely two different forms of the same equation, but with this important difference. The first equation tells us the terminal value of a given sum of money invested at interest over time; the second tells us what we would have to invest to achieve a given terminal sum over a given period of time at a specified rate of interest. It tells us, in other words what is the present value of a given sum of money received some time in the future.

EXAMPLE 7

Suppose we wish to find the present value of £50 receivable in one year's time, given the rate of interest as 10%.

The present value, P is equal to $£50(1 + 0.1)^{-1}$.

i.e. $P = \dfrac{50}{1.1} = £45.45$

If you remember the meaning of present value, the logic of this result should be obvious. £45.45 invested at 10% for 1 year would yield interest of £4.545. Thus the value of the investment after 1 year would be £45.45 + £4.545 = £50. (The marginal difference results from the tables being accurate to only four figures.)

Now suppose we wished to calculate the present value of £1000 receivable in seven year's time, assuming a 5% rate of interest. Substituting in the formula for present value we have

$$P = 1000(1 + 0.05)^{-7}$$

You will remember with relief the discount tables that give values of $(1 + r)^{-n}$. Reading from the table we have

$$(1 + 0.05)^{-7} = 0.7107$$

hence, $P = 1000 \times 0.7107 = £710.7$

What do these tables really show? In fact, they give the present value of £10,000 receivable in n year's time at a rate of interest $100r\%$ per annum. In the tutorial that follows, you will be asked to check some of these entries.

TUTORIAL THREE
1. Calculate from first principles the present value of £10,000 receivable one year hence given a rate of interest of

    (a) 3%   (b) 5%   (c) 10%   (d) 20%

    Check your answers with the first row of the table.
2. From your answer ascertain the relationship between the rate of interest and the present value of money receivable in the future.
3. If you were buying a television set would you prefer to pay

    (a) £200 now in complete settlement, or
    (b) £100 now and a second instalment of £110 in twelve months time?

    Assume a rate of interest of 5%.
4. Would your answer be any different if the rate of interest was 20%?
5. What is the present value of £1 receivable one year hence, given a rate of interest of (a) 4%, (b) 6%.
6. Use the information obtained in question 5 to ascertain the present value of £264 receivable one year hence at rates of interest of 4% and 6%.
7. What is the maximum sum you would be prepared to pay now in return for an income of £100 receivable each year for the next four years? (Assume the rate of interest to be 8%.)
8. You can buy a car either on hire purchase terms paying a deposit of £350 and £150 a year for each of the next three years or by paying cash for the full purchase price of the car. What is the maximum cash price you would expect to pay if the current rate of interest is 6%?

Present Value and Investment Decisions

In the exercise above you have calculated the present value of a future flow of income (question 7) and of costs met at different points of time. These two calculations form the basis of most investment decisions. So far as income is concerned we would prefer to undertake that project which yields revenues having the greatest present value. But considering costs we would prefer the projects where costs have the least present value. Now it is most unlikely that we can achieve both these objectives simultaneously, and we must therefore consider first what criterion we are going to use. We will assume that the objective of the firm considering investment projects is the maximisation of profit. Such an assumption may not be true at all times for all firms, but it is a good working approximation. We can say then that the aim of the firm is to maximise the difference between the present value of the future flow of income and the present value of the flow of costs.

EXAMPLE 8

A piece of equipment costing £1000 has an expected life of 5 years. It is estimated that the cash flow resulting from the use of the machine will be £400 a year. The rate of return expected from capital of this type is 15%. Is the investment worthwhile?

The appraisal would be as follows:

Year	Income	$(1 + 0.15)^{-n}$	Present Value
1	400	.8696	347.84
2	400	.7561	302.44
3	400	.6575	263.00
4	400	.5718	228.72
5	400	.4972	198.88
			1340.88

Thus, for an investment of £1000 now we will receive a return over the next five years which has a present value of £1340.88. What does this mean? It can mean on the one hand that to obtain a return of £400 a year for five years, we would have to invest £1340.88 at 15%, and hence the machine is a good investment. Alternatively we can look at it this way. Had the rate of return to capital been 15% in this case, the present value of the cash flow discounted at 15% would have been £1000. The fact that we have a surplus return of £340.88 means that the rate of return to this investment is greater than 15%. The machine is yielding a rate of return greater than capital of this type generally.

Even by trial and error it would be possible to find the actual rate of return by asking the question, 'At what rate must £400 per year for 5 years be discounted to make its present value £1000?' For present purposes, however, it is enough to note that if we are comparing two projects with different costs of purchase, both of which yield a surplus return, we can compare them by expressing the surplus return as a percentage of cost price. Thus, in this case the surplus return is £340.88, which is 34.088% of cost.

This appraisal depends on three variables, the cost of the equipment, the expected income, and the expected rate of return from capital of this type. Normally one would presume that few mistakes would be made in assessing the cost of the equipment, since the purchase price is known. It is possible, however, when the ultimate costs depend on such factors as development costs, for serious mistakes to be made in costing. This happened with Rolls Royce and the RB 211 engine a few years ago. Similarly it should be fairly easy to assess the rate of return normally obtained from this type of capital, although again expectations of industrialists can vary over time. At the end of this chapter you will find a note on what is an appropriate rate of discount in certain circumstances. The unknown factor is the expected flow of income. At best this is little more than an estimate, and many investment appraisals turn out to be false merely because industry is often too optimistic about cash flow resulting from an investment. The moral is that the value of method is no better than the accuracy with which the variables are stated.

Comparison of Investment Projects

In practice the decision is not simply to invest or not to invest. Usually there is a range of investment projects which can be undertaken, and the problem is to find the best.

EXAMPLE 9
A firm is faced with two alternative investment plans. Plan 1 will cost £750 and Plan 2 £950. Both plans involve the purchase of equipment the life of which is four years, and the current rate of return on capital is expected to be 20%. The estimated cash flows resulting from the projects are:

Year	1	2	3	4
Plan 1	£300	£400	£300	£200
Plan 2	£500	£400	£300	£300

The present value of these expected returns assuming a rate of return on capital of 20% is:

Year	Return		$(1 + 0.2)^{-n}$	Present Value	
	1	2		1	2
1	300	500	.8333	249.99	416.65
2	400	400	.6944	277.76	277.76
3	300	300	.5787	173.61	173.61
4	200	300	.4823	96.46	144.69
				797.82	1012.71

Both projects will yield a surplus. Using plan 1 we obtain a return with a present value of £797.82 for a current cost of £750 – a surplus of £47.82. Plan 2 will yield £1012.71 for a current cost of £950 – a surplus of £62.71. Both are yielding a return greater than 20%. The only way in which the two projects can be compared easily is by expressing the surplus as a percentage of the cost of the projects. We find that Project 1 has a surplus percentage return of almost 6.4%, while project 2 has a surplus percentage return of 6.6%. Thus although there is little in it, project 2 would be the preferable project to undertake.

TUTORIAL FOUR

1. A further examination of the two projects in example 9 shows that the equipment bought for project 1 will have a scrap value of £150 at the end of its life, while the equipment bought for project 2 will have a scrap value of £100. Assuming the equipment is sold for cash, will the new information affect the investment decision?

2. The Government now decide to give an investment grant of 40% of the original cost of the equipment bought, payable one year after purchase. How will this affect the decision:

    (a) Ignoring scrap values.
    (b) Allowing for scrap values as above?

3. The firm is liable to pay Corporation Tax at a rate of 40% on all receipts. Assuming that there is a delay of one year between receipts and payment of tax, which project will be undertaken,

    (a) Using the original data only,
    (b) Allowing for the investment grant of 40%,
    (c) Allowing for both an investment grant and scrap values?

4. From your calculations, what conclusions would you draw about the effect on investment projects of investment grants, corporation tax, and scrap values?

Annuities

So far we have been asking the question, 'What is the present value of future sums of money?', or what comes to the same thing, 'What sum of money invested now at $r\%$ for n years would accumulate to a given total?' Often however, it is useful to pose the question in another form. 'How much would I have to invest now at a given rate of interest $r\%$ in order to receive an income of £a per year for n years?' Some of you may have realised that in a cumbersome sort of way this is exactly what we have been doing already. When in example 5 we said that the present value of the future income flows was £797.82, we did not mean that £797.82 invested now at 20% for 4 years would equal the sum of all the income flows (£1200). (Calculate it yourself and see what it would amount to.) What we meant was that if we invested £797.82 at 20% compound interest, drawing £300 at the end of the first year, £400 at the end of the second year, and so on, by the time we drew £200 at the end of the fourth year we would just have exhausted our original deposit and the interest earned.

Now this is exactly what an annuity is — an investment which, while earning interest, enables us to draw a given sum of money for n years, by which time our capital plus interest is just exhausted. We could of course, specify the income we require and calculate the present value painstakingly by saying, 'What is the present value of £x received 20 years hence if the rate of interest is r%?', and then do the same thing for 19, 18, 17 years etc. down to 1 year. Even if we do this it will not tell us, except by trial and error, how much income we could afford to draw each year for 20 years, given that our capital is limited to £5000. How much simpler it would be if we can derive a formula from what we already know which will enable us to perform these calculations easily and rapidly. Suppose we consider an annuity yielding £a per year when the rate of interest is r%. We know that:

£a receivable one year hence has a present value of $\dfrac{£a}{1 + r}$

£a receivable two years hence has a present value of $\dfrac{£a}{(1 + r)^2}$

£a receivable n years hence has a present value of $\dfrac{£a}{(1 + r)^n}$

Thus the present value of an annuity yielding £a per year for n years is

$$\text{P.V.} = \frac{a}{1 + r} + \frac{a}{(1 + r)^2} + \frac{a}{(1 + r)^3} + \ldots \ldots \frac{a}{(1 + r)^n}$$

As you can see this is a geometric progression of which the first term is $\dfrac{a}{1 + r}$ and the common ratio $\dfrac{1}{1 + r}$. The sum to n terms of a geometric progression is, as you know, $S_n = \dfrac{a(1 - r^n)}{1 - r}$ where a is the first term and r the common ratio. (It is perhaps confusing that by convention the same letters are used in the progression as in the annuity formula.)

Thus in summing the progression we have $a = \dfrac{a}{1+r}$, and $r = \dfrac{1}{1+r}$.

The Present Value = the Sum of the geometric progression =

$$\frac{\dfrac{a}{1+r}\left[1-\left(\dfrac{1}{1+r}\right)^{n}\right]}{1-\dfrac{1}{1+r}} = \frac{a}{1+r}[1-(1+r)^{-n}]\times\frac{1+r}{r}$$

$$\therefore \ \text{P.V.} = \frac{a[1-(1+r)^{-n}]}{r}$$

You should notice that the same formula was derived in a different fashion in Tutorial II.

EXAMPLE 10

An annuity yields £100 per annum for 7 years. How much does it cost if the current market rate of interest is 6%?

$$\text{The present value of the annuity} = \frac{100\,[1-(1.06)^{-7}]}{0.06}$$

$$= \frac{100(1-.6651)}{.06} = \frac{100\times.3349}{.06} = £558.17$$

TUTORIAL FIVE

1. The formula given in the previous section enables us to calculate the present value of a given income receivable for a stated number of years. Derive from it a formula which enables you to calculate the income you could expect to receive for n years if you had £P available to invest now at a rate of interest of r%. Assume all capital to be exhausted at the end of the nth year.
2. Smith, an employee of Wein Ltd. retired from his employment at the age of 65. He received from his employers a terminal gift of £1500 or the alternative of a pension amounting to £400 per year for life. Assume a rate of interest of 5% and that a man aged 65 has a life expectancy of 7 years. Which alternative should Smith choose?
3. Had Smith adopted another course of action (using his lump sum to buy an annuity) how much income would he have gained or lost per annum?
4. A loan company grants a second mortgage on the security of house property. Jones urgently needs to borrow £400 but knows that in ten years time he will be going to live with his son in Australia.

Since he wished at that time to give the house to his daughter un-encumbered by mortgage he is worried as to whether the second mortgage will be paid off by that time. The finance company charges 15% compound interest and Jones plans to pay off the loan at £75 per annum. Can he achieve his objective? (Hint: Find $(1 + r)^{-n}$ and hence find n from the tables.)

5. You will find that Jones can not pay off the loan in less than 10 years. As his financial advisor you are asked what repayment he should make annually in order to clear the loan in nine years. What would you advise?

Sinking Funds

A related type of problem to annuities is found in the concept of a sinking fund. In its most obvious form this is a sum of money put aside at regular intervals usually monthly or annually in order to achieve a given sum at the end of a predetermined period. This concept may be found in the provision for replacement of capital, or in, say, unit trust investment of a few pounds per month by the individual saver. You will remember in Tutorial II deriving the formula

$$S = \frac{P[(1 + r)^n - 1]}{r}$$

giving us the terminal value of an annual investment of £P at r% for n years.

It can easily be seen that if we are given the required terminal sum and wish to know the annual savings we can modify the formula into:

$$P = \frac{Sr}{(1 + r)^n - 1}$$

EXAMPLE 11

A firm wishes to make provision for the replacement of certain items of capital equipment which will wear out in 8 years time. The estimated cost of replacement is £5000. If the rate of interest is 8%, what annual provision must it make to ensure funds being available?

The annual provision £$P = \dfrac{5000 \times 0.08}{(1 + r)^n - 1}$

$$(1 + r)^n = \frac{1}{(1 + r)^{-n}} = \frac{1}{.5403} = 1.851$$

$$P = \frac{£400}{1.851 - 1} = £470$$

TUTORIAL SIX

1. Suppose that in providing a sinking fund a firm invests £a at the beginning of the first year and £P per year at the end of the first year and each year thereafter for n years. The rate of interest is r%. How would you modify the formula in the previous section to take account of the initial investment of £a?

2. A man invests £100 in unit trusts and a further £100 at the end of each year thereafter for the next ten years. What will be the value of his holding at that time assuming a rate of growth in the value of the units of (a) 3% and (b) 5%.

3. In replacing capital equipment a firm wishes to allow for the effect of inflation. They therefore build up a sinking fund based on the estimated replacement cost at the time of replacement. They wish to replace equipment in ten years time which at the moment costs £2000. However, they also anticipate that the rate of inflation will be 5% per annum. Given a rate of interest of 9%,

    (a) What will be the replacement cost of the machine in 10 years time?
    (b) How much must be put aside at the end of each year to accumulate this sum?

Choosing an Appropriate Discount Rate

It should be quite obvious by now that in any of the techniques discussed the result obtained will depend largely on the rate of discount applied to future receipts. If this is not appropriate, the results can not be a guide to policy. The precise rate we choose must of course depend on the problem in hand. If we are evaluating capital investment the most generally useful rate is the rate of return generally expected from capital of this nature. What we are saying in fact is that capital investment is just worth while if when we discount cash flows the present value of future cash is just equal to the cost of the investment. If there is a surplus above this it implies that the rate of return is greater than the discount rate we have used — i.e. the capital is earning a greater percentage return than we would expect. If there is a deficit it is earning less than comparable capital elsewhere.

If on the other hand we are dealing with the lending or borrowing of cash funds we can usually find some well established market rate of interest for a loan of that type and that degree of risk which will give a more than useful guide. Anyone with a reasonable knowledge of finance will soon learn to assess comparability of loans and interest rates.

In all these techniques however, there is nothing like sound common sense. The methods outlined form the basis of an analysis which has been proved over and over again, particularly in the United States. But no technique, however good, is better than the person who uses it.

EXERCISES TO CHAPTER ELEVEN

1. A retailer's revenue and costs exactly balanced on December 31st 1971. During the following year his costs rose by £150 each month and his revenue rose by £200 each month. What were:

    (a) The costs incurred and the revenue received during the month of December 1972.
    (b) The total profits for the year ended December 31st 1972?

2. If Britain is able to raise her real national income by a constant 4% a year, how long will it take to double the standard of living?

3. A firm writes off its capital equipment at a rate of 15% of its previous year's value What is the balance sheet value of a machine costing £1000 after the expiry of five years?

4. An investment costs £2000. During the first year it costs £1600 to run and revenue from the sale of its output is £2400. During the second year costs are estimated at £1200 and revenue at £3600. Assuming a discount rate of 10%

    (a) Find the present value of profits made during this two year period.
    (b) Find the present value of total revenues received and the present value of total costs incurred. Hence check your answer to part (a).

5. A manufacturer has £5600 available to purchase a new plant. His past experience tells him that he can expect to receive a net return of 10% on this type of manufacture. His accountant informs him that he should make net profits of £600 in his first year, £800 in his second year and thereafter £1000 a year. His friends however are pressing him to purchase a plot of land which they believe will be developed and which they anticipate selling 10 years hence at a price of £14,000. They point out that his new plant will have a life of ten years only and that the gain on land speculation will

leave him with his original £5600 plus a profit of £8400, whereas his industrial venture will merely give him a profit of £9400 and plant with a negligable scrap value. What would you advise?

If the expected rate of return were 12% what is the minimum price at which you would sell the land 10 years hence?

6.   On January 1st 1970 an investor bought £5000 of 6% government stock at a price of 54. Interest is paid on December 31st each year. On January 1st 1973 he sold the stock at a price of 58. Would he have been better advised to invest the money in 9% stock which he could buy at 90 and which was redeemed at par on January 1st 1973? (Use a discount rate of 9%.)

7.   A businessman approaching retirement wishes to buy an annuity of £1000 a year payable for 15 years. Assuming a rate of interest of 7%, how much would he have to pay for such an annuity?

8.   If the same businessman had saved £1000, invested it at 7%, and then invested a further £1000 at the end of each year for 14 years what would be the cash balance in his investment account at the end of the period?

9.   If he now withdrew £5000 at the end of each year from his investment account, for how many years could he do this?

10. The problem of replacing worn out equipment is complicated by the fact that prices are not constant. A machine bought in 1960 at a cost of £1000 has an expected life of 10 years. Inflation has been at a constant annual rate of 5%. What would the cost of a replacement machine be?

How much per annum must be provided for replacement purposes if it is desired to build up a fund over a period of ten years during which time the rate of interest received is 6%? Would your answer be different if the rate of interest were 9%?

Assuming a 6% discount rate would the new machine be profitable if it yielded an income of £400 a year for five years?

Explain precisely what you mean by profitable or non-profitable in this context.

Chapter Twelve

Time Series Analysis

When we looked at the frequency distribution we took a given variable at one point in time, and analysed it at that point in time. But, of course we had no information regarding that variable a year, or even a month, later.

The time series on the other hand takes a variable and examines the way in which its magnitude has fluctuated over a period of time. We might, for example, consider the way in which output of a particular firm has varied year by year since 1955.

Why should we do this? Whenever we are interested in planning, we must make forecasts of what is likely to happen in the future. Now, however complicated the technique we use, our judgement will depend to a large extent on what has happened in the past. Thus, a reliable analysis of what has happened in relevant fields in recent years is a first step in obtaining a reliable estimate of what is likely to happen in the future. Any housewife will tell you that egg prices are likely to rise during particular months of the year, and to fall during other months. How does she know? Merely on the basis of past experience. Can we undertake a simple analysis of more complex problems to predict the future more reliably and more precisely?

The Trend

Suppose we consider the following series:

Output of A.B.C. Ltd. 1955–72 (Million Tons)

Year	Output	Year	Output	Year	Output
1955	68	1961	140	1967	250
1956	100	1962	200	1968	170
1957	120	1963	230	1969	320
1958	177	1964	180	1970	230
1959	100	1965	280	1971	210
1960	80	1966	200	1972	330

Looking at figures presented in this way will give us little information. Two things are obvious however. Figures of output show that although there have been very marked fluctuations year by year, there is a general upward rise in the figures. Moreover, certainly from 1963 to 1972, the figures tend to show a peak output every second year, followed by a fall in the intervening years.

This becomes much more apparent if you look at the graph of this time series in Figure 1. The general rise in output can be clearly seen. This general tendency of figures to move in a given direction is known as the TREND. In Figure 1, the trend has been shown as a linear, or straight line trend. In other series we examine the trend will vary its direction, at first rising and then falling, or vice versa. Such a trend is curvilinear (see diagram 3). An important characteristic of such trends is that they change direction slowly over time, and so, barring catastrophes, the continuation can be sketched in from a careful examination of the existing trend line.

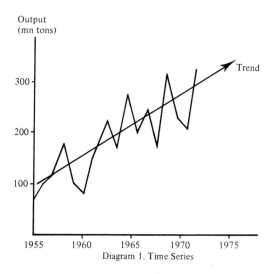

Diagram 1. Time Series

Much more obvious in Figure 1 is the succession of rises and falls in output occurring at regular intervals. Even to look at this graph suggests a certain pattern of behaviour in output in 1973. What is it?

Formal analysis of such a series can reveal much more useful data about what may happen in the future.

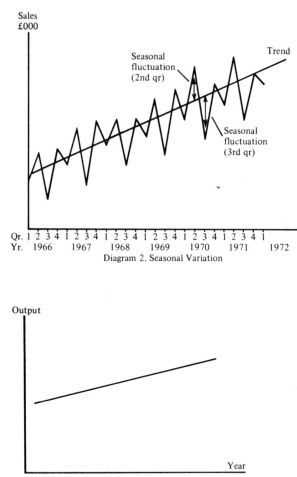

Diagram 2. Seasonal Variation

Diagram 3a. Linear Trend

Seasonal Variation

Many series are of such a nature that the figures become more significant when they are analysed on a monthly or quarterly basis rather than annually. Most economic series are like this. Sales, for example, show marked variations throughout the year. Everyone knows that sales tend to rise at Christmas or that more swimsuits are likely to be sold in summer than in winter. But if such fluctuations occur, are they sufficiently regular to be predictable? If they are, then we must try to ascertain the direction and probable magnitude of the swing. Diagram 2

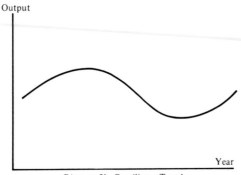

Diagram 3b. Curvilinear Trend

shows hypothetical sales data plotted quarterly over a period of years. The data show a remarkable regularity. Every year there are two sales peaks, in the second and fourth quarter, while every year sales slump in the third quarter. The pattern is regular enough for us to predict that the same thing will happen in future years.

If we take the trend as the norm, such regular fluctuations around the trend are given the name of SEASONAL VARIATIONS. It should now be apparent that if we can obtain a reliable estimate of the trend, and of the magnitude of the seasonal variation from the trend, we have gone some way towards obtaining a reliable basis for forecasting the future sales figures.

Residuals

Unfortunately for the planner, the data he has is a compound of many influences, most of which can not be foreseen and will not, in any case occur again in the same form. Exports will be affected by a dock strike, production by a shortage of some vital raw material, sales by a sudden change in taxation. Thus the figures we have can be broken down into TREND & SEASONAL VARIATION & RESIDUALS.

By their very nature residual influences are chance happenings and are not amenable to analysis. Does this mean that we should ignore residual influences? No planner can afford to ignore a factor which might make nonsense of his forecasts!

The importance of the residuals lies in this. If over the years residual influences have had a minimal effect on the figures, we can use our forecasts with some degree of confidence that they will not be unduly affected by external events. But if the data has been affected regularly and to a large degree by residual factors, then we must use our forecasts with caution.

Analysis of the Time Series

The previous paragraphs contain a large number of 'ifs'. 'If we can obtain the trend', 'if we can calculate seasonal variation' and so on. Let us now look at some simple time series and see how these things can be done. We will begin, as generations of statisticians have done, by considering the trend.

Firstly, we will take the case of a series which has an obviously linear trend i.e. which shows a constant tendency, in spite of fluctuations, for all figures to move in one direction — to rise or to fall, but not both.

The Linear Trend

A good example of figures showing a linear trend is the series showing the value of consumer durable goods bought. It is an important series for the firms producing such goods, but its importance does not end here. As all sales are valued at 1963 prices a change in standards of living can be measured by the number of cars, washing machines, refrigerators etc. we are buying.

Consumer Expenditure on Durable Goods
(£ million at 1963 prices)

1955	965	1960	1370	1965	1842
1956	849	1961	1334	1966	1821
1957	952	1962	1411	1967	1908
1958	1113	1963	1703	1968	1977
1959	1328	1964	1853	1969	1818

Source: Estimates of National Income & Balance of Payments. Cmd. 4328 April 1970.

Can you estimate the trend line? This is obviously difficult with such complex figures. How can we be sure that this is the trend line that best describes expenditure on durable goods? To be safe we should calculate the equation of the trend line from the original data rather than rely on intuition.

What do we mean by the line that best fits the data? Clearly a straight line $y = mx + c$ cannot pass through all the points and we need some criterion for calculating m and c. You will note that the points will deviate from the line, and conventionally we fix the line in such a position that the sum of the squares of the deviations from the line is a minimum (why take the sum of the squares and not just the deviations? If you remember the arguments for selecting the standard deviation in preference to the mean deviation as a measure of dispersion, the reason for taking the sum of the squares will be apparent). Armed with this criterion, we can now calculate the least squares line of best fit.

The Least – Squares Trend Line

Initially, consider the simple example shown below. Copy the points on a sheet of graph paper, and draw in what you think is the line of best fit. Later, you can draw in the line based on the least squares criterion and compare with your estimate.

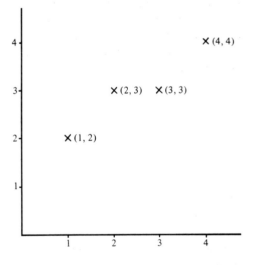

Let us first ask by how much does the point (1,2) deviate from the line you have drawn. Now its equation in general form is $y = mx + c$. This point has $x = 1$, so the value of y on your line is

$m + c$

and the deviation of the point (1,2) from $m + c$ is

$2 - (m + c) = 2 - m - c$

and the square of the deviation is

$$(2 - m - c)^2$$

i.e. $4 - 4m - 4c + m^2 + cm + c^2$

We can repeat this process for the points on the line where x equals 2, 3, and 4. It would be better to tabulate the results.

x	y	deviation	deviation squared
1	$m + c$	$2 - (m + c)$	$4 - 4m - 4c + m^2 + 2cm + c^2$
2	$2m + c$	$3 - (2m + c)$	$9 - 12m - 6c + 4m^2 + 4cm + c^2$
3	$3m + c$	$3 - (3m + c)$	$9 - 18m - 6c + 9m^2 + 6cm + c^2$
4	$4m + c$	$4 - (4m + c)$	$16 - 32m - 8c + 16m^2 + 8cm + c^2$
		adding . . .	$38 - 66m - 24c + 30m^2 + 20cm + 4c^2$

If Z is the sum of the square of the deviations, then

$$Z = 38 - 66m - 24c + 30m^2 + 20cm + 4c^2$$

To obtain the line of best fit, we wish to find values of m and c to make Z a minimum. We have solved similar problems earlier by differentiation. Suppose we deal with m first. Do you remember that to find the value of m which minimises Z we must differentiate Z with respect to m and put the derivative equal to zero?

$$\frac{dZ}{dm} = -66 + 60m + 20c$$

as

$$\frac{dZ}{dm} = 0 \text{ for a minimum value of } Z$$

then $-66 + 60m + 20c = 0$

or $30m + 10c = 33 \ldots \ldots \ldots$ (1)

However we also wish to find the value of c which minimises Z. Differentiating Z with respect to c gives

$$\frac{dZ}{dc} = -24 + 20m + 8c$$

For a minimum value of Z $\frac{dZ}{dc} = 0$

hence $-24 + 20m + 8c = 0$

or $10m + 4c = 12 \ldots \ldots \ldots$ (2)

The problem now boils down to this: if we wish to find the values m and c in the line which (according to the least square criterion) best fits the points, we must find values of m and c which fit both equations (1) and (2). In other words, we must solve the simultaneous equations

$30m + 10c = 33$
$10m + 4c = 12$
Which gives $m = 0.6$, $c = 1.5$.
The line is $y = 0.6x + 1.5$.

You can now draw this line on your graph and compare it with your estimate.

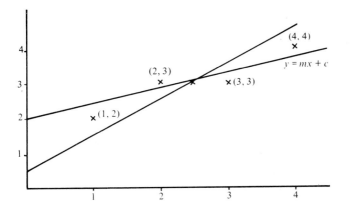

The Normal Equations

The method is not difficult, but will become tedious if we have a large number of co-ordinates, or (worse) when the co-ordinates are not integers. What we need is a general method for finding the least squares line of best fit without having to square the deviations.

Let us suppose that the line $y = mx + c$ is the best fit to the points P_1, P_2, P_3, P_n, whose co-ordinates are $(x_1 y_1), (x_2 y_2), (x_3 y_3), \ldots (x_n y_n)$. Working in the same way as before, the value of y on this line when $x = x_1$ is

$$x_1 m + c$$

and the deviation of the point $(x_1 y_1)$ from the point on the line is

$$y_1 - (x_1 m + c) = y_1 - x_1 m - c$$

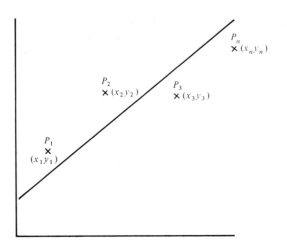

and the square of the deviation is

$$(y_1 - x_1m - c)^2$$

i.e. $y_1^2 - 2x_1y_1m - 2y_1c + x_1^2m^2 + 2x_1cm + c^2$

As before, let us tabulate the results when x equals x_2, x_3, and x_n. It will be convenient to use the summation sign (Σ) when we add the square of the deviations. You will recall from the introduction that $\Sigma x = x_1 + x_2 + x_3 \ldots + x_n$.

x	y	deviation	deviation squared
x_1	$x_1m + c$	$y_1 - (x_1m + c)$	$y_1^2 - 2x_1y_1m \; -2y_1c + x_1^2m^2 \;\; +2x_1cm + c^2$
x_2	$x_2m + c$	$y_2 - (x_2m + c)$	$y_2^2 - 2x_2y_2m \; -2y_2c + x_2^2m^2 \;\; +2x_2cm + c^2$
x_3	$x_3m + c$	$y_3 - (x_3m + c)$	$y_3^2 - 2x_3y_3m \; -2y_3c + x_3^2m^2 \;\; +2x_3cm + c^2$
x_n	$x_nm + c$	$y_n - (x_nm + c)$	$y_n^2 - 2x_ny_nm \; -2y_nc + x_n^2m^2 \;\; +2x_ncm + c^2$

Adding . . . $\Sigma y^2 - 2m\Sigma x\Sigma y - 2c\Sigma y + m^2\Sigma x^2 + 2cm\Sigma x + nc^2$

If Z is the sum of the squares, then

$$Z = \Sigma y^2 - 2m\Sigma x\Sigma y - 2c\Sigma y + m^2\Sigma x^2 + 2cm\Sigma x + nc^2$$

Differentiating Z with respect to m gives

$$\frac{dZ}{dm} = -2\Sigma x\Sigma y + 2m\Sigma x^2 + 2c\Sigma x$$

putting the derivative equal to zero

$$-2\Sigma x\Sigma y + 2m\Sigma x^2 + 2c\Sigma x = 0$$

$$\Sigma x\Sigma y = m\Sigma x^2 + c\Sigma x \ldots \ldots \ldots (1)$$

Differentiating Z with respect to c gives

$$\frac{dZ}{dc} = -2\Sigma y + 2m\Sigma x + 2nc$$

Again, putting the derivative equal to zero

$$-2\Sigma y + 2m\Sigma x + 2nc = 0$$

$$\Sigma y = m\Sigma x + nc \ldots\ldots\ldots (2)$$

We can find the values of m and c in the least squares line of best fit by solving simultaneously equations (1) and (2). They are called the Normal Equations. Earlier we found the line of best fit from first principles. Let us now find the line using the Normal Equations.

x	y	x^2	xy
1	2	1	2
2	3	4	6
3	3	9	9
4	4	16	16
$\Sigma x = 10$	$\Sigma y = 12$	$\Sigma x^2 = 30$	$\Sigma xy = 33$

Substituting these values in the Normal Equations.

$$\Sigma xy = m\Sigma x^2 + c\Sigma x$$
$$33 = 30m + 10c$$
$$\Sigma y = m\Sigma x + nc$$
$$12 = 10m + 4c$$

This method gives the same equations as previously.

Most people dislike solving simultaneous equations, but few find it difficult to substitute values in a formula. Most examples that you will meet involve solving equations far more cumbersome than the above. We could re-arrange the normal equations to give general values for m and c. This, in fact, involves solving the Normal Equations. First let us write Normal Equation (2) this way:

$$nc = \Sigma y - m\Sigma x$$

and divide by n

$$c = \frac{1}{n}(\Sigma y - m\Sigma x)$$

We can now write this value for c in normal equation (1).

$$\Sigma xy = m\Sigma x^2 + \frac{\Sigma x}{n}(\Sigma y - m\Sigma x)$$

Now remove the bracket.

$$\Sigma xy = m\Sigma x^2 + \frac{\Sigma x \Sigma y}{n} - \frac{m(\Sigma x)^2}{n}$$

$$m\Sigma x^2 - \frac{m(\Sigma x)^2}{n} = \Sigma xy - \frac{\Sigma x \Sigma y}{n}$$

thus

$$m = \frac{\Sigma xy - \dfrac{\Sigma x \Sigma y}{n}}{\Sigma x^2 - \dfrac{(\Sigma x)^2}{n}}$$

We can find a value for m by substituting in the formula above, and a value for c by substituting in the formula obtained earlier, i.e.

$$c = \frac{1}{n}(\Sigma y - m\Sigma x)$$

In the four points example we considered earlier, we found that $\Sigma x = 10$, $\Sigma y = 12$, $\Sigma x^2 = 30$, $\Sigma xy = 33$. Substituting in the formula for m

$$m = \frac{33 - \dfrac{12 \times 10}{4}}{30 - \dfrac{(10)^2}{4}}$$

$$= \frac{3}{5}$$

$$= 0.6$$

Now substituting in the formula for c.

$$c = \frac{1}{4}(12 - 0.6 \times 10)$$

$$= 1.5$$

Which agrees with the results found from first principles.

EXAMPLE 1

x	1	2	3	4	5	6	7
y	1	2	6	7	10	16	21

Find the least squares line of best fit.

x	y	x^2	xy
1	1	1	1
2	2	4	4
3	6	9	18
4	7	16	28
5	10	25	50
6	16	36	96
7	21	49	147
$\Sigma x = \underline{28}$	$\Sigma y = \underline{63}$	$\Sigma x^2 = \underline{140}$	$\Sigma xy = \underline{344}$

Hence, $\Sigma x = 28$, $\Sigma y = 63$, $\Sigma x^2 = 140$, $\Sigma xy = 344$.

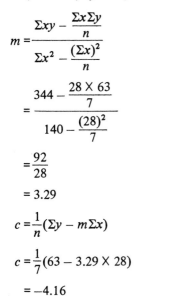

$$m = \frac{\Sigma xy - \dfrac{\Sigma x \Sigma y}{n}}{\Sigma x^2 - \dfrac{(\Sigma x)^2}{n}}$$

$$= \frac{344 - \dfrac{28 \times 63}{7}}{140 - \dfrac{(28)^2}{7}}$$

$$= \frac{92}{28}$$

$$= 3.29$$

$$c = \frac{1}{n}(\Sigma y - m\Sigma x)$$

$$c = \frac{1}{7}(63 - 3.29 \times 28)$$

$$= -4.16$$

The equation is $y = 3.29x - 4.16$.

Using an Origin

We can now use the expressions for m and c to find the straight line trend of the time series. However, we would welcome any further simplification of the method. You will remember that when calculating the mean and standard deviation of a distribution we found it convenient to select an origin and work with deviations from the origin. This considerably simplified the calculations. We can also use an origin to find the straight line of best fit. Suppose we calculate the deviations

x' from an origin 4 and y' from an origin 7, then the tabulations would look like this:

x'	y'	$(x')^2$	$x'y'$
-3	-6	9	18
-2	-5	4	10
-1	-1	1	1
0	0	0	0
1	3	1	3
2	9	4	18
3	14	9	42
$\Sigma x' = 0$	$\Sigma y' = 14$	$\Sigma(x')^2 = 28$	$\Sigma x'y' = 92$

We can now substitute the above values in the expression for m

$$m = \frac{92 - \dfrac{(0 \times 14)}{7}}{28 - \dfrac{0}{7}}$$

$$= \frac{92}{28}$$

Which is the same value that we obtained previously for m. To find c, we must convert $\Sigma x'$ into its true value Σx, and $\Sigma y'$ into its true value Σy. This presents no difficulty. Consider first $\Sigma x'$: it was obtained by subtracting 7 values from an origin 4. Thus the true value of Σx is:

$$7 \times 4 + 0 = 28$$

Likewise the true value Σy is

$$7 \times 7 + 14 = 63$$

The values Σx and Σy can now be used to calculate c as before.

You should notice that when we took 4 as an arbitrary origin for $\Sigma x'$ this made $\Sigma x' = 0$, and considerably simplified the calculation of m.

Let us again state the expression for m

$$m = \frac{\Sigma xy - \dfrac{\Sigma x \Sigma y}{n}}{\Sigma x^2 - \dfrac{(\Sigma x)^2}{n}}$$

Now if $\Sigma x' = 0$, $\dfrac{\Sigma x' \Sigma y}{n} = 0$, and $\dfrac{(\Sigma x')^2}{n} = 0$.

hence $m = \dfrac{\Sigma x'y'}{\Sigma(x')^2}$

Under what circumstances would $\Sigma x' = 0$? Only one condition must be fulfilled: we choose the arithmetic mean $\bar{x}$ as the origin. (Think back to the arithmetic mean when we proved that the sum of the deviations from a mean is zero.) Moreover, in a time series we can number the time periods as consecutive integers, which makes the mean easily obtainable by inspection. The only difficulty is that if there is an even number of time periods, then neither the mean nor the deviations will be integers. To overcome this, all examples you will have to work in this book will have an odd number of time periods.

EXAMPLE 2

We now have sufficient information to fit a trend line to the data on consumer expenditure. Let us first list the years as consecutive integers $1, 2, 3, \ldots \ldots 15$. We can take an origin 8 to simplify the calculations.

x'	y	$x'y$	$(x')^2$
-7	965	-6755	49
-6	849	-5094	36
-5	952	-4760	25
-4	1113	-4452	16
-3	1328	-3984	9
-2	1370	-2740	4
-1	1334	-1334	1
0	1411	0	0
1	1703	1703	1
2	1865	3730	4
3	1842	5526	9
4	1821	7284	16
5	1908	9540	25
6	1977	11,826	36
7	1818	12,726	49
0	22,256	23,242	280

$$m = \frac{\Sigma x'y}{\Sigma(x')^2}$$

$$= \frac{23,242}{280} = 83$$

$\Sigma x' = 0, n = 15,$

hence $\Sigma x = 15 \times 8 + 0 = 120$

$$c = \frac{1}{n}(\Sigma y - m\Sigma x)$$

$$= \frac{1}{15}(22,256 - 120 \times 83)$$

$$= 820$$

The equation of the trend line is

$$y = 83x + 820$$

y is expenditure, x is time period.

TUTORIAL ONE

1. Examine again the graph of consumer expenditure. Can you account for the declines in 1956, 1961, 1965, 1966, and 1969?
2. Use the equation of the trend line derived in example 2 to predict the position of the trend in 1970, 1971, 1972 and 1973.
3. Find from first principles the least square line $y = mx + c$, that fits the points (1,4) (3,4) (3,3) (4,2) (5,2).
4. We stated that it is possible to find a least squares line $y = mx + c$, that gives the best fit to a series of points on a graph. Such a line is called a REGRESSION LINE, and we will have more to say about this later. However, it is possible to find a second line of best fit. Why?
5. When we took deviations x' and y' we had to adjust $\Sigma x'$ and $\Sigma y'$ to their true values Σx and Σy before we calculated c. Why was no adjustment necessary to calculate m? (Hint: consider the effect on m of moving the points a uniform distance either up or down.)
6. $x =$ 1   2   3   4   5   6   7
   $y =$ 1.5   4.5   6   11   14.5   16.5   18
   Fit a straight line $y = mx + c$.
7. Find the least squares trend line of the output of ABC Ltd., on page 294. You may ignore 1955.
8. Find a linear series and omit the data for the last two years. Calculate the trend line and predict the data for the last two years. How accurate are your predictions?

Curvilinear Trends

Most of the time series you will meet in practice will probably not have a linear trend like those in the last section. This is because most aspects of our life are governed to some extent by the periodic upswing and down-swing of activity that the economist calls the trade cycle. If you own shares you will know that for some periods of time share prices will rise. But ultimately something happens and the trend turns gradually into a period of falling prices. The same sort of thing happens with employment, with output, with sales, and with most aspects of industrial life.

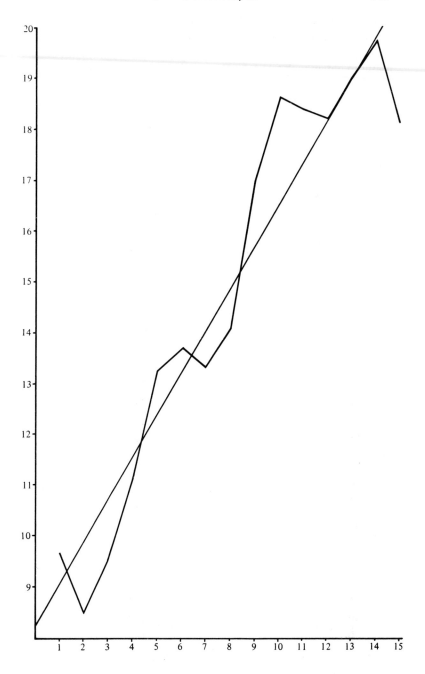

It is of course possible to derive normal equations for the curvilinear trend, but the method is complex; certainly too complex for an introductory course. You will find it far easier, and in many ways just as useful to obtain such trends by the use of what we call 'moving averages'.

What is a moving average? Quite simply it is obtained by selecting a number of consecutive values of the variable, say five, and averaging them so that the magnitude of the variations of the individual items is reduced. Then to obtain the second point on the trend line you do the same thing with the next set of five figures. But beware! The second set of five figures is not numbers 6 to 10, but numbers 2 to 6; and the third set 3 to 7 and so on. Each average we obtain then consists of four figures from the previous group plus one new one.

Let us take a simple example and calculate the five year moving average to illustrate the method.

EXAMPLE 3

Unemployment in the U.K.

Year	% Unemployed	5 Year Total	5 Year Average = Trend
1952	2.0		
1953	1.6		
1954	1.3	7.2	1.44
1955	1.1	6.6	1.32
1956	1.2	7.1	1.42
1957	1.4	8.0	1.60
1958	2.1	8.5	1.70
1959	2.2	8.8	1.76
1960	1.6	9.4	1.88
1961	1.5	9.8	1.96
1962	2.0	9.2	1.84
1963	2.5	9.0	1.80
1964	1.6	9.0	1.80
1965	1.4		
1966	1.5		

These figures are plotted in diagram 8, and you can see how the moving average smooths out the fluctuations.

You will naturally have many questions to ask about the above calculation. Why did we choose a five year moving average, and not a three,

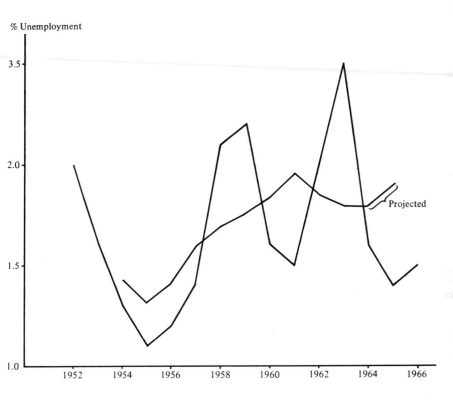

% Unemployment

or four or seven year one? There is no hard and fast rule to help you choose the right one. The correct one is the one that smooths out the fluctuations best, and only trial and error can prove this. But a good working rule is to assess the period between consecutive peaks or consecutive troughs and use this period. Do this and you will not go far wrong. We chose a five year average above because it is known that the post-war cycle in unemployment is about five years.

What about the trend figure for 1952, and 1953? A weakness of this method is that they can not be obtained from the data as given. Because it is an average, by convention we place our trend figure in the centre of the group to which it refers. But this leaves a gap at the beginning and at the end of the series. Can you see why it is easier to choose for your moving average period an odd number of years?

If you look again at the graph in diagram 8, you will notice an odd thing. Although the cycles are well marked with peaks and troughs, there is a tendency for the figures to rise. Our trend is not merely curvilinear about a static value, but itself slopes upwards. The recessions show a tendency to get more severe, and the booms to be not quite so good. This is in fact confirmed by the subsequent pattern of unemployment from 1966 to 1971.

TUTORIAL TWO

1. Graph the trend calculated in example 3. What would you infer about the trend from the fact that both in 1963 and 1964 the trend figure is 1.80?
2. Now, by projecting the graph of the trend, using your own judgement, obtain an estimate of the trend figure for 1965. (The actual figure is 1.88.)
3. Can you see any way of using this result to estimate the unemployment figure for 1967?
4. Recalculate the trend by using
    (a) a seven year moving average.
    (b) a three year moving average.
5. Graph each of the three trends we now have. Which would you say was the best suited to the data we have? Why?

The Analysis of Seasonal Variation

Analysis of long term trend has its place in business forecasting when long range planning is on the agenda. But in business, as in chess, the most important consideration is the very next move. Not what are we going to do next year, but what are we going to do next quarter, or even next month? If you think about it, most businesses show a marked variation in the level of activity within the space of a year. As an analyst you must know how to examine such data in order to find out if the pattern produced is regular enough to enable you to say that it is predictable behaviour. It is this pattern within the year that seasonal analysis examines. We will confine our attention here to a quarterly behaviour pattern, but you must remember that often a monthly pattern can be detected. Fortunately the process of analysis does not change.

Consider the expression seasonal variation. Variation implies a movement away from a point of reference or norm, and the first step in analysis is to establish that norm. It seems logical to accept the trend of a series as the reference point and accepting this we may define a seasonal variation as a movement away from the trend. Is it obvious to you that we cannot have a seasonal pattern in which all variations are above, or all are below the trend? In fact if we assess both the magnitude and direction of the seasonal fluctuations, those above the trend should cancel out those below the trend.

The Trend of a Quarterly Series

EXAMPLE 4
Let us consider the following series.

Quarterly Sales of Fertilizer 1964–68
('000 tons)

Quarter	1964	1965	1966	1967	1968
1	48	50	68	93	84
2	52	46	34	56	61
3	16	22	26	16	29
4	35	40	35	45	48

There is here a marked seasonal pattern of sales and on first inspection it appears that this is regular enough to justify analysis. So let us firstly calculate the trend line using the method of moving averages. Our first problem is that since this is a quarterly series we must use a four quarterly moving average. Now as you know, the moving average figure must be placed opposite the centre of the group of figures to which it relates. But this would place our trend between the second and third quarters, in which position we cannot use it. To overcome this problem we use a device known as 'centring'. Having totalled successive groups of four quarters we than add each of these totals in pairs and divide the result by eight. Thus the trend is centred against a quarter and can be related to it. An example will make this clear. Taking the first few figures of our table we proceed as follows.

Year	Qtr.	Sales	4 Qtr. Total	8 Qtr. Total	Moving Average
1954	1	48			
	2	52			
	3	16	151	304	38
	4	35	153	300	37.5
1965	1	50	147		
	2	46			

Our first 4 quarterly total is the sum of Quarters 1–4 of 1964; our second the sum of the last three quarters of 1964 and the first quarter of 1965, and so on. These are placed between the second and third quarters of 1964, and between the third and fourth quarters of 1964 respectively. To centre the figures we take successive pairs of totals. (Remember this means the first and second totals, the second and third totals and so on.) Thus we have a new figure, placed at the centre of the group to which it relates i.e. opposite to a particular quarter in our series. The new total is the sum of eight quarters and so to obtain the trend we must divide by eight.

The complete calculation is done below, but before looking at it try to calculate the trend of this series for yourself.

Year	Qtr.	Sales	4 Qtr. Total	8 Qtr. Total	Moving Average	Deviation $(x - \text{Trend})$
1964	1	48				
	2	52				
			151			
	3	16		304	38.0	−22.0
			153			
	4	35		300	37.5	− 2.5
			147			
1965	1	50		300	37.5	+12.5
			153			
	2	46		311	38.9	+ 7.1
			158			
	3	22		334	41.7	−19.7
			176			
	4	40		340	42.5	− 2.5
			164			
1966	1	68		332	41.5	+26.5
			168			
	2	34		331	41.4	− 7.4
			163			
	3	26		351	43.9	−17.9
			188			
	4	35		398	49.7	−14.7
			210			
1967	1	93		410	51.2	+41.8
			200			
	2	56		410	51.2	+ 4.8
			210			
	3	16		411	51.4	−35.4
			201			
	4	45		407	50.9	− 5.9
			206			
1968	1	84		425	53.1	+30.9
			219			
	2	61		441	55.1	+ 5.9
			222			
	3	29				
	4	48				

If you look at the trend you will see that it is generally rising but there is one peculiarity — a very sharp increase in the trend figure for the 4th quarter of 1966. What do you think could have caused this?

Seasonal Variation

How can we now proceed to the calculation of seasonal variation? Remembering that it is the deviation from the trend line it is a simple matter to obtain the actual deviations. But remember that in taking deviations we use $x - \bar{x}$ (original figures minus trend) and you must be sure to get the signs right.

These deviations are now listed as follows:

Quarter	1	2	3	4
1964			−22.0	− 2.5
1965	+ 12.5	+ 7.1	−19.7	− 2.5
1966	+ 26.5	− 7.4	−17.9	−14.7
1967	+ 41.8	+ 4.8	−35.4	− 5.9
1968	+ 30.9	+ 5.9		

Total	+111.7	+10.4	−95.0	−25.6	
Average	+ 27.9	+ 2.6	−23.75	− 6.4	= .35
Adjustment	− .09	− .09	− .09	− .09	
S.V.	+ 27.8	+ 2.5	−23.8	− 6.5	

The direction of the seasonal fluctuation is immediately apparent from the sign of the deviation. In the first and second quarters sales rise above trend, and in the third and fourth quarters they fall below it.

The expected magnitude of the fluctuations is the average of the deviations shown in a particular quarter. You may wonder at this but remember that our deviations include residual influences. Since these are random and may affect figures upwards or downwards it is reasonable to eliminate a large part of their impact by averaging over a period of time.

A word on the adjustment. Seasonal fluctuations from a trend are of such a nature that upswings cancel downswings exactly and so the total of our quarterly averages should be zero. In fact we are left with +0.35 (a very small figure) and this can be eliminated by subtracting 0.09 $\left(\dfrac{0.35}{4} \right)$ from each quarterly average.

Rounding to one place of decimals our results show that in quarter 1 we expect sales to be 27.8 (thousand tons) above trend, in quarter 2, 2.5 (thousand tons) above. Conversely in quarters 3 and 4 we expect sales to be 23.8 and 6.5 (thousand tons) below trend.

Knowing these expectations, if we can project an accurate trend line it is possible to forecast sales for each quarter of 1969. Draw the graph of the trend line and do this.

Trend Fitting

You will appreciate that with large figures this method can be cumbersome, it is often easier to calculate the equation of the trend line (providing it is linear) by the method you learned earlier in this chapter.

EXAMPLE 5

Let us take one more example of this. We will examine the quarterly movement of receipts from taxation on current account. Such a series will show marked fluctuations from quarter to quarter. Can you understand why? The majority of government taxation revenue is paid in the first quarter of the year, i.e. the last quarter of the fiscal year. We would expect a large positive seasonal variation in this quarter. The magnitude of the upswing becomes immediately apparent if you look at the series graphed in diagram 9.

Receipts from Taxation (Current Account)
£m

	Quarter			
	1	2	3	4
1966	3757	2276	2647	2596
1967	3995	2683	2975	2722
1968	4407	3054	3389	3338
1969	4811	3552	3781	3737
1970	5289			

Receipts from taxation – current account

Now fit a linear trend to this series by the method of least squares.

Year	Qtr.	x	x'	y	$x'y'$	$(x')^2$	Trend	Deviation
1966	1	1	−8	3757	− 30,056	64	2674	+1083
	2	2	−7	2276	− 15,932	49	2773	− 497
	3	3	−6	2647	− 15,882	36	2872	− 225
	4	4	−5	2596	− 12,980	25	2971	− 375
1967	1	5	−4	3995	− 15,980	16	3070	+ 925
	2	6	−3	2683	− 8049	9	3169	− 486

Year	Qtr	x	x'	y	$x'y'$	$(x')^2$	Trend	Deviation
	3	7	-2	2975	$-$ 5950	4	3268	$-$ 293
	4	8	-1	2722	$-$ 2722	1	3367	$-$ 645
1968	1	9	0	4407	0	0	3466	$+$ 941
	2	10	$+1$	3054	3054	1	3565	$-$ 511
	3	11	$+2$	3389	6778	4	3664	$-$ 275
	4	12	$+3$	3338	10,014	9	3763	$-$ 425
1969	1	13	$+4$	4811	19,244	16	3862	$+$ 949
	2	14	$+5$	3552	17,760	25	3961	$-$ 409
	3	15	$+6$	3781	22,686	36	4060	$-$ 279
	4	16	$+7$	3737	26,159	49	4159	$-$ 422
1970	1	17	$+8$	5289	42,312	64	4258	$+1031$
		$\underline{158}$		$\underline{59,009}$	$\underline{148,007}$	$\underline{408}$		
					$-\underline{107,551}$			
					40,456			

$$m = \frac{\Sigma x'y'}{\Sigma (x')^2} = \frac{40,456}{408} = 99.16$$

$$c = \frac{1}{n}[\Sigma y - m\Sigma x] = \frac{59,009 - 99.16 \times 153}{17}$$

$$= \frac{43770.2}{17} = 2574.17$$

$$y = 99.16x + 2574.72$$

For ease of working we will take this as

$$y = 99x + 2575$$

	1	2	3	4	
1966	$+1083$	$-$ 497	$-$ 225	$-$ 375	
1967	$+$ 925	$-$ 486	$-$ 293	$-$ 645	
1968	$+$ 941	$-$ 511	$-$ 275	$-$ 425	
1969	$+$ 949	$-$ 409	$-$ 279	$-$ 422	
1970	$+1031$				
	$+4929$	-1903	-1072	-1867	
Average	$+$ 986	$-$ 476	$-$ 268	$-$ 467	$= -225$
Adjustment	$+$ 56	$+$ 56	$+$ 56	$+$ 56	
S.V.	$+1042$	$-$ 420	$-$ 212	$-$ 411	

Forecasting

You may wonder why we stopped the series in 1970 rather than bring it up to date. The reason is that we wished to use the results for forecasting and you would not be convinced of the usefulness of the method unless you can see that the forecast is fairly close to the actual figures.

To predict the figures for the second quarter of 1970 we firstly predict the trend from our equation $y = 99x + 2575$ and then adjust for our calculated seasonal variation. Thus we have

$x \times m$ + S.V.
$19 \times 99 - 420 = 4357 - 420 = 3937$
The actual figure was 4130

We can list the predicted figures from our calculations and compare with the actual figures taken from the Bank of England Quarterly for March 1972.

	Qtr.	Trend	+	S.V.	=	Prediction	Actual
1970	2	4357		$-$ 420		3937	4130
	3	4456		$-$ 212		4244	4271
	4	4555		$-$ 411		4144	4164
1971	1	4654		$+1042$		5696	5540

The error in the forecast figures is rather high for the second quarter of 1970 but is still less than 5%, which when you remember that no allowance has been made for changes in tax rates and allowances is not a bad result.

Series with Seasonal Variation Eliminated

Most published statistics you will see are printed with seasonal variations eliminated from the figures. The underlying purpose of this is to enable us to concentrate on the general trend without being misled by seasonal influences.

To obtain such a series is quite simple. If the seasonal variation is positive, performance is raised above trend by seasonal factors and we must reduce the actual figures by the amount of the seasonal variation. If seasonal variation is negative we increase the figures by the amount of seasonal variation. Taking the last few quarters of our tax receipts series we would eliminate seasonal variation as follows:

		S.V.	Tax Receipts	Adjustment	With S.V. Eliminated
1969	1	$+1042$	4811	-1042	3769
	2	$-$ 420	3552	$+$ 420	3972
	3	$-$ 212	3781	$+$ 212	3993
	4	$-$ 411	3737	$+$ 411	4148
1970	1	$+1042$	5289	-1042	4247

You will notice that although seasonal variations are calculated as deviations from trend, eliminating them does not give us the trend figure. Try to reason out from what you know why this should be so.

EXERCISES TO CHAPTER TWELVE

1. The following table shows the growth in the use of private transport between 1952 and 1960. Calculate the least squares regression line and hence forecast the figure for 1961.

Year	1952	1953	1954	1955	1956	1957
Passenger miles (millions)	38	42	47	54	59	60

Year	1958	1959	1960
Passenger miles (millions)	73	82	89

2. The following table shows the number of passengers leaving a British airport on package holidays in recent years. Fit a straight line trend to the data by the method of least squares and hence estimate the growth of traffic during the years 1974 and 1975.

Year	1960	1961	1962	1963	1964	1965	1966
No. of passengers (thousands)	79	106	124	143	152	180	210

Year	1967	1968	1969	1970	1971	1972
No. of passengers (thousands)	242	257	309	396	453	488

3. The price of a government security and Bank Rate tend to vary inversely. The following figures show the average Bank Rate (R) for a number of years and the corresponding price of the security (C). Find the regression equation of C upon R and use it to estimate the price of the security when Bank Rate is 8%

R	C	R	C	R	C
3.7	56	4.5	56	2.0	83
4.0	57	3.9	56	3.7	58
5.0	55	3.0	67	4.6	56
4.7	55	5.5	54	4.0	55
2.0	74	2.0	76	2.4	74
2.5	67	2.0	72		

4. Sales of a particular commodity are collected over a period of years and the results summarised below. Graph the original data and superimpose the trend.

Sales — (all home branches)
(ten thousands)

| | *Quarter ended* | | | |
	Mar.	June	Sept.	Dec.
1967	295	329	344	325
1968	301	315	265	368
1969	350	386	262	405
1970	383	419	281	432
1971	393	436	302	449

5. You are informed that in the company whose sales figures are given in question 4, the need for part time employees is in the exact ratio of fluctuations in the level of sales. It is not desired to employ more than is necessary but orders must be met. What advice would you give to the personnel department regarding the employment of part time labour?

6. The annual abstract of statistics shows that until recently, there was an excess of male births over female births. The relevant data is given in the following table:

Excess of males over females per 1000 females born

Year	Excess of Males	Year	Excess of Males	Year	Excess of Males
1919	60	1926	41	1933	46
1920	52	1927	42	1934	55
1921	51	1928	44	1935	56
1922	49	1929	43	1936	54
1923	44	1930	44	1937	56
1924	47	1931	49	1938	51
1925	45	1932	50	1939	56

Fit a trend to the above data by the method of moving averages.

7. Quarterly sales of a certain car (in thousands) are as follows:

	1964	1965	1966	1967	1968
Quarter 1	66	68	86	111	102
2	70	64	52	74	79
3	34	40	44	34	47
4	53	58	53	63	66

Compute the trend by the method of moving averages, and estimate the seasonal variations in sales. Rewrite the series with Seasonal Variation eliminated.

8. Using the data of question 7 fit a trend to the data by the method of least squares and estimate the seasonal variation from this calculated trend. Compare your results with those obtained in question 7, and comment on any differences. (Omit the December qtr. of 1968 from your calculations.)

Chapter Thirteen

Correlation

We could quote many cases where it has been useful to establish a relationship between two sets of variables. In fact, the establishment of such relationships has, in some cases, changed our whole way of life. A good example of this is the relationship between the number of cigarettes smoked and the incidence of lung cancer. Another example is the relationship between the probability of infection and the degree of exposure to a certain disease. However, it can be just as useful to establish that a relationship does not exist — the statistician has been the scourge of the purveyors of quack medicine! A further use of establishing relationships is to make predictions about a variable that is difficult or costly to measure directly. For example, if we can establish a relationship between the destructive power of a bomb and the amount of explosive it contains, we can predict destructive power by measuring the explosive content. Let us now examine the form that such relationships take.

Let us suppose that we take two variables, x and y. We suspect that there is a relationship between the variables and we plot the results on a graph. The two variables which we suspect are related form a *bivariate distribution*, and the graph drawn is called a *scatter diagram*. The scatter diagram might look like this:

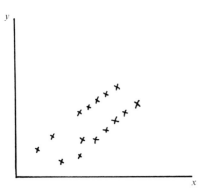

We notice that large values of *y* are associated with large values of *x*, and small values of *y* are associated with small values of *x*. We would say that in this case the variables are *positively correlated*. In other words, positive correlation exists when both variables increase together. Were we to plot the number of motor vehicles registered against the number of road accidents we would expect there to be a positive correlation. Can you think of any other examples yourself?

Now it is possible that the scatter diagram might look like this:

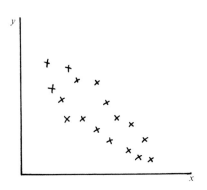

In this case small values of *y* are associated with large values of *x* and large values of *y* are associated with small values of *x*. Here, the variables are *negatively correlated*, which exists when an increase in one variable is associated with a decrease in the other. An example of negative correlation is the association between the number of T.V. licences issued and cinema admissions. You should try to compile a list of such examples.

A third case is where the scatter diagram looks like this:

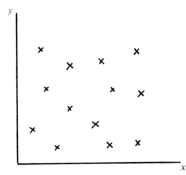

These points are scattered at random, and here we can say that there is an absence of correlation between the variables. This of course will be the most common of the three cases if we randomly select bivariate distributions.

Consider now the three diagrams below:

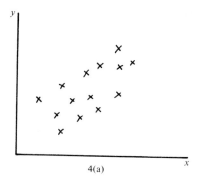

4(a)

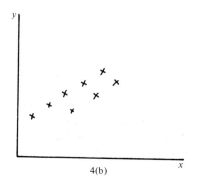

4(b)

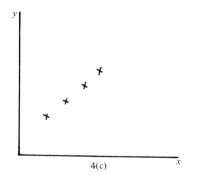

4(c)

In each case we see that the variables are positively correlated. The diagrams clearly show that association is strongest in case (c) and weakest in case (a). In case (c) all the points are on a straight line and for any value of one variable we could predict precisely the corresponding value of the other variable. Case (c) shows *perfect positive correlation*. Now in practice we will not meet perfect positive correlation, but we will meet examples that approach it quite closely. With some justification we can regard such examples as showing perfect positive correlation and any deviations of points from a straight line as resulting from experimental error.

Clearly, we need some measure of correlation, and our measure must be able to distinguish between positive and negative correlation. How can we derive such a measure? Well, you are probably thinking back to the chapter on the Time Series and realising that some connection must exist between correlation and the 'least squares' method used in that chapter. We saw that if a number of points were plotted on a graph, the least squares line of best fit could be given as $y = mx + c$, where,

$$m = \frac{\Sigma xy - \dfrac{\Sigma x \Sigma y}{n}}{\Sigma x^2 - \dfrac{(\Sigma x)^2}{n}}$$

and

$$c = \frac{1}{n}(\Sigma y - m\Sigma x)$$

If this is not clear in your mind you should read again the appropriate section.

Now the least squares line of best fit, $y = mx + c$, is called a *regression line*. Let us examine carefully how we obtained this line. We selected a point, say P_n with the coordinates $(x_n y_n)$ and said that the value of y on this line when $x = x_n$ was $x_n m + c$; and the deviation of the point $(x_n y_n)$ from the point on the line was

$$y_n - (x_n m + c) = y_n - x_n m - c$$

Now what are the implications of this? We are calculating the value of y on the line for a given value of x — which implies that we are using the values of x to predict the values of y. We are supposing that all deviations from the line occur as a result of errors in the value of y, and so we have been measuring the *vertical* deviations of the points from the line. Now must we find the regression line this way? We could have considered instead the horizontal deviations of the points from the line.

To do this we would have to consider the line

$$x = m_1 y + c_1$$

and for any value y_n the value of x on the line would be

$$y_n m_1 + c_1$$

Hence the deviation of the point $(x_n y_n)$ from this line is

$$x_n - (y_n m_1 + c_1) = x_n - y_n m_1 - c_1$$

You can see this in the diagram below:

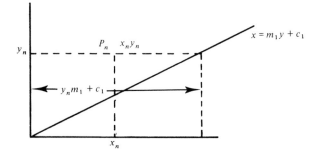

(You were asked to consider this idea in the tutorial following the analysis of least squares.)

We can use the same method to find m_1 and c_1 that we used to find m and c. This would give

$$m_1 = \frac{\Sigma xy - \dfrac{\Sigma x \Sigma y}{n}}{\Sigma y^2 - \dfrac{(\Sigma y)^2}{n}}$$

and

$$c_1 = \frac{1}{n}(\Sigma x - m_1 \Sigma y)$$

To find the regression line $y = mx + c$, we took the value of x as given and calculated the deviations from y. This line is called the regression of y on x, and is used to estimate a value of y when we are given a value of x. To find the regression line $x = m_1 y + c_1$, we took the value of y as given and calculated the deviations from x. This is the regression line of x on y and is used to estimate a value of x given a value of y. In the first case we assume that all deviations from a linear relationship are caused by differences between the observed values of y, and the values of y calculated from the regression equation.

EXAMPLE 1
When we considered the least squares method we found that for the bivariate distribution

x	1	2	3	4	5	6	7
y	1	2	6	7	10	16	21

the least squares line of best fit was

$y = 3.29x - 4.16$

This is the regression line of y on x. Let us now find the regression line of x on y.

x	y	y^2	xy
1	1	1	1
2	2	4	4
3	6	36	18
4	7	49	28
5	10	100	50
6	16	256	96
7	21	441	147
$\overline{28}$	$\overline{63}$	$\overline{887}$	$\overline{344}$

$\Sigma x = 28 \quad \Sigma y = 63 \quad \Sigma y^2 = 887 \quad \Sigma xy = 344$

$$m_1 = \frac{\Sigma xy - \dfrac{\Sigma x \Sigma y}{n}}{2y^2 - \dfrac{(\Sigma y)^2}{n}}$$

$$m_1 = \frac{344 - \dfrac{28 \times 63}{7}}{877 - \dfrac{(63)^2}{7}}$$

$m_1 = 0.288$

$$c_1 = \frac{1}{n}(\Sigma x - m_1 \Sigma y)$$

$$c_1 = \frac{1}{7}(28 - 0.28 \times 63)$$

$$c_1 = 1.43$$

The equation is $x = 0.288y + 1.43$.

The Correlation Coefficient (r)

Now let us see if we can obtain a measure of correlation, and call this the correlation coefficient (r). We will define r as the product of the coefficients m and m_1 in the regression equations, i.e.

$$r = mm_1$$

We shall consider first the case of perfect, positive correlation: the points would all lie on a straight line with a positive slope. The regression lines would coincide and the two equations $y = mx + c$ and $x = m_1y + c_1$ would be identical. Suppose we re-arrange the second equation:

$$x = m_1y + c_1$$

$$\text{so, } y = \frac{x}{m_1} - \frac{c_1}{m_1}$$

The gradient of this line is $\frac{1}{m_1}$. As the regression lines coincide, the gradient of this line is also m. Hence,

$$m = \frac{1}{m_1},$$

and the product of the coefficients will be

$$r = \frac{1}{m_1} \times m_1 = 1$$

Thus, with perfect, positive correlation, $r = 1$: a sensible value.

Now let us consider the case where no correlation exists. The regression lines will be parallel to the axes and look like the diagram on the facing page.

You may be wondering why the regression lines are parallel to the axes. If no correlation exists, then y does not depend upon x, and the regression equation is $y = c$. You will remember from the Introductory section that the line $y = c$ is parallel to the x axis. Likewise, the regression

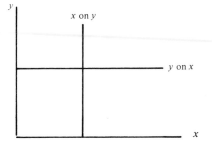

equation x on y must be $x = c_1$, which is parallel to the y axis. In this case $m = 0$ and $m_1 = 0$, so the correlation coefficient is

$0 \times 0 = 0$

When correlation is absent the value of the correlation coefficient is zero, again a sensible value.

Finally let us consider perfect negative correlation. Again the regression lines will coincide, but this time they will both have a negative slope. The regression lines will have the equations

$y = -mx + c$ and $x = -m_1 y + c_1$

Earlier we found that $m = \dfrac{1}{m_1}$ so in this case $-m = \dfrac{-1}{m_1}$ and the product

of the coefficients will be $-m_1 \times \dfrac{-1}{m_1} = 1$. Hence if perfect negative

correlation exists the value of our correlation coefficient will be one — the same as for perfect positive correlation. Now this will just not do! Our correlation coefficient must be able to distinguish between positive and negative correlation, but our coefficient $r = m\,m_1$ will not do this! What shall we do? Let us try making r the square root of the product of the coefficients, i.e.

$r^2 = m\,m_1$

Now when correlation is absent $r^2 = 0$ and $r = 0$ which is quite satisfactory. When correlation is perfect $r^2 = 1$ and $r = \pm 1$. This is much better. The positive root can signify perfect positive correlation and the negative root perfect negative correlation. But how will we know which root to take? Well, if correlation is positive, the gradient of the regression lines will be positive. Likewise if correlation is negative both m and m_1 will be negative. So if m amd m_1 are positive take the positive root of their product and if m and m_1 are negative take the negative root.

Let us summarise what we have learned so far. If we take $r^2 = m\,m_1$ as a measure of correlation, then

$$-1 < r^2 < 1$$

If $r = 0$ correlation is absent
   $r = 1$ correlation is perfect and positive
   $r = -1$ correlation is perfect and negative

So it follows that if $0 < r < 1$, correlation is positive but not perfect and if $-1 < r < 0$, correlation is negative but not perfect. The closer r approaches to its limits $(-1$ and $+1)$ the stronger is the association between the two variables.

Now let us consider the correlation coefficient for the example we considered earlier. We found that $m = 3.29$ and that $m_1 = 0.288$ so

$$r^2 = 3.29 \times 0.288 \text{ and } r = +0.97$$

A high degree of positive correlation exists. This leads us to expect that both regression lines would almost coincide. The diagram clearly shows this.

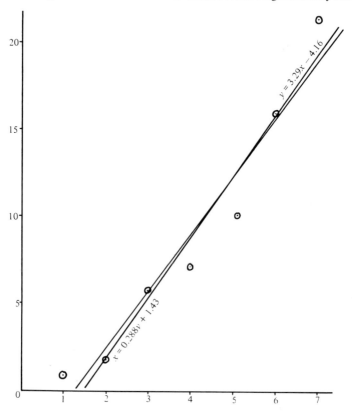

TUTORIAL ONE

1. For the following bivariate distributions which regression line would you calculate; x on y or y on x?

   (a) Index of wholesale prices (x) and index of retail prices (y).
   (b) Deaths from bronchitis (x) and mean weekly temperature (y).
   (c) The volume of a sphere (x) and the radius of the sphere (y).
   (d) Distance motored (x) and the cost of the journey (y).
   (e) Amount saved per week (x) and weekly income (y).

2. Bivariate distributions are often better described by a curve than by a straight line. One of the examples in question 1 illustrates this. Which one is it? What is the appropriate equation?

3. The annual cost of running a motor car (y) can be calculated using the equation $y = mx + c$ where x is the annual mileage. List the factors that will determine the size of the constants m and c.

4. Consider the bivariate distribution

x	0	1	2	3	4	5	6	7	8	9
y	12	15	18	21	24	27	30	33	36	39

   Find the regression lines y on x and x on y. Calculate the correlation coefficient. What would you expect to find if you drew a graph of the regression lines? Check your conclusions by drawing a graph of the regression lines and of the points.

5. Treat the following bivariate distributions in the same fashion.

(a) x	0	1	2	3	4	5	6	7	8	9
y	70	66	62	58	54	50	46	42	38	34
(b) x	3	3	3	5	5	5	7	7	7	
y	20	40	60	20	40	60	20	40	60	

   Are your suppositions about the graphs of the regression lines substantiated?

6.
x	0	1	2	3	4	5	6	7
y	15	11	12	10	5	6	4	1

   Calculate the regression lines y on x and x on y and calculate the correlation coefficient. Using an appropriate regression equation complete the following table:

x	8	9	10
y			

   Calculate the mean value for x and y and substitute these values in both regression equations. What can be deduced?

Now we have defined r^2 as $m\,m_1$, and we have seen that it is a satisfactory measure of correlation. It must follow that

$$r^2 = \frac{\Sigma xy - \dfrac{\Sigma x \Sigma y}{n}}{\Sigma x^2 - \dfrac{(\Sigma x)^2}{n}} \times \frac{\Sigma xy - \dfrac{\Sigma x \Sigma y}{n}}{\Sigma y^2 - \dfrac{(\Sigma y)^2}{n}}$$

Now suppose that we divide the top and bottom of each fraction by n the number of pairs in the bivariate distribution. Then

$$r^2 = \frac{\dfrac{1}{n}\left[\Sigma xy - \dfrac{\Sigma x \Sigma y}{n}\right]}{\dfrac{1}{n}\left[\Sigma x^2 - \dfrac{(\Sigma x)^2}{n}\right]} \times \frac{\dfrac{1}{n}\left[\Sigma xy - \dfrac{\Sigma x \Sigma y}{n}\right]}{\dfrac{1}{n}\left[\Sigma y^2 - \dfrac{(\Sigma y)^2}{n}\right]}$$

You should recognise that the denominator of the first fraction is the variance of x and that the denominator of the second fraction is the variance of y. So we can write:

$$r^2 = \frac{\dfrac{1}{n}\left[\Sigma xy - \dfrac{\Sigma x \Sigma y}{n}\right]}{\text{Var}(x)} \times \frac{\dfrac{1}{n}\left[\Sigma xy - \dfrac{\Sigma x \Sigma y}{n}\right]}{\text{Var}(y)}$$

$$\text{or } r^2 = \frac{\dfrac{1}{n^2}\left[\Sigma xy - \dfrac{\Sigma x \Sigma y}{n}\right]^2}{\text{Var}(x)\ \text{Var}(y)} \quad \text{and } r = \sqrt{\frac{\dfrac{1}{n^2}\left[\Sigma xy - \dfrac{\Sigma x \Sigma y}{n}\right]^2}{\text{Var}(x)\ \text{Var}(y)}}$$

$$r = \frac{\dfrac{1}{n}\left[\Sigma xy - \dfrac{\Sigma x \Sigma y}{n}\right]}{\sigma_x \ \sigma_y}$$

We call the numerator of this last fraction the Covariance so now we can say

$$r = \frac{\text{Covariance}(x\,y)}{\sigma_x \ \sigma_y}$$

Writing the correlation coefficient in this way we have overcome the problem of the sign of the coefficient: it is the same as the sign of the covariance.

EXAMPLE 2

Index of Earnings and Prices (1966 = 100)

	1969		1970				1971			
Quarter	3	4	1	2	3	4	1	2	3	4
Earnings	121	125	130	135	138	145	150	154	158	162
Prices	113	115	118	120	121	124	127	131	133	135

We wish to find the correlation coefficient between earnings and prices. It will simplify the arithmetic if we take deviations from 140 on the earnings index and deviations from 125 on the prices index.

x	y	x^2	y^2	xy
-19	-12	361	144	228
-15	-10	225	100	150
-10	-7	100	49	70
-5	-5	25	25	25
-2	-4	4	16	8
5	-1	25	1	-5
10	2	100	4	20
14	6	196	36	84
18	8	324	64	144
22	10	484	100	220
$\overline{18}$	$\overline{-13}$	$\overline{1844}$	$\overline{539}$	$\overline{944}$

$$\text{Covariance } (xy) = \frac{1}{n}\left[\Sigma xy - \frac{\Sigma x \Sigma y}{n} \right]$$

$$= \frac{1}{10}\left[944 - \frac{18 \times -13}{10} \right]$$

$$= \frac{1}{10}[944 + 23.4] = 96.74$$

$$\sigma_x = \sqrt{\frac{1}{n}\left[\Sigma x^2 - \frac{(\Sigma x)^2}{n} \right]} = \sqrt{\frac{1}{10}\left[1844 - \frac{18^2}{10} \right]}$$

$$= \sqrt{\frac{1}{10}[1844 - 32.4]} = \sqrt{181.16} = 13.46$$

$$\sigma_y = \sqrt{\frac{1}{n}\left[\Sigma y^2 - \frac{(\Sigma y)^2}{n} \right]} = \sqrt{\frac{1}{10}\left[539 - \frac{(-13)^2}{10} \right]}$$

$$= \sqrt{\frac{1}{10}[539 - 16.9]} \quad = \sqrt{52.21} \; = \; 7.23$$

$$r = \frac{\text{Covariance }(xy)}{\sigma_x \; \sigma_y} = \frac{96.74}{13.46 \times 7.23} = 0.994$$

The Significance of the Correlation Coefficient

Calculating the correlation coefficient has little meaning unless we can test its significance. Suppose we have a bivariate population with a zero correlation coefficient, and we draw samples of n pairs from this population. Of course, it would be unreasonable to expect all our samples to have a zero correlation coefficient: some will be greater than zero and some less, but we would expect the average value of correlation coefficients to be zero. This gives us some idea as to how we can assess the significance of r. Suppose a sample of n pairs gives a correlation coefficient r, and suppose we adopt the Null Hypothesis that the sample is drawn from an uncorrelated population, then if we can find the standard error we should be able to calculate the probability of obtaining our observed value of r.

Now we have already stated that correlation coefficients of samples all drawn from the same population are liable to sampling fluctuations. If the sample size is fairly large, then the distribution of r will be Normal. However, if n is small, then unrepresentative pairs will have a significant effect on the correlation coefficient of the sample, and the range of the correlation coefficient will be greater (the effect of an unrepresentative pair will be lost in a large sample). The distribution of the correlation coefficient for small values of n will not be Normal. This is illustrated in the diagram below.

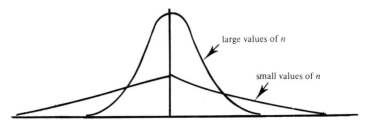

The distribution of the correlation coefficient

When n is small, the distribution of the correlation coefficient forms the so-called 't' distribution. The derivation and use of this distribution is beyond the scope of this book, and unfortunately most calculations of the correlation coefficient are from small samples. However, if you turn to the appendix, you will find that values of *r* at the 5% and 1% levels of significance are tabulated against differing sizes of *n*. Consulting this table for *n* = 10 (the number of pairs in the last example). We can state that if no correlation exists there is a 5% chance of *r* exceeding 0.58 and a 1% chance of *r* exceeding 0.71. Now the value of *r* we obtained in the last example was 0.994. What can we conclude? The probability that correlation is absent is certainly very much less than 1% and we must state that positive correlation seems highly likely.

This section on the correlation coefficient will end with a word of warning. When dealing with correlation one must never forget common sense. A significant correlation coefficient does not necessarily imply cause and effect. Many highly significant correlations are quite non-sensical. Statisticians in the past pointed to the highly significant correlation between the annual issue of broadcasting licences and the annual rate of admissions into mental institutions. The only logical conclusion you can draw from this is that quite by chance both statistics were increasing at the same rate. Even when a significant correlation seems reasonable you should not assume that cause and effect is established, as a third variable may be influencing both the others and thus explain the correlation. The high degree of positive correlation between overcrowding and infant mortality may seem to indicate clearly that high overcrowding causes a high infant mortality rate, though in fact both are indicative of income levels.

TUTORIAL TWO

1. You are attempting to assess the degree of correlation between the marriage rate and the birth rate. Given that you have the statistics of these two variables at monthly intervals which pairs would you use to try to establish a connection?

2. Obtain from the Annual abstract of statistics figures of the number of houses built in the United Kingdom over a given period of time and the volume of exports of industrial goods over the same period. Would you expect there to be any correlation between these two variables? Calculate the coefficient of correlation and if it differs

from zero attempt to assess the meaning of the value of *r* that you
have obtained.

3. Calculate the correlation coefficient for the following bivariate
distribution and test its significance. Store the result carefully as
later we will calculate the correlation coefficient using a different
method.

1971	Production of Private Cars (thousands)	Value of Exports (£ million)
Feb.	134	627
Mar.	142	771
April	138	787
May	156	755
June	176	799
Aug.	119	782
Sept.	159	834
Oct.	151	810
Nov.	145	759
Dec.	180	832

(Figures for July have been omitted because of the abnormally low
production of motor cars.)

4.
Industrial Ordinary Shares

1971	Price Index	Dividend Yield
Jan,	142	4.5
Feb.	140	4.6
Mar.	138	4.7
April	150	4.4
May	168	3.9
June	169	3.9
July	180	3.7
Aug.	182	3.6
Sept.	188	3.5
Oct.	183	3.6
Nov.	179	3.7
Dec.	190	3.5

Calculate the correlation coefficient and test its significance. Is
your conclusion what you would expect it to be? (If you do not
know what to expect, consult an economist.)

Rank Correlation

EXAMPLE 3
The correlation coefficient *r* is rather awkward to calculate and you may prefer to use instead the Rank Correlation Coefficient. This is not as accurate as the correlation coefficient *r* but does still give a fairly reliable indication as to the degree of correlation between two variables. Let us examine the following bivariate distribution

1971	Production of Private Cars (thousands)	Value of Exports (£ million)
Feb.	134	627
Mar.	142	771
April	138	787
May	156	755
June	176	799
Aug.	119	782
Sept.	159	834
Oct.	151	810
Nov.	145	759
Dec.	180	832

If we wish to calculate the rank correlation coefficient we must rank the values in each column. Let us consider the production of private cars over the ten months. The highest output was in December so we give this month a rank of one. The second greatest output was in June and we give this month a rank of two. We continue in this way ranking each month according to its output of private cars. When we have done this we rank each month according to the value of exports. Our bivariate distribution would now look like this:

Month	x Output	X Rank of Output	y Exports	Y Rank of Exports
Feb.	134	9	627	10
Mar.	142	7	771	7
April	138	8	787	5
May	156	4	755	9
June	176	2	799	4
Aug.	119	10	782	6
Sept.	159	3	834	1
Oct.	151	5	810	3
Nov.	145	6	759	8
Dec.	180	1	832	2

We can now calculate the correlation coefficient using the ranks instead of the actual values. This will considerably simplify the arithmetic.

Suppose we write p for the rank correlation coefficient. Then

$$p = \frac{\text{Covariance } (XY)}{\sigma_X \; \sigma_Y}$$

Before we actually calculate p for this bivariate distribution let us consider some preliminaries. The ranks (X) are identical to the ranks (Y) although they will not necessarily appear in the same order. It follows from this that $\Sigma x = \Sigma y$ and that $\sigma_x = \sigma_y$. If you cannot see that this must be so verify these facts for yourself using the ranks in the table above. Now we can write

$$p = \frac{\text{Covariance } (XY)}{\sigma^2_{(X)}}$$

so we need calculate only one variance rather than two standard deviations.

Let us now consider the bivariate distribution with n pairs. If we calculate p for this distribution we may well obtain an expression for p which is easier to manage than the one above. The ranks (X) in this distribution will be

1,   2,   3, ..., n

though not necessarily in that order. We will need to calculate the variance of the ranks (X).

$$\text{Variance } (X) = \frac{1}{n} \left[\Sigma x^2 - \frac{(\Sigma x)^2}{n} \right]$$

Now $\Sigma X = 1 + 2 + 3 + \ldots + n$, i.e. the sum of the first n integers. When you worked the tutorial following geometric progression (it was in the chapter on Discounted Cash Flow) you found the sum of this series and you were asked to keep the result. Find this result and check that it is $\frac{n^2 + n}{2}$. We can now write $\Sigma X = \frac{n^2 + n}{2}$. In the same way we can write

$$\Sigma X^2 = 1^2 + 2^2 + 3^2 + \ldots + n^2$$

Now it can be shown that

$$1^2 + 2^2 + 3^2 + \ldots + n^2 = \frac{n(n + 1)(2n + 1)}{6}$$

If you wish to verify this then an exercise in the next tutorial will help you to do so.

We can now say

$$\text{Variance } (X) = \frac{1}{n}\left[\frac{n(n+1)(2n+1)}{6} - \left(\frac{n^2+n}{2}\right)^2 \times \frac{1}{n}\right]$$

$$= \frac{1}{n}\left[\frac{n(n+1)(2n+1)}{6} - \frac{n^4+2n^3+n^2}{4n}\right]$$

$$= \frac{1}{n}\left[\frac{n(n+1)(2n+1)}{6} - \frac{n^3+2n^2+n}{4}\right]$$

$$= \left[\frac{(n+1)(2n+1)}{6} - \frac{n^2+2n+1}{4}\right]$$

$$= \left[\frac{(n+1)(2n+1)}{6} - \frac{(n+1)^2}{4}\right]$$

Well, this does not seem to be much of a simplification, but if we now consider the covariance the simplification will eventually occur.

$$\text{Covariance } (XY) = \frac{1}{n}\left[\Sigma XY - \frac{\Sigma X \Sigma Y}{n}\right]$$

$$= \frac{1}{n}\left[\Sigma XY - \frac{(\Sigma X)^2}{n}\right]$$

$$= \frac{\Sigma XY}{n} - \left(\frac{\Sigma X}{n}\right)^2$$

Now we have already found that the last term in the expression can be written as $\dfrac{(n+1)^2}{4}$.

We shall now see if we can find another way of writing ΣXY. To do this we shall use the fact that

$$(X - Y)^2 = X^2 - 2XY + Y^2$$
$$\text{so } \Sigma(X - Y)^2 = \Sigma X^2 - 2\Sigma XY + \Sigma Y^2$$

The quantity $(X - Y)^2$ could be found by taking the difference between the ranks, squaring them and summing the squares. If we write D^2 for this quantity

$$\Sigma D^2 = X^2 - 2XY + Y^2$$
$$\text{so } \Sigma XY = \frac{\Sigma X^2 + \Sigma Y^2 - \Sigma D^2}{2}$$

As $\Sigma X^2 = \Sigma Y^2$,

$$\Sigma XY = \frac{2\Sigma X^2 - \Sigma D^2}{2} = \Sigma X^2 - \frac{\Sigma D^2}{2}$$

We have already found an expression for ΣX^2, so

$$\Sigma XY = \frac{n(n + 1)(2n + 1)}{6} - \frac{\Sigma D^2}{2}$$

We can now write the Covariance like this

$$\text{Covariance } (XY) = \frac{n(n + 1)(2n + 1)}{6n} - \frac{\Sigma D^2}{2n} - \frac{(n + 1)^2}{4}$$

$$= \frac{(n + 1)(2n + 1)}{6} - \frac{\Sigma D^2}{2n} - \frac{(n + 1)^2}{4}$$

and the rank correlation coefficient like this:

$$p = \frac{\dfrac{(n + 1)(2n + 1)}{6} - \dfrac{\Sigma D^2}{2n} - \dfrac{(n + 1)^2}{4}}{\dfrac{(n + 1)(2n + 1)}{6} - \dfrac{(n + 1)^2}{4}}$$

$$= 1 - \frac{\dfrac{\Sigma D^2}{2n}}{\dfrac{(n + 1)(2n + 1)}{6} - \dfrac{(n + 1)^2}{4}}$$

$$= 1 - \frac{6\Sigma D^2}{2n(n + 1)(2n + 1) - 3n(n + 1)^2}$$

$$= 1 - \frac{6\Sigma D^2}{n(n^2 - 1)}$$

So the attempt at simplification has proved to be worthwhile as the rank correlation coefficient will now be simple to calculate.

X	Y	D	D^2
9	10	-1	1
7	7	0	0
8	5	3	9
4	9	-5	25
2	4	-2	4

X	Y	D	D²
10	6	4	16
3	1	2	4
5	3	2	4
6	8	-2	4
1	2	-1	1
		$\overline{0}$	$\overline{68}$

Notice that $\Sigma D = 0$. This must be so (why?) and is a useful check on the calculations performed.

$$p = 1 - \frac{6 \times 68}{10(10^2 - 1)} = 0.588$$

Again, we can test the significance of the rank correlation coefficient by consulting the table in the appendix. We see that for a sample of 10 if we adopt the null hypothesis that no correlation exists there is a 5% chance of the rank correlation coefficient exceeding 0.65. In this case it would be rather reckless to state that there is an association between the value of exports and the output of motor cars.

Tied Ranks

When variates have equal values we rank them as follows:

X =	35	89	92	92	98	99
Rank =	1	2	$3\frac{1}{2}$	$3\frac{1}{2}$	5	6

Notice that we do not rank the variates 92 as equal third. Can you see why? Can you see why we rank them both as $3\frac{1}{2}$? If we do this we will keep ΣX equal to ΣY and can use the simplified formula we used for p. Notice too that it does not matter if we assign the rank one to the lowest or the highest value of X. If, however we give a rank of one to the lowest value of X, we must also give the rank one to the lowest value of Y.

TUTORIAL THREE

1. A simple multiplication gives:

$$(n + 1)^3 = n^3 + 3n^2 + 3n + 1$$
$$\text{so } (n + 1)^3 - n^3 = 3n^2 + 3n + 1$$

Now suppose we consider the series $1, 2, 3, \ldots n - 3, n - 2, n - 1$, $n, n + 1$. We can use the expression above to obtain the following:

$(n + 1)^3 - n^3 = 3n^2 + 3n + 1$

$n^3 - (n - 1)^3 = 3(n - 1)^2 + 3(n - 1) + 1$

$(n - 1)^3 - (n - 2)^3 = 3(n - 2)^2 + 3(n - 2) + 1$

$(n - 2)^3 - (n - 3)^3 = 3(n - 3)^3 + 3(n - 3) + 1$

(if we replace n with $n - 1, n - 2$, $n - 3$ etc.)

— —

— —

$3^3 - 2^3 = 3(2)^2 + 3(2) + 1$

$2^3 - 1^3 = 3(1)^2 + 3(1) + 1$

Find the sum of the left hand side of the expressions (notice that most of the terms cancel in pairs). Now find the sum of the right hand sides using the Σ notation. You can now put the sum of the right hand side equal to the sum of the left hand side.

If you have performed these operations correctly you should notice that the left hand side contains a bracket which should be removed. The right hand side should contain $3\Sigma n$. Now we already know that $\Sigma n = \dfrac{n^2 + n}{2}$. Writing this for Σn you should now be able to show that

$$\Sigma n^2 = \frac{n(n + 1)(2n + 1)}{6}$$

2. Suppose that in a bivariate distribution X and Y were both increasing throughout the distribution though not necessarily at the same rate. What would you notice about Rank X and Rank Y for each pair. What would be the size of ΣD^2? What would be the size of the correlation coefficient.

3. Now suppose that in a bivariate distribution X was increasing and Y was decreasing throughout the distribution. Predict the size of p, bearing in mind your answer to the last question. Check your prediction by considering such a distribution with 5 pairs.

4. In many cases it is difficult or even impossible to quantify differences and in such cases ranking methods must be used. For example we can place a series of blue colours of different shades in order from light to dark, but we cannot quantify such differences. If we wish to test for consistency between judges of colour then we must use p.

Suppose three judges in a beauty contest place the ten competitors in the following order:

Competitor	A	B	C	D	E	F	G	H	I	J
Judge X	3	4	7	8	6	9	1	2	10	5
Y	6	10	1	4	9	5	7	2	8	5
Z	1	6	7	2	5	10	3	4	9	8

Use p to test whether there is any significant agreement between any pair of judges.

EXERCISES TO CHAPTER THIRTEEN
The following data taken from the records of 20 students at Liverpool Polytechnic refers to question numbers 1–6.

Examination Marks in Economics:	62	33	55	42	45	87	52	70	32	70
„ „ Accountancy:	48	49	59	31	59	89	60	55	34	68
% Attendance:	74	68	90	92	85	95	93	97	70	85

Examination Marks in Economics:	45	55	70	60	88	72	53	47	52	17
„ „ Accountancy:	58	55	73	56	78	77	42	38	50	26
% Attendance:	70	82	94	98	88	92	86	67	79	52

1. Calculate the correlation coefficient between performance in Economics and performance in accountancy.
2. Calculate the Rank correlation coefficient for the same data.
3. Test the significance of each of the above coefficients which you have calculated. Is there any connection between ability in the two subjects?
4. It is often claimed that poor performance in examinations is a function of a poor attendance record. Calculate a correlation coefficient between marks in economics and the percentage attendance.
5. Calculate the Rank coefficient of correlation between percentage attendance and the marks in accountancy.
6. What would you say to anyone who maintains that poor attendance means poor marks?
7. In a coastal area experiments have been undertaken to examine how the percentage of sand in the soil varies with distance from the coast. The following results are obtained.

Distance from
coast (yards)

| x | 0 | 600 | 1200 | 1800 | 2400 | 3000 | 3600 | 4200 | 4800 |

% sand in soil

| y | 80 | 62 | 64 | 62 | 57 | 59 | 40 | 47 | 37 |

Graph the regression lines you would use to predict the sand content in the soil at any given point and to predict your position relative to the coast given the percentage of sand in the soil.

8. Use the information so obtained to calculate a coefficient of correlation between distance from the coast and sand content.

9. Obtain the Rank correlation coefficient from the same data and assess its significance.

10. Estimate a coefficient of correlation from the following data. What conclusions can you draw?

	1972	1971	1970	1969	1968
Import of bananas into U.K. (1000 tons)	124	151	124	156	128
Suicide rate in Fiji (per 1000 of population)	13	33	12	40	12

	1967	1966	1965	1964	1963
Import of bananas into U.K. (1000 tons)	78	127	104	127	144
Suicide rate in Fiji (per 1000 of population)	7	20	4	15	26

Chapter Fourteen

Project Planning

We all know that if a project is to be undertaken efficiently, it must be well planned, especially if a number of people are engaged on the project. All too often, people work in isolation, quite ignorant of what other people working on the same project are doing. It is the task of management to co-ordinate efficiently the efforts of such people. In the past, planning has not been so important as it is today as projects were less complex — the rule of thumb method would work quite well. However, as projects have become more complex, management researchers have turned their attention increasingly to systematic planning. Consider the complexity of producing the prototype Concorde or an Apollo moonshot, and you will soon recognise the need for systematic planning. Shortly after the Second World War, researchers evolved a method called Network Analysis. The impact of this method has been quite dramatic, largely because it is applicable to such a wide variety of projects. It has been used to plan production projects, servicing projects, research projects, sales projects and military projects. This list is not intended to be exhaustive.

Network Symbols

How does the method work? Well the first stage of analysis is to divide the project into a number of different *activities*. An activity is merely a particular piece of work identifiable as an entity within the project. If, for example, the project under consideration, is the servicing of a motor car, then one of the activities would be 'check the brakes for wear'. Now an activity within a network is represented by an arrow, with the description of the activity written on it viz:

check the brakes for wear
$\xrightarrow{\hspace{5cm}}$

In addition to activities, we must also identify events. Events mark the point in time when an activity is complete and the next activity can

345

be started. Events are represented by circles.

(i) ──── check the brakes for wear ───→ (j)

The event (i) represents the point in time when the car is ready to have its brakes tested.

So far, we have not attempted to define a network and this we must now do. A network is a convenient method of showing the logical sequence of activities in a project. Suppose that in a certain project there are two activities *A* and *B*, and activity *B* cannot be started until activity *A* is completed. Using the network symbols, these activities can be represented like this.

O ──*A*──→ (x) ──*B*──→ O

The event (x) represents the point of time when activity *A* is completed, but it also represents the point of time when activity *B* can begin. We can conclude that the diagram above is a true representation of the situation when activity *B* depends upon activity *A*. Now of course it is quite likely that two or more activities are dependent upon the same activity. The situation when neither activity *D* nor *E* can start until activity *C* is complete would be represented like this:

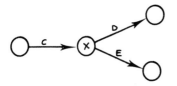

The event *X* represents the point of time when activity *C* is completed, and also the point of time when activities *D* and *E* can start, so the diagram clearly shows that *D* and *E* depend upon *C*. Also, it is likely that an activity depends upon more than one other activity. If activity *H* cannot start until activities *G* and *F* are both complete, then we would represent the situation like this:

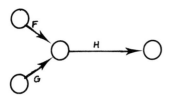

Dependence Tables

You have probably realised by now that the first task of network analysis is to sort out the logical sequence of activities. We can do this by constructing a 'dependency table'. We list all the activities, and next to them we list the activities that they depend upon. Let us suppose that our project is to understand quantitative analysis by reading this book. Now it is not absolutely necessary that you read the chapters in the order that they occur. The dependency table would look like this:

Chapter Number	Contents	Preceding Activity (Chapter No.)
0	Introductory	None
1	Inequalities	Introductory
2	Matrix Algebra	1 and Introductory
3	Simplex and L.P.	1 and 2
4	Transportation	3
5	Probability	Introductory
6	Sampling theory	5
7	Tests of Significance	5 and 6
8	Quality Control	6
9	Inventories	Introductory
10	Simulation	5 and 9
11	D.C.F.	Introductory
12	Time Series	Introductory and 11
13	Correlation	Introductory, 7, 11, and 12
14	Project Planning	None
15	Decision Theory	All

Constructing a dependence table is often the most difficult part of project analysis. Can you imagine how difficult it would be to construct a dependence table if the project were the construction of a Polaris submarine? Obviously, the construction of dependence tables is, in essence, a team activity: all the experts engaged on the project must be consulted. The last two questions of the following tutorial illustrate this point.

TUTORIAL ONE

1. Look at the dependence table for understanding quantitative analysis. It is stated in this table that understanding chapter 13 depends on your understanding of chapters 0, 7, 11, and 12. Why?

2. The dependence table quoted earlier states that this chapter does not depend on any other chapter. This is not, strictly speaking, true. Why?

3. Now imagine that you wrote this book. List as many different chapter orders as you can that still retain a logical sequence.

4. This question is reserved for female readers. Suppose that the project is to service a motor car. One of the activities is to check the brake shoes (or disc pads) for wear. What is the preceding activity?

Drawing the Network

Once we have compiled the 'dependence table', we can begin to draw the network. Suppose a certain project yielded the following 'dependence table'.

EXAMPLE 1

Activity	Preceding Activity
B, C and D	A
E and F	B
G	E
H	F
J	C
K	D
I	G and H
L	I, J and K

Notice that many relationships have been telescoped for convenience. We could represent these relationships by a series of 'sub networks'.

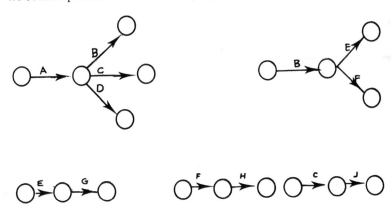

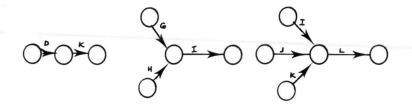

We can now attempt to join the sub-networks together and form a single network. However, one important rule must be observed: there must be just *one* start event and just *one* end event. The completed network would look like this:

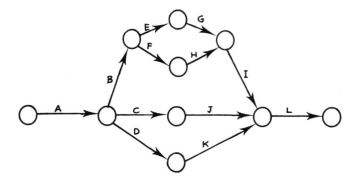

You should carefully check the sequence of activities in the network against the sub-networks and dependence table. Of course, you can omit the sub-networks if you wish and attempt to draw the complete network directly from the dependence table. Whichever way you do it, you will seldom draw the network correctly the first time you try.

The '*ij*' Event Numbering Rule

If the network above consisted of real rather than imaginary activities, then a description of each activity would be written above each arrow. Now it is convenient to use a coding system to describe a particular activity, and we do this by numbering the events according to the '*if* rule'. The rule states that the event at the end of an activity must be assigned a greater number than the event at the beginning of an activity.

There is no single way of numbering the events, for the '*ij* rule' allows considerable latitude. One numbering system that obeys the rule is:

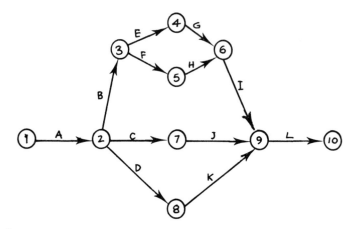

but of course there are many others. We can now describe activity *A* (remember that in fact it may have quite a lengthy description) by its '*ij*' numbers 1–2.

Dummy Activities

If activities are to be described by their '*ij*' numbers, then it is essential that no two activities have the same numbers. Now this can present some difficulty: consider the following extract from a project's dependence table.

EXAMPLE 2

Activity	Preceding Activity
B and C	A
D	B and C

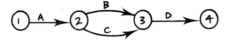

Now this network does show a logical sequence of activities, but we cannot accept it, as activities *B* and *C* both have the same '*ij*' number. We overcome this problem by introducing a *dummy activity*, which is represented by a broken line.

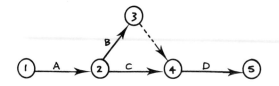

Now the dummy activity does not represent any activity as such: it is inserted to preserve the sequential numbering system of events. Such a dummy is called an *identity dummy*.

Sometimes it is necessary to insert a dummy to preserve not the sequential numbering system, but the logical sequence of events. Consider the following dependence table.

EXAMPLE 3

Activity	Preceding Activity
B and C	A
D	B
E	B and C
F	D and E

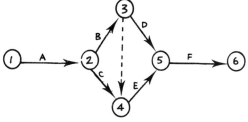

You should satisfy yourself that the network is a true representation of the logical sequence of activities, and that the dummy 3–4 is essential to preserve the logical sequence.

TUTORIAL TWO
1. It was stated in the text that usually it is possible to number the events in a number of different ways. Renumber the events in the network of example 1 in two different ways.
2. Is it possible to re-number the events in example 3 differently, and still obey the '*ij*' rule?

3. Re-examine the network for example 2. It need not be drawn pre-
    cisely this way. Can you redraw it in three other ways? Is it possible
    to redraw example 3?
4. Draw the network of the following 'dependence table'.

Activity	Preceding Activity
B and C	A
E	B
F	C and D
G	E and F

    Remember that there must be one start event and one end event.
5. Examine the dependence table for the above exercise, and also the
    dependence table for example 3 in the text. Can you find rules for
    when logical dummies are needed and where they will be needed?
6. A businessman decides to form a team of mathematicians from
    certain members of his staff, and (wisely) chooses this book as his
    training manual. You are asked to investigate the best method of
    training the team on the understanding that the businessman would
    be happy to have a team of specialists.
    Consider the dependence table for this book. Clearly it can be
    telescoped, i.e.

Chapter	Preceding Chapter
1, 5, 9, 11	Introduction

    Find the telescoped dependence table. (You should remember
    from the previous tutorial that an understanding of all aspects of
    this chapter assumes that all other chapters have been read.) Draw
    the network. What is the minimum number of specialists on the
    team? Assume the team is to be assembled as quickly as possible.

Total Project Time: Critical Path

In exercise 5 of the last tutorial you were asked to find rules for when
and where logical dummies would be needed. Now these rules are most
convenient to use, so they will be stated here.

1. If an activity occurs in the right hand column of a dependence table
    but not in the left-hand column, it cannot depend upon another
    activity having been completed. Hence, it must be a 'starting activity'.
    If there is more than one start activity you *may* need dummies.

2. If any activity occurs more than once in the right-hand column then you *will* need to introduce dummies. If the activity occurs n times, then $(n-1)$ dummies will have to be drawn from its end event.

EXAMPLE 2

The dependence table of a certain project looks like this:

Activity	Preceding Activity
B, C and D	A
E and F	B
G	E
H	F
I	G and H
J	G, H, C and D
K	D
L	I, J and K

Firstly, we note that there is only one starting activity. Both G and H occur twice, so the dummy will be needed from the end event of both these activities. The network will look like this.

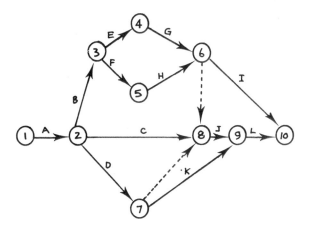

Let us suppose that the network describes the activities in a manufacturing process. The network clearly shows that activities B, C and D can all be started together as long as we have sufficient resources to do so. Let us assume that we can easily obtain all the resources that we need. Later we can drop this assumption. Now let us assume that the project is such that the only resource needed is labour, and that all the labour available is equally capable of performing any activity. The time taken to complete each activity is known

Activity	A	B	C	D	E	F	G	H	I	J	K	L
Time (hrs.)	3	4	5	6	2	1	7	4	3	5	6	2

How long will the project take? To answer this, we must examine all the routes through the network. Can you see that there are seven possible routes? Let us list each and find the time taken.

Route	*Time*	
A B E G I L	$3 + 4 + 2 + 7 + 3 + 2$	= 21 hrs.
A B E G Dummy J L	$3 + 4 + 2 + 7 + 0 + 5 + 2$ = 23 hrs.	
A B F H I L	$3 + 4 + 1 + 4 + 3 + 2$	= 17 hrs.
A B F H Dummy J L	$3 + 4 + 1 + 4 + 0 + 5 + 2$ = 19 hrs.	
A C J L	$3 + 5 + 5 + 2$	= 15 hrs.
A D Dummy I L	$3 + 6 + 0 + 5 + 2$	= 16 hrs.
A D K L	$3 + 6 + 6 + 2$	= 17 hrs.

The project cannot be completed in less than 23 hours. This is determined by the longest route through the network — called the *critical path*. Activities on this route must be completed on time otherwise the total project time will lengthen; (i.e. the activities have critical times). You should realise that it would be very difficult to calculate the total project time without first drawing the network. Try to find the total project time just using the dependence table and you will realise how true this is!

Most networks will be much more complicated than the one we have examined, and it will be tedious to identify all the routes. In fact it is more than likely that some routes will be overlooked. A more efficient method is to use *the earliest event times* i.e. the earliest time that each event can occur. We divide the circle showing the event into three parts like this:

A is the event number,
B is the earliest event time.

We start at event 1 and arbitrarily assign to it a start time of zero. Event two occurs when activity *A* is complete, so the earliest event time is $0 + 3 = $ third hour.

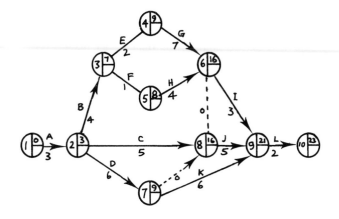

Likewise we can deduce that

the earliest event time for event 3 is 7 hours (activity *B* complete)
the earliest event time for event 4 is 9 hours (activity *E* complete)
the earliest event time for event 5 is 8 hours (activity *F* complete)
the earliest event time for event 7 is 9 hours (activity *D* complete)

Now let us consider event 6. This cannot occur until both activity *G* and activity *H* are complete.

Activity *G* is complete after 9 + 7 = 16 hours at the earliest.
Activity *H* is complete after 8 + 4 = 12 hours at the earliest.

So the earliest time for event 6 is the 16th hour.

For event 8, the earliest time would be either 16 + 0 = 16 (dummy activity 6–8 complete – remember that it is not an activity as such and so cannot occupy any time) or 3 + 5 = 8 (activity *C* complete) or 9 + 0 = 9 (dummy activity 7–8 complete). Clearly, the earliest time for event 8 is the 16th hour. Continuing in this way for all the other events, we finally reach the end event 10 which has an earliest time of 23 hours. Clearly, the earliest time for the end event must be the same as the total project time. Using this method to find the total project time is more efficient and (as we shall see later) is essential to further analysis of the network.

TUTORIAL THREE

1. The question refers to the last network of the text. Suppose each activity was programmed to start at the earliest possible time.

Deduce the effects on the network if

    (a) Activity J takes 3 hours longer than expected.
    (b) Activity I takes 3 hours longer than expected.
    (c) Activity K takes 6 hours longer than expected.
    (d) Activity F takes 1 hour longer than expected.
    (e) Activity H takes 2 hours longer than expected.
    (f) Activity J takes 3 hours less than expected.

The Critical Path

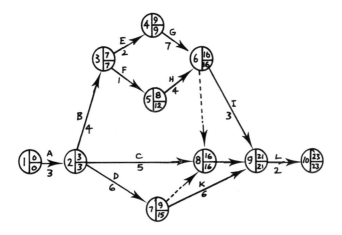

Finding the earliest event times has certainly helped to determine the total project time, but it has not helped us to isolate the critical path. To do this, we must use latest event times i.e. the latest time that each event can occur if the network is to be completed on time. Obviously, the latest time for event 10 is the 23rd hour, and the latest time for event 9 is the $23 - 2 = 21$st hour. Again, the latest time for event 8 is quite straightforward: it is $21 - 5 = 16$th hour. For event 6, we must consider the two following activities 6–8 and 6–9. If activity 6–8 is to be complete on time, then event 6 has a latest time of $16 - 0 = 16$th hour. If activity 6–9 is to be complete on time then event 6 has a latest time of $21 - 3 = 18$th hour. Can you see that if the project is to be complete on time the event 6 has the 16th hour as its latest time? Continuing in this way, we can obtain the latest times for all the events.

    Think back to critical activities — they must be started on time, otherwise the total project time will lengthen. What does this imply? Each event on the critical path must have the same earliest and latest times. Using this fact we can easily identify the critical path in the above network — but beware! The method is not infallable. Consider the network below.

EXAMPLE 3

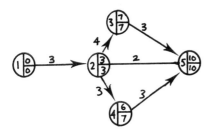

If we state the critical path contains those event which have the same earliest and latest times, then we would isolate two paths:

$$1 - 2 - 3 - 5$$
$$\text{and } 1 - 2 - 5$$

Are they both critical paths?

Float

What can we deduce about non-critical activities? Well firstly, it is not necessary for them to start at a particular, specified time. Secondly, they can take longer than the time specified. Let us extract some non-critical activities from the last network.

Consider first activity 3–5. The earliest time it can start is the 7th hour, and it must be completed by the 12th hour. It follows that we have $12 - 7 = 5$ hours available to complete this task. Now as the activity should only take 1 hour, we could expand the time spent on this task by $5 - 1 = 4$ hours without affecting the total project time. Alternatively, this activity could start 4 hours late without affecting the total project time. This 4 hours latitude we have on activity 3–5 is called its *total float*. We can define total float as time available for an activity *minus expected activity duration*, or using *i* as the start event and *j* as the end event.

latest time for event *j*
minus earliest time for event *i*
minus activity duration

Critical activities will have a zero total float – this is the only reliable method of extracting the critical path.

The total float of activity 5–6 is 16 − 8 − 4 = 4 hours. Now suppose we start activity 3–5 on the 11th hour (i.e. take advantage of its total float), then this task will be completed on the 11 + 1 = 12th hour. Activity 5–6 now cannot start until the 12th hour, and its total float is 16 − 12 − 4 = 0. Thus, the total float of 4 hours refers to the path 3–6 rather than the activities on that path.

Now suppose we take up all the float on activity 5–6, then this activity must end on the sixteenth hour. Now no activity starting with event 6 can begin before this time, so taking up the float on 5–6 does not effect the following activities. The float on 5–6 then, is essentially different to the float on 3–5. We say that activity 5–6 has a *free float* of 4 hours. Thus, total float affects following activities, whereas free float does not.

Event 5–6 has a total float of 4 hours, all of which is also free float. However, if free float exists, it does not always follow that total float will be the same as the free float. Consider the example below.

EXAMPLE 4

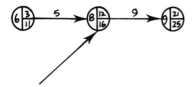

Activity 6–8 has a total float of 16 − 3 − 5 = 8 hours, and as long as this activity is completed by the twelfth hour, then activity 8–9 will be unaffected. We can calculate free float as follows.

Earliest time for event *j*
Minus earliest time for event *i*
Minus time taken by activity *ij*

TUTORIAL FOUR

Examine again the network showing earliest and latest event times for example 11. The times for each activity assumes a 'one man, one job' allocation of labour.

1. Assume that each activity starts at its earliest time. How many men would be required if the project is to be completed on time?
2. It is usual to summarise a network in a chart like this:

Activity	Duration	Earliest		Latest		Float	
		Start	End	Start	End	Total	Free
1–2	3	0	3	0	3	0	0
2–3	4	3	7	3	7	0	0
2–7	6	3	9	9	15	6	0
: :	:	:	:	:	:	:	:
7–9	6	9	15	15	21	6	0

The activity duration, earliest start and latest end are extracted from the network. So,

Earliest start + duration = earliest end
Latest end − duration = latest start

Can you find a simple rule for calculating total float? Complete the chart.
3. Would it be possible to complete this project on time if (a) three men, (b) two men, were available?

Gantt Charts: 'Loading' the Network

So far, our analysis has ignored the supply of resources that are available to work on a project. Let us continue to analyse the same project, and assume that we require one man per activity. Allocating resources to the project is called 'loading the network', and this is usually done with a *Gantt Chart*. On a Gantt chart, the activities are represented by lines having lengths proportional to the duration of each activity. The Gantt chart for our project would look like the diagram overleaf.

Each activity on the Gantt chart is identified by its *ij* number. Can you see the assumption on which the chart has been drawn? It has been assumed that each activity starts at its earliest time. The line representing activity 5–6 shows that the activity starts at the end of the eighth hour and ends at the end of the twelfth hour. The dotted line 6–6 shows that activity 5–6 must be completed at the end of the 16th hour at the latest. The dotted line shows the total float of each activity. At the foot

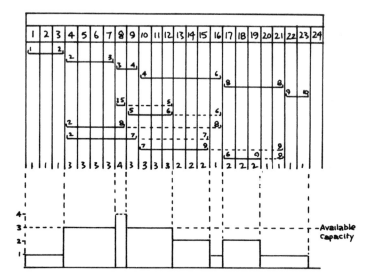

of each column, the number of men required for that hour is shown. This is obtained by counting the number of solid horizontal lines in each column. The histogram shows the amount of labour required on an hourly basis.

The histogram shows that the labour scheduled to this project is un-evenly distributed. It also shows that there are 3 men available for this project. For most of the time, the labour required is less than the labour available, and the project is said to be *underloaded*. However in the 8th hour the project is overloaded, and this overload must be removed. It would also seem sensible to smooth the histogram as much as possible.

Obviously, we must concentrate on activities that have float, and as activities with free float do not affect other activities, we will consider them first. Activity 7–9 has a free float, and if we put this forward one day, the gaps in the histogram in the 16th day will be filled. This, how-ever, will leave another gap at the 10th hour. We can fill this gap and remove the overload in the eighth hour by advancing activities 3–5 and 5–6 by two hours. This removes the overload, but does not smooth the histogram as the demand for labour is now like this:

Hour	Labour required
0–3	1 man
4–8	3 men
9–10	2 men

Hour	Labour required
11—14	3 men
15—19	2 men
20—23	1 man

The unevenness in the 9th and 10th hour can be removed if we advance activities 2—8 and 2—7 by an hour. The Gantt chart and demand for labour now looks like this:

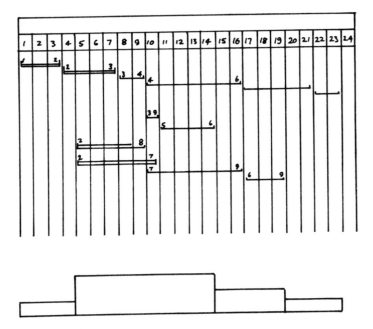

We can now state exactly how labour will be scheduled to the project. One man can be instructed to work on the critical path. A second man will be required from the 5th to the 14th hour, working on activity 2—8, 3—5 and 5—6.

Progress Charts

The Gantt Chart can be used effectively to show how a project is actually progressing, and we do this by drawing a line above the activity symbol which is proportional to the amount of work completed. Suppose, for example that the Gantt Chart above represents the state of the project at the end of the 8th hour. We can conclude that activity 3—4 should be 50% complete but that it has not yet been started.

Activity 2–7 should be 83.33% complete – in fact it has been completed. This interpretation begs the question as to whether it is possible to transfer labour to the critical path, as the project time is in danger of lengthening. Used in this way, the Gantt chart is an important element of project control.

TUTORIAL FIVE

1. Suppose that the labour available to work on the project discussed in the text was as follows:

   > two men up to the end of the 11th hour.
   > three men from the 11th to the 23rd hour.

   Redraw the Gantt Chart to take account of the supply of labour.

2. Now suppose only two men are available: the project time will have to increase. Schedule resources to the activities so that the increased total project time is a minimum.

3.

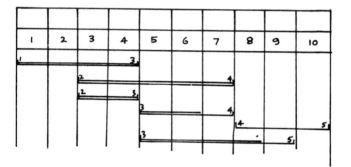

The progress chart shows the state of a project at the end of the 7th day. What conclusions would you draw? Have you any recommendations?

EXERCISES TO CHAPTER FOURTEEN

1. You have identified that in a particular project there are 15 activities, and have deduced the following dependency table.

Activity	Preceding Activity
A	None
B	A
C	B
D	C
E	D
F	A
G	F
H	D, G, I, + J
I	A
J	A
K	A
L	A
M	L
N	M
O	E, H, K, + N

Telescope this table into a more manageable form.

2. Look again at the telescoped dependence table you have derived in question 1. How many dummies are you likely to need? Draw the network of this project and number the events.

3. The activities referred to in question 1 have the following estimated durations

Activity	A	B	C	D	E	F	G	H	I	J	K	L	M	N	O
Days	5	10	5	2	3	8	1	12	5	5	10	2	10	6	5

Insert these times on your network and so deduce the earliest and latest event times. What is the total project time? What is the critical path?

4. Prepare a summary chart for the project, and find the total and free float for each activity. What would be the effect of activity E taking 9 days longer than expected?

5. Each activity requires one man, except activity B. This activity requires one man for one day at its start, and one man for one day to complete the activity. Draw the Gantt Chart on the understanding that each activity starts at its earliest possible time, and obtain the resource profile.

6. The labour availability for this project is — three men to the end of day 18, and two men thereafter. Can the project be completed on time?

7. At the end of day 14, activity *L* is just complete (it has taken one day longer than expected). If activity *K* is also just complete, what would you recommend?

8. The '*ij*' numbers and durations of activities in a certain project are

ij No.	*Duration*
1–2	3
2–3	4
2–4	5
2–6	6
3–4	3
3–5	2
4–7	4
5–7	0
5–8	5
6–7	0
6–8	1
7–8	3

Draw the network and deduce the total project time.

9. Label the activities in the last question from *A* to *J* (in the order they occur in the last question). Do *not* label the dummies. Now derive a dependence table for the project.

10. Your assistant has been given the task of analysing a project. He derives the following 'dependence table'.

Activity	*Preceding Activity*
B, *C* and *N*	*A*
E	*B*
F	*C* and *G*
G	*D* and *H*
H	*F*
I	*E*
J	*F*
K	*D* and *H*

He is having difficulty with this task. Identify the difficulty and advise him where the fault lies.

Chapter Fifteen

Conclusion – Decision Taking in Industry

Those of you who have worked through the examples and exercises in this book should by now have some knowledge of the analytical techniques used in the taking of decisions in industry and commerce. We would now like you to consider seriously the following simple situation.

A manufacturer engaged in a highly competitive business finds that a competitor has developed a new technique of production which enables him to undercut the current market price. Sales fall off and our manufacturer is faced with a situation in which there are a number of alternative courses of action.

(a) He can cut his own prices to the level of those of his competitor. If he does so his accountant estimates that he will be selling at a loss which in total will amount to £1500 a year. Meanwhile his own research team will be working on the problem of cost reduction and they believe that ultimately they can come up with an answer that will enable him to compete effectively once more.

(b) He can temporarily close the unit which can not make a profit until such time as his research team come up with the answers. If he does so, however, he will still have certain charges to meet, such as the wages of caretaking and maintenance staff, interest on debentures and other loans, rent of the factory premises etc. In total it is estimated that these will amount to £1300 a year.

(c) He can cut his losses and close his productive unit completely, abandoning all hope of being able to meet the new competition at any time in the future.

There are probably many other courses open to him but these will suffice to illustrate a crucial factor in decision taking.

We will eliminate the third alternative immediately. Why? Although economists assume, and probably correctly, that people are in business for money, other factors are often as important. The pride of 'running one's own show' may be important; the sense of power given by 'being the boss' can be a critical consideration; an unwillingness to admit defeat may influence the decision; or perhaps our business man is imbued with an all pervading sense of optimism that often seems to take control of the businessman. As Frank Livesey argues in his textbook, 'Economics', sheer survival can be a major industrial objective.[1]

Thus, on personal and psychological grounds, we reject the extreme course of action, and concentrate our attention on the other two.

Of the two, on purely financial grounds the second might appear to be the better alternative, since the annual cash outflow is less. How long we can stand such an outflow depends on the state of our financial reserves, and we might ultimately be forced out of business. For the time being however would we not be wise to husband our resources for as long as possible and adopt the alternative which 'costs' us least. Or would we? A moment's consideration will convince you that there are problems other than the simple accounting cost of a particular course of action. If we make our workers redundant will we be able to persuade them to rejoin us when the firm is once more a viable proposition? Or will they have found other employment? Even if they do come back to us, what will be the state of industrial relations within the factory? Will a decision to close result in a sit-in, or possibly a strike in other units or even other firms? Will we recover the markets we are losing? Has the firm a social purpose which may at times be more important than the purely economic or financial?

The list of questions we could ask is long, and before the 'best' decision is made, all would require an answer. Anyone who has been faced with this type of problem will readily appreciate that the ability to take decisions involves more than the mere application of analytical techniques. This is not to say, of course, that for several months now you have been wasting your time. We are firmly convinced that a good industrial decision requires two types of information on which it can be based:

(a) Concrete measurable information which can be obtained by the use of the techniques examined in this book. Price/output policy is of little use unless based on a reliable estimate of future demand; investment policy is less than sound without the use of discounted cash flow techniques; production planning would lose much of its validity without critical path analysis. But, as you must be aware, only inanimate behaviour is capable of quantitative analysis. Any policy decided upon must, if it is to yield results, be put into practice — and it must be put into practice by human beings.

(b) Once we begin to take human beings into account we are faced

---

[1] F. Livesey, Economics pages 19–20.

by a set of data which by its very nature is intangible and not capable of mathematical measurement. As individuals, and as a working group we will react to given circumstances in a unique way. Yet the way we react is of crucial importance in determining whether a given course of action is the right one or not. The successful decision taker must not be merely a mathematician, but also a psychologist, a human relations expert, a marketing specialist and so on.

Now, obviously, no one man can be such a walking encyclopedia of expertise. It is more true to say that the successful decision taker is the man who can listen to experts, integrate their opinions, assess their worth, and on the basis of the combined knowledge of a team make a decision.

In the past too many industries have fallen into the error of not using the mass of data available to them, or the results of statistical analysis at all. Decisions have been based purely on hunches or value judgements. But just as great a sin in our opinion is that of placing entire reliance on statistical analysis and excluding from the decision taking process those value judgements which must form a part of it. In recent years when industry has been inflicted by unrest over such things as decisions to close (Upper Clyde, Courtaulds), by wage claims based on the fact that bigger or more complex equipment is being used (Jumbo jets, high speed passenger trains), and by oppositions to policies which appear to threaten job security or working conditions, it is becoming all too apparent that decision taking must take into account all aspects of the problem.

One of the more recent development in technique which promises to have a great future is the technique employed in the decision to build the Victoria line and the decision on the siting of the third London airport – Cost Benefit Analysis. One cannot do justice to this new approach to decision taking without extending this book to an unacceptable length, and those who are expert in this technique may feel that anything other than a full treatment should be ignored. It is, we think, important that all who are concerned with decisions should be aware that the technique exists, and that, in addition to the quantitative analysis we have been considering, it takes into account many intangible factors which previously were assessed only minimally. Examples of the type of intangible which has been taken into account are, the impact of noise, the effect of atmospheric pollution on health, the long term effect of unemployment, the spin-off that results from providing a certain number of employment opportunities. Obviously such factors are not amenable to strict quantitative analysis, and yet any accurate measurement of the full cost and the full benefit of a project must be couched in quantitative terms. You may not consider it very satisfactory to give a monetary value to the individual whose health suffers as a result of large jet airliners screaming overhead; perhaps you will not accept that we can quantify the gain from living in a quiet rural area or the benefit to our nerves resulting from a more traffic free city centre. Yet ultimately such things must be quantitatively assessed, and the technique of Cost-Benefit Analysis, although as yet in its infancy, goes farther than most techniques in this direction. The more lusty the infant becomes, the more there will be a need for a generalist to assess the data collected by and the opinions of the teams of specialists who will emerge.

It is our belief that British industry needs, perhaps more than ever before, the 'General' manager, and 'General' management to co-ordinate the work of the experts, and it is our hope that in writing this book we have been able to make a useful contribution to that need.

ANSWERS TO THE EXERCISES (Mainly odd numbered questions)

Introduction

1. (a) $F = \dfrac{C}{50} + 17$.

    (b) Fee on sale price of £4000 = £97.
        Fee on sale price of £10,000 = £217.

    (c) £77; £167; £257.

3. $X = 5$.

5. $P = \dfrac{a[1 - (1 + r)^{-n}]}{r}$

7. Class intervals: 30–39 etc.
    $\bar{x} = 52.25$
    $\sigma = 10.1$
    Mean of original data is 52.55.

Chapter 1

1. Graphical question.
3. Produce 18 bats and 4 rackets at a profit of £49⅓.
5. $x = 10$, $y = 30$, Profit = £130.
7. $x = 4$, $y = 5$, Cost is £9 per 1000 gallons.

Chapter 2

1. $B = A^{-1}C$
    $B = \begin{bmatrix} 1 & 2 \\ 3 & 1 \end{bmatrix}$
    $A = CB^{-1}$
    $A = \begin{bmatrix} 2 & 1 \\ 1 & 3 \end{bmatrix}$

3. New position is $\begin{bmatrix} 1 & 2 \\ -1 & -1 \end{bmatrix}$

    Transformation matrix is $\begin{bmatrix} 0 & 1 \\ -1 & 0 \end{bmatrix}$

    New position of second line is $\begin{bmatrix} 1 & 2 \\ -1 & -3 \end{bmatrix}$

5. $P = \begin{bmatrix} 240 & 0 \\ 0 & 300 \end{bmatrix}$ $P' = \begin{bmatrix} 160 & 80 \\ 0 & 300 \end{bmatrix}$ $Q = \begin{bmatrix} \frac{2}{3} & \frac{4}{15} \\ 0 & 1 \end{bmatrix}$

$P'' = \begin{bmatrix} 200 & 40 \\ 150 & 150 \end{bmatrix}$ $R = \begin{bmatrix} \frac{5}{6} & \frac{2}{15} \\ \frac{5}{8} & \frac{1}{2} \end{bmatrix}$

Chapter 3

2. $\left[\begin{array}{ccccc|c} -\frac{13}{16} & 0 & 1 & \frac{15}{8} & -\frac{25}{16} & 75 \\ \frac{33}{24} & 1 & 0 & -\frac{5}{4} & \frac{45}{24} & 150 \\ \hline \frac{3}{24} & 0 & 0 & \frac{5}{4} & \frac{15}{24} & 450 \end{array} \right]$

4. $\left[\begin{array}{ccccc|c} 0 & 1 & \frac{1}{6} & -\frac{1}{6} & 0 & 100 \\ 1 & 0 & \frac{2}{3} & \frac{1}{3} & 0 & 400 \\ 0 & 0 & \frac{8}{3} & \frac{7}{3} & 1 & 1600 \\ \hline 0 & 0 & \frac{5}{6} & \frac{1}{6} & 0 & 500 \end{array} \right]$

5. $\left[\begin{array}{cccccccc|c} 0 & 0 & 0 & 1 & -\frac{1}{3} & -\frac{4}{3} & 0 & 0 & -\frac{11}{3} & 10 \\ 0 & 0 & 0 & 0 & \frac{2}{3} & -\frac{1}{3} & 0 & 1 & \frac{1}{3} & 26 \\ 0 & 0 & 0 & 0 & -\frac{1}{3} & \frac{2}{3} & 1 & 0 & \frac{4}{3} & 32 \\ 1 & 0 & 0 & 0 & -\frac{1}{3} & \frac{2}{3} & 0 & 0 & \frac{4}{3} & 42 \\ 0 & 1 & 0 & 0 & \frac{2}{3} & -\frac{1}{3} & 0 & 0 & \frac{1}{3} & 36 \\ 0 & 0 & 1 & 0 & 0 & 0 & 0 & 0 & -1 & 10 \\ \hline 0 & 0 & 0 & 0 & \frac{8}{3} & \frac{2}{3} & 0 & 0 & \frac{7}{3} & 434 \end{array} \right]$

7. $3A - 3B - 2C \leqslant 0$
$2B + C - 4A \leqslant 0$
$5B - 7A - 5C \leqslant 0$
$2C - A - 2B \leqslant 0$
$A \leqslant 20 \quad B \leqslant 34 \quad C \leqslant 32$
Maximise $A + B + C$

9. Material y has a scarcity value of £$\frac{1}{7}$, range of -35 to $+7$.
Material z has a scarcity value of £$\frac{3}{7}$, range of -7 to $+21$.

Chapter 4

1. Minimise $7x_{11} + 5x_{12} + 9x_{13} + \ldots + 5x_{53} + 7x_{54} + 9x_{55}$

Subject to $x_{11} + x_{12} + x_{13} + x_{14} + x_{15} = 71$

$x_{21} + x_{22} + x_{23} + x_{24} + x_{25} = 26$

$x_{31} + x_{32} + x_{33} + x_{34} + x_{35} = 50$

$x_{41} + x_{42} + x_{43} + x_{44} + x_{45} = 133$

$$x_{51} + x_{52} + x_{53} + x_{54} + x_{55} = 76$$
$$x_{11} + x_{21} + x_{31} + x_{41} + x_{51} = 110$$
$$x_{12} + x_{22} + x_{32} + x_{42} + x_{52} = 38$$
$$x_{13} + x_{23} + x_{33} + x_{43} + x_{53} = 52$$
$$x_{14} + x_{24} + x_{34} + x_{44} + x_{54} = 130$$
$$x_{15} + x_{25} + x_{35} + x_{45} + x_{55} = 26$$

3.

			Distributor		
		A	*B*	*C*	*D*	*E*
	A				71	
	B					26
Factory	*C*	12	38			
	D	22		52	59	
	E	76				

Loss of profit is £14,382.

5.

	1000 cc.	1250 cc.	1500 cc.
Saloon	25		20
Van	35	10	
Convertible	10		

Profit is £9550.

7. Flight commander : *F*
   Pilot : *A*
   Navigator : *E*
   Engineer : *C*
   Back-up man : *D*
   Communications : *B*

Chapter 5

1. $\dfrac{1}{26^3 \times 999}$

3. (a) 85 ways
   (b) 35 ways

5. Expand $(\frac{1}{2} + \frac{1}{2})^6$

   $P(2, 3, \text{ or } 4 \text{ girls}) = \dfrac{25}{32}$

7. Expand $(\frac{3}{4} + \frac{1}{4})^6$

$$P(3) = \frac{347}{2048}$$

9. $P(6 \text{ or more defectives}) = 0.017$
$P(\text{Two defective batches}) = (0.017)^2$

11. $P(\text{No strikes}) = 0.4066$
$P(x \geqslant 4 \text{ new strikes}) = 0.013$

Chapter 6

1. 15.87%, 30.85%, 78.88%, 12.93%.
3. 7640 hours.
5. Annual saving is £1231.
7. 99.8% of sample means will be within the range 32 ± 0.0618.
   Machine needs resetting.
9. $(3.335 \pm 0.0081) \times 10^{-3}$ cms.
11. Actual increase could be as low as 0.642%. A larger sample is needed.

Chapter 7

1. $2\frac{1}{2}\%$.
3. Standard error is 0.28. Significant at 99.9% level.
5. Standard error is 0.081. Significant at 99.9% level.
7. Expected frequencies: 2 5 3 12 30 18 6 15 9
   Chi – squared = 25.28, highly significant.
9. Mean = 2, var = 1.058, $q = \dfrac{1.058}{2} = 0.53$.
   $240(0.53 + 0.47)^4 = 18.94 + 67.16 + 89.33 + 52.81 + 11.71$.
   Chi – squared = 4.627. Binomial distribution fits.
11. Mean = 0.75.
    Expected frequencies: 47 35 18 (for 2 or more).
    Chi – squared = 1.535. Poisson distribution fits.

Chapter 8

1. $P(1 \text{ defective}) = 26.4\%$, so this proportion of batches would be returned.

3. Expanding $e^{-1}\left(\dfrac{a}{x}\right)^x$
   Only one sample mean in 20 should have more than 3 defectives and only 1 in 1000 more than 5 defectives. Process is under control.

5. Sample means   Warning limit: 25.02 ± 0.023
                            Action limit: 25.02 ± 0.036
     Sample ranges   Warning limit: 0.169
                            Action limit: 0.202
    Specifications cannot be met.
7. 25 ± 1.336.
9. Reset the machine to 25 thousandths.

Chapter 9

1. (a) $9x^2 + 8x + 5$ (b) $\dfrac{x^2 + x + 1}{2}$

   (c) $-2\left[\dfrac{6}{x^4} + \dfrac{3}{x^3} + \dfrac{1}{x^2}\right]$ (d) $\dfrac{1}{4(4\sqrt{x})^3} + \dfrac{1}{3(3\sqrt{x})^2} + \dfrac{1}{2\sqrt{x}}$

3. Range is 1110 miles, average velocity is $138\frac{3}{4}$ m.p.m. orbital velocity is 256 m.p.m.
$r = 256t - 948, t > 8$
Length of orbit is 36,342 miles.
5. $x = 74$. Total revenue = £5476. Price = £74.
7. M.R. $= 148 - 2x$. M.C. $= 36$. $x = 56$.
9. $q = \sqrt{\dfrac{2RC_2 D}{C_1(R-D)}} = 5000$

Run length = 1 week. Inventory cycle = 5 weeks.

Chapter 10

1. Poisson frequencies: 13.5, 27.1, 18, 9, etc.
Random numbers: 000–134, 135–405, 406–676 etc.
3. 14.3 (using Poisson frequencies), 29 (using a simulated sequence).
5. Waiting times: 0 0 3 2 2 0 1 0 0 0 3 1 3 4 6 2 4 5 6 9 11 11 12 13
7. $Q = 200$

Chapter 11

1. Costs £$(x + 1800)$. Revenue £$(x + 2400)$. Profit £3900.
3. $a = 1000$. $r = 0.85$. $n = 6$.
Balance sheet value = £443.6.

5. P.V. of industrial profits £5615.58.
   P.V. of land speculation return £5397. Do not speculate.
   Maximum selling price £17,391.
7. Cost £9108.57.
9. $n = 6$. Small balance remains in the account.

Chapter 12

1. Equation is $y = 6.367x + 28.61$. Forecast for 1961 is 92.28.
3. Equation is $C = 88.97 - 7.42R$. Estimated price 29.6
4. Trend: 299, 298, 299, 307, 318, 333, 342, 346, 355, 363, 370, 375, 380, 383, 388, 393.
7. Trend: 56.0, 55.5, 55.5, 56.9, 59.8, 60.5, 59.5, 59.4, 61.9, 67.8, 69.3, 69.3, 69.4, 68.9, 71.1, 73.1.
   Seasonal variation: +28, +2.5, −24, −6.5.

Chapter 13

1. Taking 50 as origin for both x and y, $\Sigma x = 107$, $\Sigma y = 105$, $\Sigma x^2 = 6633$, $\Sigma y^2 = 5705$, $\Sigma xy = 5323$, $r = 0.888$.
3. Both r and p are significant.
5. $D^2 = 570$, $p = 0.57$.
7. Using 24 as origin for x, and 57 as origin for y; $\Sigma xy = -1608$, $\Sigma x^2 = 2160$, $\Sigma y^2 = 1421$. Equations are: $y = 75 - 0.75x$, $x = 88 - 1.13y$, $r = -0.92$.
9. $D^2 = 231.5$, $p = -0.93$.

Chapter 14

1.

Activity	Preceding Activity
B, F, I, J, K, L	A
C	B
D	C
E	D
G	F
M	L
N	M
H	D, G, I, J
O	F, H, K, N

3. Total project time is 39 days. Critical path A, B, C, D, H, O.

5. Resource profile.

Day	1–5	6	7–10	11–15	16–22	23	24–25	26–39
Labour required	1	6	5	3	2	3	2	1

7. Divert activity G from man A to man C. Man A can now complete B on time and man C can complete G on any day up to day 22.

9. | Activity | Preceding Activity |
    |---|---|
    | B, C, D | A |
    | E, F | B |
    | G | C, E |
    | H | F |
    | I | D |
    | J | D, F, G |

Appendices

Logarithms
Antilogarithms
Squares
Square Roots of Numbers

e^{-x}

The Normal Distribution Function [Area under the Normal Curve $N(0,1)$]
Percentage Points of the χ^2 Distribution
Discount Tables – Values of $(1 + r)^{-n}$
Reciprocals
Random Sampling Numbers
Significance of the Correlation Coefficient
Minimum Values of p which are significant at the 5% and 1% level

COMMON LOGARITHMS   log₁₀x

$\log_{10} x$

x	0	1	2	3	4	5	6	7	8	9	Δ_m +	1 2 3	4 5 6	7 8 9 ADD
10	·0000	0043	0086	0128	0170	0212					42	4 8 13	17 21 25	29 34 38
						0212	0253	0294	0334	0374	40	4 8 12	16 20 24	28 32 36
11	·0414	0453	0492	0531	0569	0607					39	4 8 12	16 19 23	27 31 35
						0607	0645	0682	0719	0755	37	4 7 11	15 19 22	26 30 33
12	·0792	0828	0864	0899	0934	0969					35	4 7 11	14 18 21	25 28 32
						0969	1004	1038	1072	1106	34	3 7 10	14 17 20	24 27 31
13	·1139	1173	1206	1239	1271	1303					33	3 7 10	13 16 20	23 26 30
						1303	1335	1367	1399	1430	32	3 6 10	13 16 19	22 26 29
14	·1461	1492	1523	1553	1584	1614	1644	1673	1703	1732	30	3 6 9	12 15 18	21 24 27
15	·1761	1790	1818	1847	1875	1903	1931	1959	1987	2014	28	3 6 8	11 14 17	20 22 25
16	·2041	2068	2095	2122	2148	2175	2201	2227	2253	2279	26	3 5 8	10 13 16	18 21 23
17	·2304	2330	2355	2380	2405	2430	2455	2480	2504	2529	25	2 5 7	10 12 15	17 20 22
18	·2553	2577	2601	2625	2648	2672	2695	2718	2742	2765	24	2 5 7	10 12 14	17 19 22
19	·2788	2810	2833	2856	2878	2900	2923	2945	2967	2989	22	2 4 7	9 11 13	15 18 20
20	·3010	3032	3054	3075	3096	3118	3139	3160	3181	3201	21	2 4 6	8 11 13	15 17 19
21	·3222	3243	3263	3284	3304	3324	3345	3365	3385	3404	20	2 4 6	8 10 12	14 16 18
22	·3424	3444	3464	3483	3502	3522	3541	3560	3579	3598	19	2 4 6	8 10 11	13 15 17
23	·3617	3636	3655	3674	3692	3711	3729	3747	3766	3784	18	2 4 5	7 9 11	13 14 16
24	·3802	3820	3838	3856	3874	3892	3909	3927	3945	3962	18	2 4 5	7 9 11	13 14 16
25	·3979	3997	4014	4031	4048	4065	4082	4099	4116	4133	17	2 3 5	7 9 10	12 14 15
26	·4150	4166	4183	4200	4216	4232	4249	4265	4281	4298	16	2 3 5	6 8 10	11 13 14
27	·4314	4330	4346	4362	4378	4393	4409	4425	4440	4456	16	2 3 5	6 8 10	11 13 14
28	·4472	4487	4502	4518	4533	4548	4564	4579	4594	4609	15	2 3 5	6 8 9	11 12 14
29	·4624	4639	4654	4669	4683	4698	4713	4728	4742	4757	15	1 3 4	6 7 9	10 12 13
30	·4771	4786	4800	4814	4829	4843	4857	4871	4886	4900	14	1 3 4	6 7 8	10 11 13
31	·4914	4928	4942	4955	4969	4983	4997	5011	5024	5038	14	1 3 4	6 7 8	10 11 13
32	·5051	5065	5079	5092	5105	5119	5132	5145	5159	5172	13	1 3 4	5 7 8	9 10 12
33	·5185	5198	5211	5224	5237	5250	5263	5276	5289	5302	13	1 3 4	5 6 8	9 10 12
34	·5315	5328	5340	5353	5366	5378	5391	5403	5416	5428	13	1 3 4	5 6 8	9 10 12
35	·5441	5453	5465	5478	5490	5502	5514	5527	5539	5551	12	1 2 4	5 6 7	8 10 11
36	·5563	5575	5587	5599	5611	5623	5635	5647	5658	5670	12	1 2 4	5 6 7	8 10 11
37	·5682	5694	5705	5717	5729	5740	5752	5763	5775	5786	12	1 2 4	5 6 7	8 10 11
38	·5798	5809	5821	5832	5843	5855	5866	5877	5888	5899	11	1 2 3	4 6 7	8 9 10
39	·5911	5922	5933	5944	5955	5966	5977	5988	5999	6010	11	1 2 3	4 6 7	8 9 10
40	·6021	6031	6042	6053	6064	6075	6085	6096	6107	6117	11	1 2 3	4 5 7	8 9 10
41	·6128	6138	6149	6160	6170	6180	6191	6201	6212	6222	10	1 2 3	4 5 6	7 8 9
42	·6232	6243	6253	6263	6274	6284	6294	6304	6314	6325	10	1 2 3	4 5 6	7 8 9
43	·6335	6345	6355	6365	6375	6385	6395	6405	6415	6425	10	1 2 3	4 5 6	7 8 9
44	·6435	6444	6454	6464	6474	6484	6493	6503	6513	6522	10	1 2 3	4 5 6	7 8 9
45	·6532	6542	6551	6561	6571	6580	6590	6599	6609	6618	10	1 2 3	4 5 6	7 8 9
46	·6628	6637	6646	6656	6665	6675	6684	6693	6702	6712	9	1 2 3	4 5 5	6 7 8
47	·6721	6730	6739	6749	6758	6767	6776	6785	6794	6803	9	1 2 3	4 5 5	6 7 8
48	·6812	6821	6830	6839	6848	6857	6866	6875	6884	6893	9	1 2 3	4 4 5	6 7 8
49	·6902	6911	6920	6928	6937	6946	6955	6964	6972	6981	9	1 2 3	4 4 5	6 7 8

COMMON LOGARITHMS $\log_{10}x$

x	0	1	2	3	4	5	6	7	8	9	Δ_m +	1 2 3	4 5 6	7 8 9
													ADD	
50	·6990	6998 7007 7016			7024 7033 7042			7050 7059 7067			9	1 2 3	4 4 5	6 7 8
51	·7076	7084 7093 7101			7110 7118 7126			7135 7143 7152			8	1 2 2	3 4 5	6 6 7
52	·7160	7168 7177 7185			7193 7202 7210			7218 7226 7235			8	1 2 2	3 4 5	6 6 7
53	·7243	7251 7259 7267			7275 7284 7292			7300 7308 7316			8	1 2 2	3 4 5	6 6 7
54	·7324	7332 7340 7348			7356 7364 7372			7380 7388 7396			8	1 2 2	3 4 5	6 6 7
55	·7404	7412 7419 7427			7435 7443 7451			7459 7466 7474			8	1 2 2	3 4 5	6 6 7
56	·7482	7490 7497 7505			7513 7520 7528			7536 7543 7551			8	1 2 2	3 4 5	6 6 7
57	·7559	7566 7574 7582			7589 7597 7604			7612 7619 7627			8	1 2 2	3 4 5	6 6 7
58	·7634	7642 7649 7657			7664 7672 7679			7686 7694 7701			8	1 2 2	3 4 5	6 6 7
59	·7709	7716 7723 7731			7738 7745 7752			7760 7767 7774			7	1 1 2	3 4 4	5 6 6
60	·7782	7789 7796 7803			7810 7818 7825			7832 7839 7846			7	1 1 2	3 4 4	5 6 6
61	·7853	7860 7868 7875			7882 7889 7896			7903 7910 7917			7	1 1 2	3 4 4	5 6 6
62	·7924	7931 7938 7945			7952 7959 7966			7973 7980 7987			7	1 1 2	3 3 4	5 6 6
63	·7993	8000 8007 8014			8021 8028 8035			8041 8048 8055			7	1 1 2	3 3 4	5 6 6
64	·8062	8069 8075 8082			8089 8096 8102			8109 8116 8122			7	1 1 2	3 3 4	5 6 6
65	·8129	8136 8142 8149			8156 8162 8169			8176 8182 8189			7	1 1 2	3 3 4	5 6 6
66	·8195	8202 8209 8215			8222 8228 8235			8241 8248 8254			7	1 1 2	3 3 4	5 6 6
67	·8261	8267 8274 8280			8287 8293 8299			8306 8312 8319			6	1 1 2	2 3 4	4 5 5
68	·8325	8331 8338 8344			8351 8357 8363			8370 8376 8382			6	1 1 2	2 3 4	4 5 5
69	·8388	8395 8401 8407			8414 8420 8426			8432 8439 8445			6	1 1 2	2 3 4	4 5 5
70	·8451	8457 8463 8470			8476 8482 8488			8494 8500 8506			6	1 1 2	2 3 4	4 5 5
71	·8513	8519 8525 8531			8537 8543 8549			8555 8561 8567			6	1 1 2	2 3 4	4 5 5
72	·8573	8579 8585 8591			8597 8603 8609			8615 8621 8627			6	1 1 2	2 3 4	4 5 5
73	·8633	8639 8645 8651			8657 8663 8669			8675 8681 8686			6	1 1 2	2 3 4	4 5 5
74	·8692	8698 8704 8710			8716 8722 8727			8733 8739 8745			6	1 1 2	2 3 4	4 5 5
75	·8751	8756 8762 8768			8774 8779 8785			8791 8797 8802			6	1 1 2	2 3 4	4 5 5
76	·8808	8814 8820 8825			8831 8837 8842			8848 8854 8859			6	1 1 2	2 3 4	4 5 5
77	·8865	8871 8876 8882			8887 8893 8899			8904 8910 8915			6	1 1 2	2 3 4	4 5 5
78	·8921	8927 8932 8938			8943 8949 8954			8960 8965 8971			6	1 1 2	2 3 4	4 5 5
79	·8976	8982 8987 8993			8998 9004 9009			9015 9020 9025			6	1 1 2	2 3 4	4 5 5
80	·9031	9036 9042 9047			9053 9058 9063			9069 9074 9079			5	1 1 2	2 3 3	4 4 5
81	·9085	9090 9096 9101			9106 9112 9117			9122 9128 9133			5	1 1 2	2 3 3	4 4 5
82	·9138	9143 9149 9154			9159 9165 9170			9175 9180 9186			5	1 1 2	2 3 3	4 4 5
83	·9191	9196 9201 9206			9212 9217 9222			9227 9232 9238			5	1 1 2	2 3 3	4 4 5
84	·9243	9248 9253 9258			9263 9269 9274			9279 9284 9289			5	1 1 2	2 3 3	4 4 5
85	·9294	9299 9304 9309			9315 9320 9325			9330 9335 9340			5	1 1 2	2 3 3	4 4 5
86	·9345	9350 9355 9360			9365 9370 9375			9380 9385 9390			5	1 1 2	2 3 3	4 4 5
87	·9395	9400 9405 9410			9415 9420 9425			9430 9435 9440			5	0 1 1	2 2 3	3 4 4
88	·9445	9450 9455 9460			9465 9469 9474			9479 9484 9489			5	0 1 1	2 2 3	3 4 4
89	·9494	9499 9504 9509			9513 9518 9523			9528 9533 9538			5	0 1 1	2 2 3	3 4 4
90	·9542	9547 9552 9557			9562 9566 9571			9576 9581 9586			5	0 1 1	2 2 3	3 4 4
91	·9590	9595 9600 9605			9609 9614 9619			9624 9628 9633			5	0 1 1	2 2 3	3 4 4
92	·9638	9643 9647 9652			9657 9661 9666			9671 9675 9680			5	0 1 1	2 2 3	3 4 4
93	·9685	9689 9694 9699			9703 9708 9713			9717 9722 9727			5	0 1 1	2 2 3	3 4 4
94	·9731	9736 9741 9745			9750 9754 9759			9763 9768 9773			5	0 1 1	2 2 3	3 4 4
95	·9777	9782 9786 9791			9795 9800 9805			9809 9814 9818			5	0 1 1	2 2 3	3 4 4
96	·9823	9827 9832 9836			9841 9845 9850			9854 9859 9863			4	0 1 1	2 2 2	3 3 4
97	·9868	9872 9877 9881			9886 9890 9894			9899 9903 9908			4	0 1 1	2 2 2	3 3 4
98	·9912	9917 9921 9926			9930 9934 9939			9943 9948 9952			4	0 1 1	2 2 2	3 3 4
99	·9956	9961 9965 9969			9974 9978 9983			9987 9991 9996			4	0 1 1	2 2 2	3 3 4

ANTILOGARITHMS 10^x

x	0	1	2	3	4	5	6	7	8	9	Δ_m +	1 2 3	4 5 6	7 8 9
													ADD	
·00	1000	1002	1005	1007	1009	1012	1014	1016	1019	1021	2	0 0 1	1 1 1	1 2 2
·01	1023	1026	1028	1030	1033	1035	1038	1040	1042	1045	2	0 0 1	1 1 1	1 2 2
·02	1047	1050	1052	1054	1057	1059	1062	1064	1067	1069	2	0 0 1	1 1 1	1 2 2
·03	1072	1074	1076	1079	1081	1084	1086	1089	1091	1094	2	0 0 1	1 1 1	1 2 2
·04	1096	1099	1102	1104	1107	1109	1112	1114	1117	1119	3	0 1 1	1 1 2	2 2 3
·05	1122	1125	1127	1130	1132	1135	1138	1140	1143	1146	3	0 1 1	1 1 2	2 2 3
·06	1148	1151	1153	1156	1159	1161	1164	1167	1169	1172	3	0 1 1	1 1 2	2 2 3
·07	1175	1178	1180	1183	1186	1189	1191	1194	1197	1199	3	0 1 1	1 1 2	2 2 3
·08	1202	1205	1208	1211	1213	1216	1219	1222	1225	1227	3	0 1 1	1 1 2	2 2 3
·09	1230	1233	1236	1239	1242	1245	1247	1250	1253	1256	3	0 1 1	1 1 2	2 2 3
·10	1259	1262	1265	1268	1271	1274	1276	1279	1282	1285	3	0 1 1	1 1 2	2 2 3
·11	1288	1291	1294	1297	1300	1303	1306	1309	1312	1315	3	0 1 1	1 2 2	2 2 3
·12	1318	1321	1324	1327	1330	1334	1337	1340	1343	1346	3	0 1 1	1 2 2	2 2 3
·13	1349	1352	1355	1358	1361	1365	1368	1371	1374	1377	3	0 1 1	1 2 2	2 2 3
·14	1380	1384	1387	1390	1393	1396	1400	1403	1406	1409	3	0 1 1	1 2 2	2 2 3
·15	1413	1416	1419	1422	1426	1429	1432	1435	1439	1442	3	0 1 1	1 2 2	2 2 3
·16	1445	1449	1452	1455	1459	1462	1466	1469	1472	1476	3	0 1 1	1 2 2	2 2 3
·17	1479	1483	1486	1489	1493	1496	1500	1503	1507	1510	4	0 1 1	2 2 2	3 3 4
·18	1514	1517	1521	1524	1528	1531	1535	1538	1542	1545	4	0 1 1	2 2 2	3 3 4
·19	1549	1552	1556	1560	1563	1567	1570	1574	1578	1581	4	0 1 1	2 2 2	3 3 4
·20	1585	1589	1592	1596	1600	1603	1607	1611	1614	1618	4	0 1 1	2 2 2	3 3 4
·21	1622	1626	1629	1633	1637	1641	1644	1648	1652	1656	4	0 1 1	2 2 2	3 3 4
·22	1660	1663	1667	1671	1675	1679	1683	1687	1690	1694	4	0 1 1	2 2 2	3 3 4
·23	1698	1702	1706	1710	1714	1718	1722	1726	1730	1734	4	0 1 1	2 2 2	3 3 4
·24	1738	1742	1746	1750	1754	1758	1762	1766	1770	1774	4	0 1 1	2 2 2	3 3 4
·25	1778	1782	1786	1791	1795	1799	1803	1807	1811	1816	4	0 1 1	2 2 2	3 3 4
·26	1820	1824	1828	1832	1837	1841	1845	1849	1854	1858	4	0 1 1	2 2 2	3 3 4
·27	1862	1866	1871	1875	1879	1884	1888	1892	1897	1901	4	0 1 1	2 2 2	3 3 4
·28	1905	1910	1914	1919	1923	1928	1932	1936	1941	1945	4	0 1 1	2 2 2	3 3 4
·29	1950	1954	1959	1963	1968	1972	1977	1982	1986	1991	4	0 1 1	2 2 2	3 3 4
·30	1995	2000	2004	2009	2014	2018	2023	2028	2032	2037	5	0 1 1	2 2 3	3 4 4
·31	2042	2046	2051	2056	2061	2065	2070	2075	2080	2084	5	0 1 1	2 2 3	3 4 4
·32	2089	2094	2099	2104	2109	2113	2118	2123	2128	2133	5	0 1 1	2 2 3	3 4 4
·33	2138	2143	2148	2153	2158	2163	2168	2173	2178	2183	5	1 1 2	2 3 3	4 4 5
·34	2188	2193	2198	2203	2208	2213	2218	2223	2228	2234	5	1 1 2	2 3 3	4 4 5
·35	2239	2244	2249	2254	2259	2265	2270	2275	2280	2286	5	1 1 2	2 3 3	4 4 5
·36	2291	2296	2301	2307	2312	2317	2323	2328	2333	2339	5	1 1 2	2 3 3	4 4 5
·37	2344	2350	2355	2360	2366	2371	2377	2382	2388	2393	6	1 1 2	2 3 4	4 5 5
·38	2399	2404	2410	2415	2421	2427	2432	2438	2443	2449	6	1 1 2	2 3 4	4 5 5
·39	2455	2460	2466	2472	2477	2483	2489	2495	2500	2506	6	1 1 2	2 3 4	4 5 5
·40	2512	2518	2523	2529	2535	2541	2547	2553	2559	2564	6	1 1 2	2 3 4	4 5 5
·41	2570	2576	2582	2588	2594	2600	2606	2612	2618	2624	6	1 1 2	2 3 4	4 5 5
·42	2630	2636	2642	2649	2655	2661	2667	2673	2679	2685	6	1 1 2	2 3 4	4 5 5
·43	2692	2698	2704	2710	2716	2723	2729	2735	2742	2748	6	1 1 2	2 3 4	4 5 5
·44	2754	2761	2767	2773	2780	2786	2793	2799	2805	2812	6	1 1 2	2 3 4	4 5 5
·45	2818	2825	2831	2838	2844	2851	2858	2864	2871	2877	7	1 1 2	3 3 4	5 6 6
·46	2884	2891	2897	2904	2911	2917	2924	2931	2938	2944	7	1 1 2	3 3 4	5 6 6
·47	2951	2958	2965	2972	2979	2985	2992	2999	3006	3013	7	1 1 2	3 3 4	5 6 6
·48	3020	3027	3034	3041	3048	3055	3062	3069	3076	3083	7	1 1 2	3 4 4	5 6 6
·49	3090	3097	3105	3112	3119	3126	3133	3141	3148	3155	7	1 1 2	3 4 4	5 6 6

ANTILOGARITHMS 10ˣ

x	0	1	2	3	4	5	6	7	8	9	Δ_m +	1 2 3	4 5 6	7 8 9 ADD
·50	3162	3170	3177	3184	3192	3199	3206	3214	3221	3228	7	1 1 2	3 4 4	5 6 6
·51	3236	3243	3251	3258	3266	3273	3281	3289	3296	3304	8	1 2 2	3 4 5	6 6 7
·52	3311	3319	3327	3334	3342	3350	3357	3365	3373	3381	8	1 2 2	3 4 5	6 6 7
·53	3388	3396	3404	3412	3420	3428	3436	3443	3451	3459	8	1 2 2	3 4 5	6 6 7
·54	3467	3475	3483	3491	3499	3508	3516	3524	3532	3540	8	1 2 2	3 4 5	6 6 7
·55	3548	3556	3565	3573	3581	3589	3597	3606	3614	3622	8	1 2 2	3 4 5	6 6 7
·56	3631	3639	3648	3656	3664	3673	3681	3690	3698	3707	8	1 2 2	3 4 5	6 6 7
·57	3715	3724	3733	3741	3750	3758	3767	3776	3784	3793	9	1 2 3	4 4 5	6 7 8
·58	3802	3811	3819	3828	3837	3846	3855	3864	3873	3882	9	1 2 3	4 4 5	6 7 8
·59	3890	3899	3908	3917	3926	3936	3945	3954	3963	3972	9	1 2 3	4 5 5	6 7 8
·60	3981	3990	3999	4009	4018	4027	4036	4046	4055	4064	9	1 2 3	4 5 5	6 7 8
·61	4074	4083	4093	4102	4111	4121	4130	4140	4150	4159	10	1 2 3	4 5 6	7 8 9
·62	4169	4178	4188	4198	4207	4217	4227	4236	4246	4256	10	1 2 3	4 5 6	7 8 9
·63	4266	4276	4285	4295	4305	4315	4325	4335	4345	4355	10	1 2 3	4 5 6	7 8 9
·64	4365	4375	4385	4395	4406	4416	4426	4436	4446	4457	10	1 2 3	4 5 6	7 8 9
·65	4467	4477	4487	4498	4508	4519	4529	4539	4550	4560	10	1 2 3	4 5 6	7 8 9
·66	4571	4581	4592	4603	4613	4624	4634	4645	4656	4667	11	1 2 3	4 5 7	8 9 10
·67	4677	4688	4699	4710	4721	4732	4742	4753	4764	4775	11	1 2 3	4 5 7	8 9 10
·68	4786	4797	4808	4819	4831	4842	4853	4864	4875	4887	11	1 2 3	4 6 7	8 9 10
·69	4898	4909	4920	4932	4943	4955	4966	4977	4989	5000	11	1 2 3	4 6 7	8 9 10
·70	5012	5023	5035	5047	5058	5070	5082	5093	5105	5117	12	1 2 4	5 6 7	8 10 11
·71	5129	5140	5152	5164	5176	5188	5200	5212	5224	5236	12	1 2 4	5 6 7	8 10 11
·72	5248	5260	5272	5284	5297	5309	5321	5333	5346	5358	12	1 2 4	5 6 7	8 10 11
·73	5370	5383	5395	5408	5420	5433	5445	5458	5470	5483	12	1 2 4	5 6 7	8 10 11
·74	5495	5508	5521	5534	5546	5559	5572	5585	5598	5610	13	1 3 4	5 6 8	9 10 12
·75	5623	5636	5649	5662	5675	5689	5702	5715	5728	5741	13	1 3 4	5 7 8	9 10 12
·76	5754	5768	5781	5794	5808	5821	5834	5848	5861	5875	13	1 3 4	5 7 8	9 10 12
·77	5888	5902	5916	5929	5943	5957	5970	5984	5998	6012	14	1 3 4	6 7 8	10 11 13
·78	6026	6039	6053	6067	6081	6095	6109	6124	6138	6152	14	1 3 4	6 7 8	10 11 13
·79	6166	6180	6194	6209	6223	6237	6252	6266	6281	6295	14	1 3 4	6 7 8	10 11 13
·80	6310	6324	6339	6353	6368	6383	6397	6412	6427	6442	15	1 3 4	6 7 9	10 12 13
·81	6457	6471	6486	6501	6516	6531	6546	6561	6577	6592	15	2 3 5	6 8 9	11 12 14
·82	6607	6622	6637	6653	6668	6683	6699	6714	6730	6745	15	2 3 5	6 8 9	11 12 14
·83	6761	6776	6792	6808	6823	6839	6855	6871	6887	6902	16	2 3 5	6 8 10	11 13 14
·84	6918	6934	6950	6966	6982	6998	7015	7031	7047	7063	16	2 3 5	6 8 10	11 13 14
·85	7079	7096	7112	7129	7145	7161	7178	7194	7211	7228	16	2 3 5	6 8 10	11 13 14
·86	7244	7261	7278	7295	7311	7328	7345	7362	7379	7396	17	2 3 5	7 8 10	12 14 15
·87	7413	7430	7447	7464	7482	7499	7516	7534	7551	7568	17	2 3 5	7 9 10	12 14 15
·88	7586	7603	7621	7638	7656	7674	7691	7709	7727	7745	18	2 4 5	7 9 11	13 14 16
·89	7762	7780	7798	7816	7834	7852	7870	7889	7907	7925	18	2 4 5	7 9 11	13 14 16
·90	7943	7962	7980	7998	8017	8035	8054	8072	8091	8110	18	2 4 5	7 9 11	13 15 17
·91	8128	8147	8166	8185	8204	8222	8241	8260	8279	8299	19	2 4 6	8 10 11	13 15 17
·92	8318	8337	8356	8375	8395	8414	8433	8453	8472	8492	19	2 4 6	8 10 11	13 15 17
·93	8511	8531	8551	8570	8590	8610	8630	8650	8670	8690	20	2 4 6	8 10 12	14 16 18
·94	8710	8730	8750	8770	8790	8810	8831	8851	8872	8892	20	2 4 6	8 10 12	14 16 18
·95	8913	8933	8954	8974	8995	9016	9036	9057	9078	9099	21	2 4 6	8 10 13	15 17 19
·96	9120	9141	9162	9183	9204	9226	9247	9268	9290	9311	21	2 4 6	8 11 13	15 17 19
·97	9333	9354	9376	9397	9419	9441	9462	9484	9506	9528	22	2 4 7	9 11 13	15 18 20
·98	9550	9572	9594	9616	9638	9661	9683	9705	9727	9750	22	2 4 7	9 11 13	15 18 20
·99	9772	9795	9817	9840	9863	9886	9908	9931	9954	9977	23	2 5 7	9 11 14	16 18 21

SQUARES　　　x^2

x	0	1	2	3	4	5	6	7	8	9	Δ_m +	1	2	3	4	5	6	7	8	9
															ADD					
10	1000	1020	1040	1061	1082	1103	1124	1145	1166	1188	21	2	4	6	8	10	13	15	17	19
11	1210	1232	1254	1277	1300	1323	1346	1369	1392	1416	23	2	5	7	9	11	14	16	18	21
12	1440	1464	1488	1513	1538	1563	1588	1613	1638	1664	25	2	5	7	10	12	15	17	20	22
13	1690	1716	1742	1769	1796	1823	1850	1877	1904	1932	27	3	5	8	11	13	16	19	22	24
14	1960	1988	2016	2045	2074	2103	2132	2161	2190	2220	29	3	6	9	12	14	17	20	23	26
15	2250	2280	2310	2341	2372	2403	2434	2465	2496	2528	31	3	6	9	12	15	19	22	25	28
16	2560	2592	2624	2657	2690	2723	2756	2789	2822	2856	33	3	7	10	13	16	20	23	26	30
17	2890	2924	2958	2993	3028	3063	3098	3133	3168	3204	35	3	7	10	14	17	21	24	28	31
18	3240	3276	3312	3349	3386	3423	3460	3497	3534	3572	37	4	7	11	15	18	22	26	30	33
19	3610	3648	3686	3725	3764	3803	3842	3881	3920	3960	39	4	8	12	16	19	23	27	31	35
20	4000	4040	4080	4121	4162	4203	4244	4285	4326	4368	41	4	8	12	16	20	25	29	33	37
21	4410	4452	4494	4537	4580	4623	4666	4709	4752	4796	43	4	9	13	17	21	26	30	34	39
22	4840	4884	4928	4973	5018	5063	5108	5153	5198	5244	45	4	9	13	18	22	27	31	36	40
23	5290	5336	5382	5429	5476	5523	5570	5617	5664	5712	47	5	9	14	19	23	28	33	38	42
24	5760	5808	5856	5905	5954	6003	6052	6101	6150	6200	49	5	10	15	20	24	29	34	39	44
25	6250	6300	6350	6401	6452	6503	6554	6605	6656	6708	51	5	10	15	20	25	31	36	41	46
26	6760	6812	6864	6917	6970	7023	7076	7129	7182	7236	53	5	11	16	21	26	32	37	42	48
27	7290	7344	7398	7453	7508	7563	7618	7673	7728	7784	55	5	11	16	22	27	33	38	44	49
28	7840	7896	7952	8009	8066	8123	8180	8237	8294	8352	57	6	11	17	23	28	34	40	46	51
29	8410	8468	8526	8585	8644	8703	8762	8821	8880	8940	59	6	12	18	24	29	35	41	47	53
30	9000	9060	9120	9181	9242	9303	9364	9425	9486	9548	61	6	12	18	24	30	37	43	49	55
31	9610	9672	9734	9797	9860	9923	9986				63	6	13	19	25	31	38	44	50	57
							999	1005	1011	1018	6	1	1	2	2	3	4	4	5	5
32	1024	1030	1037	1043	1050	1056	1063	1069	1076	1082	6	1	1	2	2	3	4	4	5	5
33	1089	1096	1102	1109	1116	1122	1129	1136	1142	1149	7	1	1	2	3	3	4	5	6	6
34	1156	1163	1170	1176	1183	1190	1197	1204	1211	1218	7	1	1	2	3	3	4	5	6	6
35	1225	1232	1239	1246	1253	1260	1267	1274	1282	1289	7	1	1	2	3	4	4	5	6	6
36	1296	1303	1310	1318	1325	1332	1340	1347	1354	1362	7	1	1	2	3	4	4	5	6	6
37	1369	1376	1384	1391	1399	1406	1414	1421	1429	1436	8	1	2	2	3	4	5	6	6	7
38	1444	1452	1459	1467	1475	1482	1490	1498	1505	1513	8	1	2	2	3	4	5	6	6	7
39	1521	1529	1537	1544	1552	1560	1568	1576	1584	1592	8	1	2	2	3	4	5	6	6	7
40	1600	1608	1616	1624	1632	1640	1648	1656	1665	1673	8	1	2	2	3	4	5	6	6	7
41	1681	1689	1697	1706	1714	1722	1731	1739	1747	1756	8	1	2	2	3	4	5	6	6	7
42	1764	1772	1781	1789	1798	1806	1815	1823	1832	1840	8	1	2	2	3	4	5	6	6	7
43	1849	1858	1866	1875	1884	1892	1901	1910	1918	1927	9	1	2	3	4	4	5	6	7	8
44	1936	1945	1954	1962	1971	1980	1989	1998	2007	2016	9	1	2	3	4	4	5	6	7	8
45	2025	2034	2043	2052	2061	2070	2079	2088	2098	2107	9	1	2	3	4	5	5	6	7	8
46	2116	2125	2134	2144	2153	2162	2172	2181	2190	2200	9	1	2	3	4	5	5	6	7	8
47	2209	2218	2228	2237	2247	2256	2266	2275	2285	2294	10	1	2	3	4	5	6	7	8	9
48	2304	2314	2323	2333	2343	2352	2362	2372	2381	2391	10	1	2	3	4	5	6	7	8	9
49	2401	2411	2421	2430	2440	2450	2460	2470	2480	2490	10	1	2	3	4	5	6	7	8	9
50	2500	2510	2520	2530	2540	2550	2560	2570	2581	2591	10	1	2	3	4	5	6	7	8	9
51	2601	2611	2621	2632	2642	2652	2663	2673	2683	2694	10	1	2	3	4	5	6	7	8	9
52	2704	2714	2725	2735	2746	2756	2767	2777	2788	2798	10	1	2	3	4	5	6	7	8	9
53	2809	2820	2830	2841	2852	2862	2873	2884	2894	2905	11	1	2	3	4	5	7	8	9	10
54	2916	2927	2938	2948	2959	2970	2981	2992	3003	3014	11	1	2	3	4	5	7	8	9	10
55	3025	3036	3047	3058	3069	3080	3091	3102	3114	3125	11	1	2	3	4	6	7	8	9	10
56	3136	3147	3158	3170	3181	3192	3204	3215	3226	3238	11	1	2	3	4	6	7	8	9	10
57	3249	3260	3272	3283	3295	3306	3318	3329	3341	3352	12	1	2	4	5	6	7	8	10	11
58	3364	3376	3387	3399	3411	3422	3434	3446	3457	3469	12	1	2	4	5	6	7	8	10	11
59	3481	3493	3505	3516	3528	3540	3552	3564	3576	3588	12	1	2	4	5	6	7	8	10	11

SQUARES x^2

x	0	1	2	3	4	5	6	7	8	9	Δ_m	1	2	3	4	5	6	7	8	9
											+	ADD								
60	3600	3612	3624	3636	3648	3660	3672	3684	3697	3709	12	1	2	4	5	6	7	8	10	11
61	3721	3733	3745	3758	3770	3782	3795	3807	3819	3832	12	1	2	4	5	6	7	8	10	11
62	3844	3856	3869	3881	3894	3906	3919	3931	3944	3956	12	1	2	4	5	6	7	8	10	11
63	3969	3982	3994	4007	4020	4032	4045	4058	4070	4083	13	1	3	4	5	6	8	9	10	12
64	4096	4109	4122	4134	4147	4160	4173	4186	4199	4212	13	1	3	4	5	6	8	9	10	12
65	4225	4238	4251	4264	4277	4290	4303	4316	4330	4343	13	1	3	4	5	7	8	9	10	12
66	4356	4369	4382	4396	4409	4422	4436	4449	4462	4476	13	1	3	4	5	7	8	9	10	12
67	4489	4502	4516	4529	4543	4556	4570	4583	4597	4610	14	1	3	4	6	7	8	10	11	13
68	4624	4638	4651	4665	4679	4692	4706	4720	4733	4747	14	1	3	4	6	7	8	10	11	13
69	4761	4775	4789	4802	4816	4830	4844	4858	4872	4886	14	1	3	4	6	7	8	10	11	13
70	4900	4914	4928	4942	4956	4970	4984	4998	5013	5027	14	1	3	4	6	7	8	10	11	13
71	5041	5055	5069	5084	5098	5112	5127	5141	5155	5170	14	1	3	4	6	7	8	10	11	13
72	5184	5198	5213	5227	5242	5256	5271	5285	5300	5314	14	1	3	4	6	7	8	10	11	13
73	5329	5344	5358	5373	5388	5402	5417	5432	5446	5461	15	1	3	4	6	7	9	10	12	13
74	5476	5491	5506	5520	5535	5550	5565	5580	5595	5610	15	1	3	4	6	7	9	10	12	13
75	5625	5640	5655	5670	5685	5700	5715	5730	5746	5761	15	2	3	5	6	8	9	11	12	14
76	5776	5791	5806	5822	5837	5852	5868	5883	5898	5914	15	2	3	5	6	8	9	11	12	14
77	5929	5944	5960	5975	5991	6006	6022	6037	6053	6068	16	2	3	5	6	8	10	11	13	14
78	6084	6100	6115	6131	6147	6162	6178	6194	6209	6225	16	2	3	5	6	8	10	11	13	14
79	6241	6257	6273	6288	6304	6320	6336	6352	6368	6384	16	2	3	5	6	8	10	11	13	14
80	6400	6416	6432	6448	6464	6480	6496	6512	6529	6545	16	2	3	5	6	8	10	11	13	14
81	6561	6577	6593	6610	6626	6642	6659	6675	6691	6708	16	2	3	5	6	8	10	11	13	14
82	6724	6740	6757	6773	6790	6806	6823	6839	6856	6872	16	2	3	5	6	8	10	11	13	14
83	6889	6906	6922	6939	6956	6972	6989	7006	7022	7039	17	2	3	5	7	8	10	12	14	15
84	7056	7073	7090	7106	7123	7140	7157	7174	7191	7208	17	2	3	5	7	8	10	12	14	15
85	7225	7242	7259	7276	7293	7310	7327	7344	7362	7379	17	2	3	5	7	9	10	12	14	15
86	7396	7413	7430	7448	7465	7482	7500	7517	7534	7552	17	2	3	5	7	9	10	12	14	15
87	7569	7586	7604	7621	7639	7656	7674	7691	7709	7726	18	2	4	5	7	9	11	13	14	16
88	7744	7762	7779	7797	7815	7832	7850	7868	7885	7903	18	2	4	5	7	9	11	13	14	16
89	7921	7939	7957	7974	7992	8010	8028	8046	8064	8082	18	2	4	5	7	9	11	13	14	16
90	8100	8118	8136	8154	8172	8190	8208	8226	8245	8263	18	2	4	5	7	9	11	13	14	16
91	8281	8299	8317	8336	8354	8372	8391	8409	8427	8446	18	2	4	5	7	9	11	13	14	16
92	8464	8482	8501	8519	8538	8556	8575	8593	8612	8630	18	2	4	5	7	9	11	13	14	16
93	8649	8668	8686	8705	8724	8742	8761	8780	8798	8817	19	2	4	6	8	9	11	13	15	17
94	8836	8855	8874	8892	8911	8930	8949	8968	8987	9006	19	2	4	6	8	9	11	13	15	17
95	9025	9044	9063	9082	9101	9120	9139	9158	9178	9197	19	2	4	6	8	10	11	13	15	17
96	9216	9235	9254	9274	9293	9312	9332	9351	9370	9390	19	2	4	6	8	10	11	13	15	17
97	9409	9428	9448	9467	9487	9506	9526	9545	9565	9584	20	2	4	6	8	10	12	14	16	18
98	9604	9624	9643	9663	9683	9702	9722	9742	9761	9781	20	2	4	6	8	10	12	14	16	18
99	9801	9821	9841	9860	9880	9900	9920	9940	9960	9980	20	2	4	6	8	10	12	14	16	18

The decimal point must be inserted by inspection.

Examples: $(1{\cdot}43)^2 \doteqdot 2{\cdot}045$ $(6{\cdot}935)^2 \doteqdot 48{\cdot}09$

$(232{\cdot}8)^2 = (2{\cdot}328 \times 10^2)^2 \doteqdot 5{\cdot}420 \times 10^4$

$(0{\cdot}007035)^2 = (7{\cdot}035 \times 10^{-3})^2 \doteqdot 49{\cdot}49 \times 10^{-6} = 0{\cdot}00004949$

SQUARE ROOTS   √x or x^½

x	0	1	2	3	4	5	6	7	8	9	Δ_m +	1 2 3	4 5 6	7 8 9 ADD
10	1000	1005	1010	1015	1020	1025	1030	1034	1039	1044	5	0 1 1	2 2 3	3 4 4
	3162	3178	3194	3209	3225	3240	3256	3271	3286	3302	16	2 3 5	6 8 10	11 13 14
11	1049	1054	1058	1063	1068	1072	1077	1082	1086	1091	5	0 1 1	2 2 3	3 4 4
	3317	3332	3347	3362	3376	3391	3406	3421	3435	3450	15	1 3 4	6 7 9	10 12 13
12	1095	1100	1105	1109	1114	1118	1122	1127	1131	1136	4	0 1 1	2 2 2	3 3 4
	3464	3479	3493	3507	3521	3536	3550	3564	3578	3592	14	1 3 4	6 7 8	10 11 13
13	1140	1145	1149	1153	1158	1162	1166	1170	1175	1179	4	0 1 1	2 2 2	3 3 4
	3606	3619	3633	3647	3661	3674	3688	3701	3715	3728	14	1 3 4	6 7 8	10 11 13
14	1183	1187	1192	1196	1200	1204	1208	1212	1217	1221	4	0 1 1	2 2 2	3 3 4
	3742	3755	3768	3782	3795	3808	3821	3834	3847	3860	13	1 3 4	5 7 8	9 10 12
15	1225	1229	1233	1237	1241	1245	1249	1253	1257	1261	4	0 1 1	2 2 2	3 3 4
	3873	3886	3899	3912	3924	3937	3950	3962	3975	3987	13	1 3 4	5 6 8	9 10 12
16	1265	1269	1273	1277	1281	1285	1288	1292	1296	1300	4	0 1 1	2 2 2	3 3 4
	4000	4012	4025	4037	4050	4062	4074	4087	4099	4111	12	1 2 4	5 6 7	8 10 11
17	1304	1308	1311	1315	1319	1323	1327	1330	1334	1338	4	0 1 1	2 2 2	3 3 4
	4123	4135	4147	4159	4171	4183	4195	4207	4219	4231	12	1 2 4	5 6 7	8 10 11
18	1342	1345	1349	1353	1356	1360	1364	1367	1371	1375	4	0 1 1	2 2 2	3 3 4
	4243	4254	4266	4278	4290	4301	4313	4324	4336	4347	12	1 2 4	5 6 7	8 10 11
19	1378	1382	1386	1389	1393	1396	1400	1404	1407	1411	4	0 1 1	2 2 2	3 3 4
	4359	4370	4382	4393	4405	4416	4427	4438	4450	4461	11	1 2 3	4 6 7	8 9 10
20	1414	1418	1421	1425	1428	1432	1435	1439	1442	1446	4	0 1 1	2 2 2	3 3 4
	4472	4483	4494	4506	4517	4528	4539	4550	4561	4572	11	1 2 3	4 6 7	8 9 10
21	1449	1453	1456	1459	1463	1466	1470	1473	1476	1480	3	0 1 1	1 2 2	2 2 3
	4583	4593	4604	4615	4626	4637	4648	4658	4669	4680	11	1 2 3	4 5 7	8 9 10
22	1483	1487	1490	1493	1497	1500	1503	1507	1510	1513	3	0 1 1	1 2 2	2 2 3
	4690	4701	4712	4722	4733	4743	4754	4764	4775	4785	11	1 2 3	4 5 7	8 9 10
23	1517	1520	1523	1526	1530	1533	1536	1539	1543	1546	3	0 1 1	1 2 2	2 2 3
	4796	4806	4817	4827	4837	4848	4858	4868	4879	4889	10	1 2 3	4 5 6	7 8 9
24	1549	1552	1556	1559	1562	1565	1568	1572	1575	1578	3	0 1 1	1 2 2	2 2 3
	4899	4909	4919	4930	4940	4950	4960	4970	4980	4990	10	1 2 3	4 5 6	7 8 9
25	1581	1584	1587	1591	1594	1597	1600	1603	1606	1609	3	0 1 1	1 2 2	2 2 3
	5000	5010	5020	5030	5040	5050	5060	5070	5079	5089	10	1 2 3	4 5 6	7 8 9
26	1612	1616	1619	1622	1625	1628	1631	1634	1637	1640	3	0 1 1	1 2 2	2 2 3
	5099	5109	5119	5128	5138	5148	5158	5167	5177	5187	10	1 2 3	4 5 6	7 8 9
27	1643	1646	1649	1652	1655	1658	1661	1664	1667	1670	3	0 1 1	1 2 2	2 2 3
	5196	5206	5215	5225	5235	5244	5254	5263	5273	5282	10	1 2 3	4 5 6	7 8 9
28	1673	1676	1679	1682	1685	1688	1691	1694	1697	1700	3	0 1 1	1 2 2	2 2 3
	5292	5301	5310	5320	5329	5339	5348	5357	5367	5376	9	1 2 3	4 5 5	6 7 8
29	1703	1706	1709	1712	1715	1718	1720	1723	1726	1729	3	0 1 1	1 1 2	2 2 3
	5385	5394	5404	5413	5422	5431	5441	5450	5459	5468	9	1 2 3	4 5 5	6 7 8
30	1732	1735	1738	1741	1744	1746	1749	1752	1755	1758	3	0 1 1	1 1 2	2 2 3
	5477	5486	5495	5505	5514	5523	5532	5541	5550	5559	9	1 2 3	4 5 5	6 7 8

The decimal point must be inserted by inspection.

Examples:

$$\sqrt{1 \cdot 856} \doteqdot 1 \cdot 362 \qquad \sqrt{217 \cdot 3} \doteqdot 14 \cdot 74$$
$$\sqrt{27 \cdot 12} \doteqdot 5 \cdot 208 \qquad \sqrt{2930} \doteqdot 54 \cdot 13$$
$$\sqrt{0 \cdot 236} \doteqdot 0 \cdot 4858 \qquad \sqrt{0 \cdot 0306} \doteqdot 0 \cdot 1749$$

SQUARE ROOTS $\quad \sqrt{x}$ OR $\quad x^{\frac{1}{2}}$

x	0	1	2	3	4	5	6	7	8	9	Δ_m +	1 2 3	4 5 6	7 8 9 ADD
31	1761	1764	1766	1769	1772	1775	1778	1780	1783	1786	3	0 1 1	1 1 2	2 2 3
	5568	5577	5586	5595	5604	5612	5621	5630	5639	5648	9	1 2 3	4 4 5	6 7 8
32	1789	1792	1794	1797	1800	1803	1806	1808	1811	1814	3	0 1 1	1 1 2	2 2 3
	5657	5666	5675	5683	5692	5701	5710	5718	5727	5736	9	1 2 3	4 4 5	6 7 8
33	1817	1819	1822	1825	1828	1830	1833	1836	1838	1841	3	0 1 1	1 1 2	2 2 3
	5745	5753	5762	5771	5779	5788	5797	5805	5814	5822	9	1 2 3	4 4 5	6 7 8
34	1844	1847	1849	1852	1855	1857	1860	1863	1865	1868	3	0 1 1	1 1 2	2 2 3
	5831	5840	5848	5857	5865	5874	5882	5891	5899	5908	9	1 2 3	4 4 5	6 7 8
35	1871	1873	1876	1879	1881	1884	1887	1889	1892	1895	3	0 1 1	1 1 2	2 2 3
	5916	5925	5933	5941	5950	5958	5967	5975	5983	5992	8	1 2 2	3 4 5	6 6 7
36	1897	1900	1903	1905	1908	1910	1913	1916	1918	1921	3	0 1 1	1 1 2	2 2 3
	6000	6008	6017	6025	6033	6042	6050	6058	6066	6075	8	1 2 2	3 4 5	6 6 7
37	1924	1926	1929	1931	1934	1936	1939	1942	1944	1947	3	0 1 1	1 1 2	2 2 3
	6083	6091	6099	6107	6116	6124	6132	6140	6148	6156	8	1 2 2	3 4 5	6 6 7
38	1949	1952	1954	1957	1960	1962	1965	1967	1970	1972	3	0 1 1	1 1 2	2 2 3
	6164	6173	6181	6189	6197	6205	6213	6221	6229	6237	8	1 2 2	3 4 5	6 6 7
39	1975	1977	1980	1982	1985	1987	1990	1992	1995	1997	2	0 0 1	1 1 1	1 2 2
	6245	6253	6261	6269	6277	6285	6293	6301	6309	6317	8	1 2 2	3 4 5	6 6 7
40	2000	2002	2005	2007	2010	2012	2015	2017	2020	2022	2	0 0 1	1 1 1	1 2 2
	6325	6332	6340	6348	6356	6364	6372	6380	6387	6395	8	1 2 2	3 4 5	6 6 7
41	2025	2027	2030	2032	2035	2037	2040	2042	2045	2047	2	0 0 1	1 1 1	1 2 2
	6403	6411	6419	6427	6434	6442	6450	6458	6465	6473	8	1 2 2	3 4 5	6 6 7
42	2049	2052	2054	2057	2059	2062	2064	2066	2069	2071	2	0 0 1	1 1 1	1 2 2
	6481	6488	6496	6504	6512	6519	6527	6535	6542	6550	8	1 2 2	3 4 5	6 6 7
43	2074	2076	2078	2081	2083	2086	2088	2090	2093	2095	2	0 0 1	1 1 1	1 2 2
	6557	6565	6573	6580	6588	6595	6603	6611	6618	6626	8	1 2 2	3 4 5	6 6 7
44	2098	2100	2102	2105	2107	2110	2112	2114	2117	2119	2	0 0 1	1 1 1	1 2 2
	6633	6641	6648	6656	6663	6671	6678	6686	6693	6701	8	1 2 2	3 4 5	6 6 7
45	2121	2124	2126	2128	2131	2133	2135	2138	2140	2142	2	0 0 1	1 1 1	1 2 2
	6708	6716	6723	6731	6738	6745	6753	6760	6768	6775	7	1 1 2	3 4 4	5 6 6
46	2145	2147	2149	2152	2154	2156	2159	2161	2163	2166	2	0 0 1	1 1 1	1 2 2
	6782	6790	6797	6804	6812	6819	6826	6834	6841	6848	7	1 1 2	3 4 4	5 6 6
47	2168	2170	2173	2175	2177	2179	2182	2184	2186	2189	2	0 0 1	1 1 1	1 2 2
	6856	6863	6870	6877	6885	6892	6899	6907	6914	6921	7	1 1 2	3 4 4	5 6 6
48	2191	2193	2195	2198	2200	2202	2205	2207	2209	2211	2	0 0 1	1 1 1	1 2 2
	6928	6935	6943	6950	6957	6964	6971	6979	6986	6993	7	1 1 2	3 4 4	5 6 6
49	2214	2216	2218	2220	2223	2225	2227	2229	2232	2234	2	0 0 1	1 1 1	1 2 2
	7000	7007	7014	7021	7029	7036	7043	7050	7057	7064	7	1 1 2	3 4 4	5 6 6
50	2236	2238	2241	2243	2245	2247	2249	2252	2254	2256	2	0 0 1	1 1 1	1 2 2
	7071	7078	7085	7092	7099	7106	7113	7120	7127	7134	7	1 1 2	3 4 4	5 6 6
51	2258	2261	2263	2265	2267	2269	2272	2274	2276	2278	2	0 0 1	1 1 1	1 2 2
	7141	7148	7155	7162	7169	7176	7183	7190	7197	7204	7	1 1 2	3 4 4	5 6 6
52	2280	2283	2285	2287	2289	2291	2293	2296	2298	2300	2	0 0 1	1 1 1	1 2 2
	7211	7218	7225	7232	7239	7246	7253	7259	7266	7273	7	1 1 2	3 3 4	5 6 6
53	2302	2304	2307	2309	2311	2313	2315	2317	2319	2322	2	0 0 1	1 1 1	1 2 2
	7280	7287	7294	7301	7308	7314	7321	7328	7335	7342	7	1 1 2	3 3 4	5 6 6

Examples: $\quad \sqrt{37450000} = \sqrt{37\cdot45 \times 10^6} \doteqdot 6\cdot120 \times 10^3 = 6120$

$\sqrt{0\cdot0005328} = \sqrt{5\cdot328 \times 10^{-4}} \doteqdot 2\cdot309 \times 10^{-2} = 0\cdot02309$

Note that the power of 10 extracted under the root sign must always be even. If this leaves **one** figure before the decimal point, use the **upper** line of a pair; if it leaves **two** figures before the decimal point, use the **lower** line.

SQUARE ROOTS $\sqrt{x}$ OR $x^{\frac{1}{2}}$

x	0	1	2	3	4	5	6	7	8	9	Δ_m +	1 2 3	4 5 6	7 8 9 ADD
54	2324	2326	2328	2330	2332	2335	2337	2339	2341	2343	2	0 0 1	1 1 1	1 2 2
	7348	7355	7362	7369	7376	7382	7389	7396	7403	7409	7	1 1 2	3 3 4	5 6 6
55	2345	2347	2349	2352	2354	2356	2358	2360	2362	2364	2	0 0 1	1 1 1	1 2 2
	7416	7423	7430	7436	7443	7450	7457	7463	7470	7477	7	1 1 2	3 3 4	5 6 6
56	2366	2369	2371	2373	2375	2377	2379	2381	2383	2385	2	0 0 1	1 1 1	1 2 2
	7483	7490	7497	7503	7510	7517	7523	7530	7537	7543	7	1 1 2	3 3 4	5 6 6
57	2387	2390	2392	2394	2396	2398	2400	2402	2404	2406	2	0 0 1	1 1 1	1 2 2
	7550	7556	7563	7570	7576	7583	7589	7596	7603	7609	7	1 1 2	3 3 4	5 6 6
58	2408	2410	2412	2415	2417	2419	2421	2423	2425	2427	2	0 0 1	1 1 1	1 2 2
	7616	7622	7629	7635	7642	7649	7655	7662	7668	7675	6	1 1 2	2 3 4	4 5 5
59	2429	2431	2433	2435	2437	2439	2441	2443	2445	2447	2	0 0 1	1 1 1	1 2 2
	7681	7688	7694	7701	7707	7714	7720	7727	7733	7740	6	1 1 2	2 3 4	4 5 5
60	2449	2452	2454	2456	2458	2460	2462	2464	2466	2468	2	0 0 1	1 1 1	1 2 2
	7746	7752	7759	7765	7772	7778	7785	7791	7797	7804	6	1 1 2	2 3 4	4 5 5
61	2470	2472	2474	2476	2478	2480	2482	2484	2486	2488	2	0 0 1	1 1 1	1 2 2
	7810	7817	7823	7829	7836	7842	7849	7855	7861	7868	6	1 1 2	2 3 4	4 5 5
62	2490	2492	2494	2496	2498	2500	2502	2504	2506	2508	2	0 0 1	1 1 1	1 2 2
	7874	7880	7887	7893	7899	7906	7912	7918	7925	7931	6	1 1 2	2 3 4	4 5 5
63	2510	2512	2514	2516	2518	2520	2522	2524	2526	2528	2	0 0 1	1 1 1	1 2 2
	7937	7944	7950	7956	7962	7969	7975	7981	7987	7994	6	1 1 2	2 3 4	4 5 5
64	2530	2532	2534	2536	2538	2540	2542	2544	2546	2548	2	0 0 1	1 1 1	1 2 2
	8000	8006	8012	8019	8025	8031	8037	8044	8050	8056	6	1 1 2	2 3 4	4 5 5
65	2550	2551	2553	2555	2557	2559	2561	2563	2565	2567	2	0 0 1	1 1 1	1 2 2
	8062	8068	8075	8081	8087	8093	8099	8106	8112	8118	6	1 1 2	2 3 4	4 5 5
66	2569	2571	2573	2575	2577	2579	2581	2583	2585	2587	2	0 0 1	1 1 1	1 2 2
	8124	8130	8136	8142	8149	8155	8161	8167	8173	8179	6	1 1 2	2 3 4	4 5 5
67	2588	2590	2592	2594	2596	2598	2600	2602	2604	2606	2	0 0 1	1 1 1	1 2 2
	8185	8191	8198	8204	8210	8216	8222	8228	8234	8240	6	1 1 2	2 3 4	4 5 5
68	2608	2610	2612	2613	2615	2617	2619	2621	2623	2625	2	0 0 1	1 1 1	1 2 2
	8246	8252	8258	8264	8270	8276	8283	8289	8295	8301	6	1 1 2	2 3 4	4 5 5
69	2627	2629	2631	2632	2634	2636	2638	2640	2642	2644	2	0 0 1	1 1 1	1 2 2
	8307	8313	8319	8325	8331	8337	8343	8349	8355	8361	6	1 1 2	2 3 4	4 5 5
70	2646	2648	2650	2651	2653	2655	2657	2659	2661	2663	2	0 0 1	1 1 1	1 2 2
	8367	8373	8379	8385	8390	8396	8402	8408	8414	8420	6	1 1 2	2 3 4	4 5 5
71	2665	2666	2668	2670	2672	2674	2676	2678	2680	2681	2	0 0 1	1 1 1	1 2 2
	8426	8432	8438	8444	8450	8456	8462	8468	8473	8479	6	1 1 2	2 3 4	4 5 5
72	2683	2685	2687	2689	2691	2693	2694	2696	2698	2700	2	0 0 1	1 1 1	1 2 2
	8485	8491	8497	8503	8509	8515	8521	8526	8532	8538	6	1 1 2	2 3 4	4 5 5
73	2702	2704	2706	2707	2709	2711	2713	2715	2717	2718	2	0 0 1	1 1 1	1 2 2
	8544	8550	8556	8562	8567	8573	8579	8585	8591	8597	6	1 1 2	2 3 4	4 5 5
74	2720	2722	2724	2726	2728	2729	2731	2733	2735	2737	2	0 0 1	1 1 1	1 2 2
	8602	8608	8614	8620	8626	8631	8637	8643	8649	8654	6	1 1 2	2 3 4	4 5 5
75	2739	2740	2742	2744	2746	2748	2750	2751	2753	2755	2	0 0 1	1 1 1	1 2 2
	8660	8666	8672	8678	8683	8689	8695	8701	8706	8712	6	1 1 2	2 3 4	4 5 5
76	2757	2759	2760	2762	2764	2766	2768	2769	2771	2773	2	0 0 1	1 1 1	1 2 2
	8718	8724	8729	8735	8741	8746	8752	8758	8764	8769	6	1 1 2	2 3 4	4 5 5

The decimal point must be inserted by inspection.

Examples: $\sqrt{5\cdot978} \doteq 2\cdot445$ $\sqrt{67\cdot42} \doteq 8\cdot211$

$\sqrt{723\cdot1} \doteq 26\cdot89$ $\sqrt{0\cdot7591} \doteq 0\cdot8713$

SQUARE ROOTS $\sqrt{x}$ OR $x^{\frac{1}{2}}$

x	0	1	2	3	4	5	6	7	8	9	Δ_m +	1 2 3	4 5 6	7 8 9 (ADD)
77	2775	2777	2778	2780	2782	2784	2786	2787	2789	2791	2	0 0 1	1 1 1	1 2 2
	8775	8781	8786	8792	8798	8803	8809	8815	8820	8826	6	1 1 2	2 3 4	4 5 5
78	2793	2795	2796	2798	2800	2802	2804	2805	2807	2809	2	0 0 1	1 1 1	1 2 2
	8832	8837	8843	8849	8854	8860	8866	8871	8877	8883	6	1 1 2	2 3 4	4 5 5
79	2811	2812	2814	2816	2818	2820	2821	2823	2825	2827	2	0 0 1	1 1 1	1 2 2
	8888	8894	8899	8905	8911	8916	8922	8927	8933	8939	6	1 1 2	2 3 4	4 5 5
80	2828	2830	2832	2834	2835	2837	2839	2841	2843	2844	2	0 0 1	1 1 1	1 2 2
	8944	8950	8955	8961	8967	8972	8978	8983	8989	8994	6	1 1 2	2 3 4	4 5 5
81	2846	2848	2850	2851	2853	2855	2857	2858	2860	2862	2	0 0 1	1 1 1	1 2 2
	9000	9006	9011	9017	9022	9028	9033	9039	9044	9050	6	1 1 2	2 3 4	4 5 5
82	2864	2865	2867	2869	2871	2872	2874	2876	2877	2879	2	0 0 1	1 1 1	1 2 2
	9055	9061	9066	9072	9077	9083	9088	9094	9099	9105	6	1 1 2	2 3 4	4 5 5
83	2881	2883	2884	2886	2888	2890	2891	2893	2895	2897	2	0 0 1	1 1 1	1 2 2
	9110	9116	9121	9127	9132	9138	9143	9149	9154	9160	5	1 1 2	2 3 3	4 4 5
84	2898	2900	2902	2903	2905	2907	2909	2910	2912	2914	2	0 0 1	1 1 1	1 2 2
	9165	9171	9176	9182	9187	9192	9198	9203	9209	9214	5	1 1 2	2 3 3	4 4 5
85	2915	2917	2919	2921	2922	2924	2926	2927	2929	2931	2	0 0 1	1 1 1	1 2 2
	9220	9225	9230	9236	9241	9247	9252	9257	9263	9268	5	1 1 2	2 3 3	4 4 5
86	2933	2934	2936	2938	2939	2941	2943	2944	2946	2948	2	0 0 1	1 1 1	1 2 2
	9274	9279	9284	9290	9295	9301	9306	9311	9317	9322	5	1 1 2	2 3 3	4 4 5
87	2950	2951	2953	2955	2956	2958	2960	2961	2963	2965	2	0 0 1	1 1 1	1 2 2
	9327	9333	9338	9343	9349	9354	9359	9365	9370	9375	5	1 1 2	2 3 3	4 4 5
88	2966	2968	2970	2972	2973	2975	2977	2978	2980	2982	2	0 0 1	1 1 1	1 2 2
	9381	9386	9391	9397	9402	9407	9413	9418	9423	9429	5	1 1 2	2 3 3	4 4 5
89	2983	2985	2987	2988	2990	2992	2993	2995	2997	2998	2	0 0 1	1 1 1	1 2 2
	9434	9439	9445	9450	9455	9460	9466	9471	9476	9482	5	1 1 2	2 3 3	4 4 5
90	3000	3002	3003	3005	3007	3008	3010	3012	3013	3015	2	0 0 1	1 1 1	1 2 2
	9487	9492	9497	9503	9508	9513	9518	9524	9529	9534	5	1 1 2	2 3 3	4 4 5
91	3017	3018	3020	3022	3023	3025	3027	3028	3030	3032	2	0 0 1	1 1 1	1 2 2
	9539	9545	9550	9555	9560	9566	9571	9576	9581	9586	5	1 1 2	2 3 3	4 4 5
92	3033	3035	3036	3038	3040	3041	3043	3045	3046	3048	2	0 0 1	1 1 1	1 2 2
	9592	9597	9602	9607	9612	9618	9623	9628	9633	9638	5	1 1 2	2 3 3	4 4 5
93	3050	3051	3053	3055	3056	3058	3059	3061	3063	3064	2	0 0 1	1 1 1	1 2 2
	9644	9649	9654	9659	9664	9670	9675	9680	9685	9690	5	1 1 2	2 3 3	4 4 5
94	3066	3068	3069	3071	3072	3074	3076	3077	3079	3081	2	0 0 1	1 1 1	1 2 2
	9695	9701	9706	9711	9716	9721	9726	9731	9737	9742	5	1 1 2	2 3 3	4 4 5
95	3082	3084	3085	3087	3089	3090	3092	3094	3095	3097	2	0 0 1	1 1 1	1 2 2
	9747	9752	9757	9762	9767	9772	9778	9783	9788	9793	5	1 1 2	2 3 3	4 4 5
96	3098	3100	3102	3103	3105	3106	3108	3110	3111	3113	2	0 0 1	1 1 1	1 2 2
	9798	9803	9808	9813	9818	9823	9829	9834	9839	9844	5	1 1 2	2 3 3	4 4 5
97	3114	3116	3118	3119	3121	3122	3124	3126	3127	3129	2	0 0 1	1 1 1	1 2 2
	9849	9854	9859	9864	9869	9874	9879	9884	9889	9894	5	1 1 2	2 3 3	4 4 5
98	3130	3132	3134	3135	3137	3138	3140	3142	3143	3145	2	0 0 1	1 1 1	1 2 2
	9899	9905	9910	9915	9920	9925	9930	9935	9940	9945	5	1 1 2	2 3 3	4 4 5
99	3146	3148	3150	3151	3153	3154	3156	3158	3159	3161	2	0 0 1	1 1 1	1 2 2
	9950	9955	9960	9965	9970	9975	9980	9985	9990	9995	5	1 1 2	2 3 3	4 4 5

Examples:

$$\sqrt{862300} = \sqrt{86.230 \times 10^4} \doteq 9.286 \times 10^2 = 928.6$$
$$\sqrt{0.0927} = \sqrt{9.27 \times 10^{-2}} \doteq 3.045 \times 10^{-1} = 0.3045$$

Note that the power of 10 extracted under the root sign must always be even. If this leaves **one** figure before the decimal point, use the **upper** line of a pair; if it leaves **two** figures before the decimal point, use the **lower** line.

$$e^{-x}$$

x	.00	.01	.02	.03	.04	.05	.06	.07	.08	.09
0.0	1.0000	.9900	.9802	.9704	.9608	.9512	.9418	.9324	.9231	.9139
0.1	0.9048	.8958	.8869	.8781	.8694	.8607	.8521	.8437	.8353	.8270
.2	.8187	.8106	.8025	.7945	.7866	.7788	.7711	.7634	.7558	.7483
.3	.7408	.7334	.7261	.7189	.7118	.7047	.6977	.6907	.6839	.6771
.4	.6703	.6637	.6570	.6505	.6440	.6376	.6313	.6250	.6188	.6126
.5	.6065	.6005	.5945	.5886	.5827	.5769	.5712	.5655	.5599	.5543
.6	.5488	.5434	.5379	.5326	.5273	.5220	.5169	.5117	.5066	.5016
.7	.4966	.4916	.4868	.4819	.4771	.4724	.4677	.4630	.4584	.4538
.8	.4493	.4449	.4404	.4360	.4317	.4274	.4232	.4190	.4148	.4107
.9	.4066	.4025	.3985	.3946	.3906	.3867	.3829	.3791	.3753	.3716
1.0	0.3679	.3642	.3606	.3570	.3535	.3499	.3465	.3430	.3396	.3362
1.1	.3329	.3296	.3263	.3230	.3198	.3166	.3135	.3104	.3073	.3042
.2	.3012	.2982	.2952	.2923	.2894	.2865	.2837	.2808	.2780	.2753
.3	.2725	.2698	.2671	.2645	.2618	.2592	.2567	.2541	.2516	.2491
.4	.2466	.2441	.2417	.2393	.2369	.2346	.2322	.2299	.2276	.2254
.5	.2231	.2209	.2187	.2165	.2144	.2122	.2101	.2080	.2060	.2039
.6	.2019	.1999	.1979	.1959	.1940	.1920	.1901	.1882	.1864	.1845
.7	.1827	.1809	.1791	.1773	.1755	.1738	.1720	.1703	.1686	.1670
.8	.1653	.1637	.1620	.1604	.1588	.1572	.1557	.1541	.1526	.1511
.9	.1496	.1481	.1466	.1451	.1437	.1423	.1409	.1395	.1381	.1367
2.0	0.1353	.1340	.1327	.1313	.1300	.1287	.1275	.1262	.1249	.1237
2.1	0.1225	.1212	.1200	.1188	.1177	.1165	.1153	.1142	.1130	.1119
.2	.1108	.1097	.1086	.1075	.1065	.1054	.1044	.1033	.1023	.1013
.3	.1003	.0993	.0983	.0973	.0963	.0954	.0944	.0935	.0925	.0916
.4	.0907	.0898	.0889	.0880	.0872	.0863	.0854	.0846	.0837	.0829
.5	.0821	.0813	.0805	.0797	.0789	.0781	.0773	.0765	.0758	.0750
.6	.0743	.0735	.0728	.0721	.0714	.0707	.0699	.0693	.0686	.0679
.7	.0672	.0665	.0659	.0652	.0646	.0639	.0633	.0627	.0620	.0614
.8	.0608	.0602	.0596	.0590	.0584	.0578	.0573	.0567	.0561	.0556
9	.0550	.0545	.0539	.0534	.0529	.0523	.0518	.0513	.0508	.0503
3.0	0.0498	.0493	.0488	.0483	.0478	.0474	.0469	.0464	.0460	.0455
3.1	.0450	.0446	.0442	.0437	.0433	.0429	.0424	.0420	.0416	.0412
.2	.0408	.0404	.0400	.0396	.0392	.0388	.0384	.0380	.0376	.0373
.3	.0369	.0365	.0362	.0358	.0354	.0351	.0347	.0344	.0340	.0337
.4	.0334	.0330	.0327	.0324	.0321	.0317	.0314	.0311	.0308	.0305
.5	.0302	.0299	.0296	.0293	.0290	.0287	.0284	.0282	.0279	.0276
.6	.0273	.0271	.0268	.0265	.0260	.0257	.0257	.0255	.0252	.0250
.7	.0247	.0245	.0242	.0240	.0238	.0235	.0233	.0231	.0228	.0226
.8	.0224	.0221	.0219	.0217	.0215	.0213	.0211	.0209	.0207	.0204
.9	.0202	.0200	.0198	.0196	.0194	.0193	.0191	.0189	.0187	.0185
4.0	0.0183									
x	.00	.01	.02	.03	.04	.05	.06	.07	.08	.09

THE NORMAL DISTRIBUTION FUNCTION

x	$\Phi(x)$	x	$\Phi(x)$	x	$\Phi(x)$	x	$\Phi(x)$	x	$\Phi(x)$
0·00	0·5000 ^{40}	0·50	0·6915 ^{35}	1·00	0·8413 ^{25}	1·50	0·9332 ^{13}	2·00	0·97725 ^{53}
·01	·5040 ^{40}	·51	·6950 ^{35}	·01	·8438 ^{23}	·51	·9345 ^{12}	·01	·97778 ^{53}
·02	·5080 ^{40}	·52	·6985 ^{34}	·02	·8461 ^{24}	·52	·9357 ^{13}	·02	·97831 ^{51}
·03	·5120 ^{40}	·53	·7019 ^{35}	·03	·8485 ^{23}	·53	·9370 ^{12}	·03	·97882 ^{50}
·04	·5160 ^{39}	·54	·7054 ^{34}	·04	·8508 ^{23}	·54	·9382 ^{12}	·04	·97932 ^{50}
0·05	0·5199 ^{40}	0·55	0·7088 ^{35}	1·05	0·8531 ^{23}	1·55	0·9394 ^{12}	2·05	0·97982 ^{48}
·06	·5239 ^{40}	·56	·7123 ^{34}	·06	·8554 ^{23}	·56	·9406 ^{12}	·06	·98030 ^{47}
·07	·5279 ^{40}	·57	·7157 ^{33}	·07	·8577 ^{22}	·57	·9418 ^{11}	·07	·98077 ^{47}
·08	·5319 ^{40}	·58	·7190 ^{34}	·08	·8599 ^{22}	·58	·9429 ^{12}	·08	·98124 ^{45}
·09	·5359 ^{39}	·59	·7224 ^{33}	·09	·8621 ^{22}	·59	·9441 ^{11}	·09	·98169 ^{45}
0·10	0·5398 ^{40}	0·60	0·7257 ^{34}	1·10	0·8643 ^{22}	1·60	0·9452 ^{11}	2·10	0·98214 ^{43}
·11	·5438 ^{40}	·61	·7291 ^{33}	·11	·8665 ^{21}	·61	·9463 ^{11}	·11	·98257 ^{43}
·12	·5478 ^{39}	·62	·7324 ^{33}	·12	·8686 ^{22}	·62	·9474 ^{10}	·12	·98300 ^{41}
·13	·5517 ^{40}	·63	·7357 ^{32}	·13	·8708 ^{21}	·63	·9484 ^{11}	·13	·98341 ^{41}
·14	·5557 ^{39}	·64	·7389 ^{33}	·14	·8729 ^{20}	·64	·9495 ^{10}	·14	·98382 ^{40}
0·15	0·5596 ^{40}	0·65	0·7422 ^{32}	1·15	0·8749 ^{21}	1·65	0·9505 ^{10}	2·15	0·98422 ^{39}
·16	·5636 ^{39}	·66	·7454 ^{32}	·16	·8770 ^{20}	·66	·9515 ^{10}	·16	·98461 ^{39}
·17	·5675 ^{39}	·67	·7486 ^{31}	·17	·8790 ^{20}	·67	·9525 ^{10}	·17	·98500 ^{37}
·18	·5714 ^{39}	·68	·7517 ^{32}	·18	·8810 ^{20}	·68	·9535 ^{10}	·18	·98537 ^{37}
·19	·5753 ^{40}	·69	·7549 ^{31}	·19	·8830 ^{19}	·69	·9545 ^{9}	·19	·98574 ^{36}
0·20	0·5793 ^{39}	0·70	0·7580 ^{31}	1·20	0·8849 ^{20}	1·70	0·9554 ^{10}	2·20	0·98610 ^{35}
·21	·5832 ^{39}	·71	·7611 ^{31}	·21	·8869 ^{19}	·71	·9564 ^{9}	·21	·98645 ^{34}
·22	·5871 ^{39}	·72	·7642 ^{31}	·22	·8888 ^{19}	·72	·9573 ^{9}	·22	·98679 ^{34}
·23	·5910 ^{38}	·73	·7673 ^{31}	·23	·8907 ^{18}	·73	·9582 ^{9}	·23	·98713 ^{32}
·24	·5948 ^{39}	·74	·7704 ^{30}	·24	·8925 ^{19}	·74	·9591 ^{8}	·24	·98745 ^{33}
0·25	0·5987 ^{39}	0·75	0·7734 ^{30}	1·25	0·8944 ^{18}	1·75	0·9599 ^{9}	2·25	0·98778 ^{31}
·26	·6026 ^{38}	·76	·7764 ^{30}	·26	·8962 ^{18}	·76	·9608 ^{8}	·26	·98809 ^{31}
·27	·6064 ^{39}	·77	·7794 ^{29}	·27	·8980 ^{17}	·77	·9616 ^{9}	·27	·98840 ^{30}
·28	·6103 ^{38}	·78	·7823 ^{29}	·28	·8997 ^{18}	·78	·9625 ^{8}	·28	·98870 ^{29}
·29	·6141 ^{38}	·79	·7852 ^{29}	·29	·9015 ^{17}	·79	·9633 ^{8}	·29	·98899 ^{29}
0·30	0·6179 ^{38}	0·80	0·7881 ^{29}	1·30	0·9032 ^{17}	1·80	0·9641 ^{8}	2·30	0·98928 ^{28}
·31	·6217 ^{38}	·81	·7910 ^{29}	·31	·9049 ^{17}	·81	·9649 ^{7}	·31	·98956 ^{27}
·32	·6255 ^{38}	·82	·7939 ^{28}	·32	·9066 ^{16}	·82	·9656 ^{8}	·32	·98983 ^{27}
·33	·6293 ^{38}	·83	·7967 ^{28}	·33	·9082 ^{17}	·83	·9664 ^{7}	·33	·99010 ^{26}
·34	·6331 ^{37}	·84	·7995 ^{28}	·34	·9099 ^{16}	·84	·9671 ^{7}	·34	·99036 ^{25}
0·35	0·6368 ^{38}	0·85	0·8023 ^{28}	1·35	0·9115 ^{16}	1·85	0·9678 ^{8}	2·35	0·99061 ^{25}
·36	·6406 ^{37}	·86	·8051 ^{27}	·36	·9131 ^{16}	·86	·9686 ^{7}	·36	·99086 ^{25}
·37	·6443 ^{37}	·87	·8078 ^{28}	·37	·9147 ^{15}	·87	·9693 ^{6}	·37	·99111 ^{23}
·38	·6480 ^{37}	·88	·8106 ^{27}	·38	·9162 ^{15}	·88	·9699 ^{7}	·38	·99134 ^{24}
·39	·6517 ^{37}	·89	·8133 ^{26}	·39	·9177 ^{15}	·89	·9706 ^{7}	·39	·99158 ^{22}
0·40	0·6554 ^{37}	0·90	0·8159 ^{27}	1·40	0·9192 ^{15}	1·90	0·9713 ^{6}	2·40	0·99180 ^{22}
·41	·6591 ^{37}	·91	·8186 ^{26}	·41	·9207 ^{15}	·91	·9719 ^{7}	·41	·99202 ^{22}
·42	·6628 ^{36}	·92	·8212 ^{26}	·42	·9222 ^{14}	·92	·9726 ^{6}	·42	·99224 ^{21}
·43	·6664 ^{36}	·93	·8238 ^{26}	·43	·9236 ^{15}	·93	·9732 ^{6}	·43	·99245 ^{21}
·44	·6700 ^{36}	·94	·8264 ^{25}	·44	·9251 ^{14}	·94	·9738 ^{6}	·44	·99266 ^{20}
0·45	0·6736 ^{36}	0·95	0·8289 ^{26}	1·45	0·9265 ^{14}	1·95	0·9744 ^{6}	2·45	0·99286 ^{19}
·46	·6772 ^{36}	·96	·8315 ^{25}	·46	·9279 ^{13}	·96	·9750 ^{6}	·46	·99305 ^{19}
·47	·6808 ^{36}	·97	·8340 ^{25}	·47	·9292 ^{14}	·97	·9756 ^{5}	·47	·99324 ^{19}
·48	·6844 ^{35}	·98	·8365 ^{24}	·48	·9306 ^{13}	·98	·9761 ^{6}	·48	·99343 ^{18}
·49	·6879 ^{36}	·99	·8389 ^{24}	·49	·9319 ^{13}	·99	·9767 ^{5}	·49	·99361 ^{18}
0·50	0·6915	1·00	0·8413	1·50	0·9332	2·00	0·9772	2·50	0·99379

x	Φ(x)		x	Φ(x)		x	Φ(x)	
2·50	0·99379	17	2·70	0·99653	11	2·90	0·99813	6
·51	·99396	17	·71	·99664	10	·91	·99819	6
·52	·99413	17	·72	·99674	9	·92	·99825	6
·53	·99430	16	·73	·99683	10	·93	·99831	5
·54	·99446	15	·74	·99693	9	·94	·99836	5
2·55	0·99461	16	2·75	0·99702	9	2·95	0·99841	5
·56	·99477	15	·76	·99711	9	·96	·99846	5
·57	·99492	14	·77	·99720	8	·97	·99851	5
·58	·99506	14	·78	·99728	8	·98	·99856	5
·59	·99520	14	·79	·99736	8	·99	·99861	4
2·60	0·99534	13	2·80	0·99744	8	3·0	0·99865	38
·61	·99547	13	·81	·99752	8	3·1	·99903	28
·62	·99560	13	·82	·99760	7	3·2	·99931	21
·63	·99573	12	·83	·99767	7	3·3	·99952	14
·64	·99585	13	·84	·99774	7	3·4	·99966	11
2·65	0·99598	11	2·85	0·99781	7	3·5	0·99977	7
·66	·99609	12	·86	·99788	7	3·6	·99984	5
·67	·99621	11	·87	·99795	6	3·7	·99989	4
·68	·99632	11	·88	·99801	6	3·8	·99993	2
·69	·99643	10	·89	·99807	6	3·9	·99995	2
2·70	0·99653		2·90	0·99813		4·0	0·99997	

PERCENTAGE POINTS OF THE χ^2-DISTRIBUTION

P	99.5	99	97.5	95	10	5	2.5	1	0.5	0.1
$\nu = 1$	0.0⁴393	0.0³157	0.0³982	0.00393	2.71	3.84	5.02	6.63	7.88	10.83
2	0.0100	0.0201	0.0506	0.103	4.61	5.99	7.38	9.21	10.60	13.81
3	0.0717	0.115	0.216	0.352	6.25	7.81	9.35	11.34	12.84	16.27
4	0.207	0.297	0.484	0.711	7.78	9.49	11.14	13.28	14.86	18.47
5	0.412	0.554	0.831	1.15	9.24	11.07	12.83	15.09	16.75	20.52
6	0.676	0.872	1.24	1.64	10.64	12.59	14.45	16.81	18.55	22.46
7	0.989	1.24	1.69	2.17	12.02	14.07	16.01	18.48	20.28	24.32
8	1.34	1.65	2.18	2.73	13.36	15.51	17.53	20.09	21.95	26.12
9	1.73	2.09	2.70	3.33	14.68	16.92	19.02	21.67	23.59	27.88
10	2.16	2.56	3.25	3.94	15.99	18.31	20.48	23.21	25.19	29.59
11	2.60	3.05	3.82	4.57	17.28	19.68	21.92	24.73	26.76	31.26
12	3.07	3.57	4.40	5.23	18.55	21.03	23.34	26.22	28.30	32.91
13	3.57	4.11	5.01	5.89	19.81	22.36	24.74	27.69	29.82	34.53
14	4.07	4.66	5.63	6.57	21.06	23.68	26.12	29.14	31.32	36.12
15	4.60	5.23	6.26	7.26	22.31	25.00	27.49	30.58	32.80	37.70
16	5.14	5.81	6.91	7.96	23.54	26.30	28.85	32.00	34.27	39.25
17	5.70	6.41	7.56	8.67	24.77	27.59	30.19	33.41	35.72	40.79
18	6.26	7.01	8.23	9.39	25.99	28.87	31.53	34.81	37.16	42.31
19	6.84	7.63	8.91	10.12	27.20	30.14	32.85	36.19	38.58	43.82
20	7.43	8.26	9.59	10.85	28.41	31.41	34.17	37.57	40.00	45.31
21	8.03	8.90	10.28	11.59	29.62	32.67	35.48	38.93	41.40	46.80
22	8.64	9.54	10.98	12.34	30.81	33.92	36.78	40.29	42.80	48.27
23	9.26	10.20	11.69	13.09	32.01	35.17	38.08	41.64	44.18	49.73
24	9.89	10.86	12.40	13.85	33.20	36.42	39.36	42.98	45.56	51.18
25	10.52	11.52	13.12	14.61	34.38	37.65	40.65	44.31	46.93	52.62
26	11.16	12.20	13.84	15.38	35.56	38.89	41.92	45.64	48.29	54.05
27	11.81	12.88	14.57	16.15	36.74	40.11	43.19	46.96	49.64	55.48
28	12.46	13.56	15.31	16.93	37.92	41.34	44.46	48.28	50.99	56.89
29	13.12	14.26	16.05	17.71	39.09	42.56	45.72	49.59	52.34	58.30
30	13.79	14.95	16.79	18.49	40.26	43.77	46.98	50.89	53.67	59.70
40	20.71	22.16	24.43	26.51	51.81	55.76	59.34	63.69	66.77	73.40
50	27.99	29.71	32.36	34.76	63.17	67.50	71.42	76.15	79.49	86.66
60	35.53	37.48	40.48	43.19	74.40	79.08	83.30	88.38	91.95	99.61
70	43.28	45.44	48.76	51.74	85.53	90.53	95.02	100.4	104.2	112.3
80	51.17	53.54	57.15	60.39	96.58	101.9	106.6	112.3	116.3	124.8
90	59.20	61.75	65.65	69.13	107.6	113.1	118.1	124.1	128.3	137.2
100	67.33	70.06	74.22	77.93	118.5	124.3	129.6	135.8	140.2	149.4

DISCOUNT TABLES
VALUES OF $(1 + r)^{-n}$

n	0.02	0.03	0.04	0.05	0.06	0.07	0.08	0.09	0.10	0.11	0.12	0.13	0.14	0.15	0.16	0.20
1	9804	9709	9615	9524	9434	9346	9259	9174	9091	9009	8929	8850	8772	8696	8621	8333
2	9612	9426	9246	9070	8900	8734	8573	8417	8264	8116	7972	7831	7695	7561	7432	6944
3	9423	9151	8890	8638	8396	8163	7938	7722	7513	7312	7118	6931	6750	6575	6407	5787
4	9238	8885	8548	8227	7921	7629	7350	7084	6830	6587	6355	6133	5921	5718	5523	4823
5	9057	8626	8219	7835	7473	7130	6806	6499	6209	5935	5674	5428	5194	4972	4761	4019
6	8880	8375	7903	7462	7050	6663	6302	5963	5645	5346	5066	4803	4556	4323	4104	3349
7	8706	8131	7599	7107	6651	6227	5835	5470	5132	4817	4523	4251	3996	3759	3538	2791
8	8535	7894	7307	6768	6274	5820	5403	5019	4665	4339	4039	3762	3506	3269	3050	2326
9	8368	7664	7026	6446	5919	5439	5002	4604	4241	3909	3606	3329	3075	2843	2630	1938
10	8203	7441	6756	6139	5584	5083	4632	4224	3855	3522	3220	2946	2697	2472	2267	1615

Use with reciprocal tables for values of $(1 + r)^n$

RECIPROCALS 1/x OR x⁻¹

x	0	1	2	3	4	5	6	7	8	9	Δ_m	1 2 3	4 5 6	7 8 9
											−	SUBTRACT		
10	10000	9901	9804								98	10 20 29	39 49 59	69 78 88
			9804	9709	9615						94	9 19 28	38 47 56	66 75 85
					9615	9524	9434	9346	9259		89*	9 18 27	36 44 53	62 71 80
									9259	9174	84	8 17 25	34 42 50	59 67 76
11	9091	9009	8929	8850	8772						80*	8 16 24	32 40 48	56 64 72
					8772	8696	8621	8547			75	8 15 23	30 38 45	53 60 68
								8547	8475	8403	71	7 14 21	28 36 43	50 57 64
12	8333	8264	8197	8130	8065	8000					67*	7 13 20	27 33 40	47 54 60
						8000	7937	7874	7812	7752	62*	6 12 19	25 31 37	43 50 56
13	7692	7634	7576	7519	7463	7407					57	6 11 17	23 28 34	40 46 51
						7407	7353	7299	7246	7194	53*	5 11 16	21 26 32	37 42 48
14	7143	7092	7042	6993	6944	6897	6849				49*	5 10 15	20 24 29	34 39 44
							6849	6803	6757	6711	45	5 9 14	18 23 27	32 36 41
15	6667	6623	6579	6536	6494	6452	6410				43	4 9 13	17 21 26	30 34 39
							6410	6369	6329	6289	40	4 8 12	16 20 24	28 32 36
16	6250	6211	6173	6135	6098	6061	6024	5988	5952	5917	37*	4 7 11	15 18 22	26 30 33
17	5882	5848	5814	5780	5747	5714	5682	5650	5618	5587	33*	3 7 10	13 16 20	23 26 30
18	5556	5525	5495	5464	5435	5405	5376	5348	5319	5291	29*	3 6 9	12 15 17	20 23 26
19	5263	5236	5208	5181	5155	5128	5102	5076	5051	5025	26*	3 5 8	10 13 16	18 21 23
20	5000	4975	4950	4926	4902	4878	4854	4831	4808	4785	24	2 5 7	10 12 14	17 19 22
21	4762	4739	4717	4695	4673	4651	4630	4608	4587	4566	22	2 4 7	9 11 13	15 18 20
22	4545	4525	4505	4484	4464	4444	4425	4405	4386	4367	20	2 4 6	8 10 12	14 16 18
23	4348	4329	4310	4292	4274	4255	4237	4219	4202	4184	18	2 4 5	7 9 11	13 14 16
24	4167	4149	4132	4115	4098	4082	4065	4049	4032	4016	17	2 3 5	7 8 10	12 14 15
25	4000	3984	3968	3953	3937	3922	3906	3891	3876	3861	15	2 3 5	6 8 9	11 12 14
26	3846	3831	3817	3802	3788	3774	3759	3745	3731	3717	14	1 3 4	6 7 8	10 11 13
27	3704	3690	3676	3663	3650	3636	3623	3610	3597	3584	13	1 3 4	5 7 8	9 10 12
28	3571	3559	3546	3534	3521	3509	3497	3484	3472	3460	12	1 2 4	5 6 7	8 10 11
29	3448	3436	3425	3413	3401	3390	3378	3367	3356	3344	12	1 2 4	5 6 7	8 10 11
30	3333	3322	3311	3300	3289	3279	3268	3257	3247	3236	11	1 2 3	4 5 7	8 9 10
31	3226	3215	3205	3195	3185	3175	3165	3155	3145	3135	10	1 2 3	4 5 6	7 8 9
32	3125	3115	3106	3096	3086	3077	3067	3058	3049	3040	10	1 2 3	4 5 6	7 8 9
33	3030	3021	3012	3003	2994	2985	2976	2967	2959	2950	9	1 2 3	4 4 5	6 7 8
34	2941	2933	2924	2915	2907	2899	2890	2882	2874	2865	8	1 2 2	3 4 5	6 6 7
35	2857	2849	2841	2833	2825	2817	2809	2801	2793	2786	8	1 2 2	3 4 5	6 6 7
36	2778	2770	2762	2755	2747	2740	2732	2725	2717	2710	8	1 2 2	3 4 5	6 6 7
37	2703	2695	2688	2681	2674	2667	2660	2653	2646	2639	7	1 1 2	3 4 4	5 6 6
38	2632	2625	2618	2611	2604	2597	2591	2584	2577	2571	7	1 1 2	3 3 4	5 6 6
39	2564	2558	2551	2545	2538	2532	2525	2519	2513	2506	6	1 1 2	2 3 4	4 5 5
40	2500	2494	2488	2481	2475	2469	2463	2457	2451	2445	6	1 1 2	2 3 4	4 5 5
41	2439	2433	2427	2421	2415	2410	2404	2398	2392	2387	6	1 1 2	2 3 4	4 5 5
42	2381	2375	2370	2364	2358	2353	2347	2342	2336	2331	6	1 1 2	2 3 4	4 5 5
43	2326	2320	2315	2309	2304	2299	2294	2288	2283	2278	5	1 1 2	2 3 3	4 4 5
44	2273	2268	2262	2257	2252	2247	2242	2237	2232	2227	5	1 1 2	2 3 3	4 4 5
45	2222	2217	2212	2208	2203	2198	2193	2188	2183	2179	5	0 1 1	2 2 3	3 4 4
46	2174	2169	2165	2160	2155	2151	2146	2141	2137	2132	5	0 1 1	2 2 3	3 4 4
47	2128	2123	2119	2114	2110	2105	2101	2096	2092	2088	4	0 1 1	2 2 2	3 3 4
48	2083	2079	2075	2070	2066	2062	2058	2053	2049	2045	4	0 1 1	2 2 2	3 3 4
49	2041	2037	2033	2028	2024	2020	2016	2012	2008	2004	4	0 1 1	2 2 2	3 3 4

RECIPROCALS $1/x$ OR x^{-1}

x	0	1	2	3	4	5	6	7	8	9	Δ_m	1 2 3	4 5 6	7 8 9
											−		SUBTRACT	
50	2000	1996	1992	1988	1984	1980	1976	1972	1969	1965	4	0 1 1	2 2 2	3 3 4
51	1961	1957	1953	1949	1946	1942	1938	1934	1931	1927	4	0 1 1	2 2 2	3 3 4
52	1923	1919	1916	1912	1908	1905	1901	1898	1894	1890	4	0 1 1	2 2 2	3 3 4
53	1887	1883	1880	1876	1873	1869	1866	1862	1859	1855	4	0 1 1	2 2 2	3 3 4
54	1852	1848	1845	1842	1838	1835	1832	1828	1825	1821	3	0 1 1	1 2 2	2 2 3
55	1818	1815	1812	1808	1805	1802	1799	1795	1792	1789	3	0 1 1	1 2 2	2 2 3
56	1786	1783	1779	1776	1773	1770	1767	1764	1761	1757	3	0 1 1	1 2 2	2 2 3
57	1754	1751	1748	1745	1742	1739	1736	1733	1730	1727	3	0 1 1	1 2 2	2 2 3
58	1724	1721	1718	1715	1712	1709	1706	1704	1701	1698	3	0 1 1	1 1 2	2 2 3
59	1695	1692	1689	1686	1684	1681	1678	1675	1672	1669	3	0 1 1	1 1 2	2 2 3
60	1667	1664	1661	1658	1656	1653	1650	1647	1645	1642	3	0 1 1	1 1 2	2 2 3
61	1639	1637	1634	1631	1629	1626	1623	1621	1618	1616	3	0 1 1	1 1 2	2 2 3
62	1613	1610	1608	1605	1603	1600	1597	1595	1592	1590	3	0 1 1	1 1 2	2 2 3
63	1587	1585	1582	1580	1577	1575	1572	1570	1567	1565	2	0 0 1	1 1 1	1 2 2
64	1562	1560	1558	1555	1553	1550	1548	1546	1543	1541	2	0 0 1	1 1 1	1 2 2
65	1538	1536	1534	1531	1529	1527	1524	1522	1520	1517	2	0 0 1	1 1 1	1 2 2
66	1515	1513	1511	1508	1506	1504	1502	1499	1497	1495	2	0 0 1	1 1 1	1 2 2
67	1493	1490	1488	1486	1484	1481	1479	1477	1475	1473	2	0 0 1	1 1 1	1 2 2
68	1471	1468	1466	1464	1462	1460	1458	1456	1453	1451	2	0 0 1	1 1 1	1 2 2
69	1449	1447	1445	1443	1441	1439	1437	1435	1433	1431	2	0 0 1	1 1 1	1 2 2
70	1429	1427	1425	1422	1420	1418	1416	1414	1412	1410	2	0 0 1	1 1 1	1 2 2
71	1408	1406	1404	1403	1401	1399	1397	1395	1393	1391	2	0 0 1	1 1 1	1 2 2
72	1389	1387	1385	1383	1381	1379	1377	1376	1374	1372	2	0 0 1	1 1 1	1 2 2
73	1370	1368	1366	1364	1362	1361	1359	1357	1355	1353	2	0 0 1	1 1 1	1 2 2
74	1351	1350	1348	1346	1344	1342	1340	1339	1337	1335	2	0 0 1	1 1 1	1 2 2
75	1333	1332	1330	1328	1326	1325	1323	1321	1319	1318	2	0 0 1	1 1 1	1 2 2
76	1316	1314	1312	1311	1309	1307	1305	1304	1302	1300	2	0 0 1	1 1 1	1 2 2
77	1299	1297	1295	1294	1292	1290	1289	1287	1285	1284	2	0 0 1	1 1 1	1 2 2
78	1282	1280	1279	1277	1276	1274	1272	1271	1269	1267	2	0 0 1	1 1 1	1 2 2
79	1266	1264	1263	1261	1259	1258	1256	1255	1253	1252	2	0 0 1	1 1 1	1 2 2
80	1250	1248	1247	1245	1244	1242	1241	1239	1238	1236	2	0 0 1	1 1 1	1 2 2
81	1235	1233	1232	1230	1229	1227	1225	1224	1222	1221				
82	1220	1218	1217	1215	1214	1212	1211	1209	1208	1206				
83	1205	1203	1202	1200	1199	1198	1196	1195	1193	1192				
84	1190	1189	1188	1186	1185	1183	1182	1181	1179	1178				
85	1176	1175	1174	1172	1171	1170	1168	1167	1166	1164				
86	1163	1161	1160	1159	1157	1156	1155	1153	1152	1151				
87	1149	1148	1147	1145	1144	1143	1142	1140	1139	1138				
88	1136	1135	1134	1133	1131	1130	1129	1127	1126	1125				
89	1124	1122	1121	1120	1119	1117	1116	1115	1114	1112				
90	1111	1110	1109	1107	1106	1105	1104	1103	1101	1100				
91	1099	1098	1096	1095	1094	1093	1092	1091	1089	1088				
92	1087	1086	1085	1083	1082	1081	1080	1079	1078	1076				
93	1075	1074	1073	1072	1071	1070	1068	1067	1066	1065				
94	1064	1063	1062	1060	1059	1058	1057	1056	1055	1054				
95	1053	1052	1050	1049	1048	1047	1046	1045	1044	1043				
96	1042	1041	1040	1038	1037	1036	1035	1034	1033	1032				
97	1031	1030	1029	1028	1027	1026	1025	1024	1022	1021				
98	1020	1019	1018	1017	1016	1015	1014	1013	1012	1011				
99	1010	1009	1008	1007	1006	1005	1004	1003	1002	1001				

RANDOM SAMPLING NUMBERS

20 17	42 28	23 17	59 66	38 61	02 10	86 10	51 55	92 52	44 25
74 49	04 49	03 04	10 33	53 70	11 54	48 63	94 60	94 49	57 38
94 70	49 31	38 67	23 42	29 65	40 88	78 71	37 18	48 64	06 57
22 15	78 15	69 84	32 52	32 54	15 12	54 02	01 37	38 37	12 93
93 29	12 18	27 30	30 55	91 87	50 57	58 51	49 36	12 53	96 40
45 04	77 97	36 14	99 45	52 95	69 85	03 83	51 87	85 56	22 37
44 91	99 49	89 39	94 60	48 49	06 77	64 72	59 26	08 51	25 57
16 23	91 02	19 96	47 59	89 65	27 84	30 92	63 37	26 24	23 66
04 50	65 04	65 65	82 42	70 51	55 04	61 47	88 83	99 34	82 37
32 70	17 72	03 61	66 26	24 71	22 77	88 33	17 78	08 92	73 49
03 64	59 07	42 95	81 39	06 41	20 81	92 34	51 90	39 08	21 42
62 49	00 90	67 86	93 48	31 83	19 07	67 68	49 03	27 47	52 03
61 00	95 86	98 36	14 03	48 88	51 07	33 40	06 86	33 76	68 57
89 03	90 49	28 74	21 04	09 96	60 45	22 03	52 80	01 79	33 81
01 72	33 85	52 40	60 07	06 71	89 27	14 29	55 24	85 79	31 96
27 56	49 79	34 34	32 22	60 53	91 17	33 26	44 70	93 14	99 70
49 05	74 48	10 55	35 25	24 28	20 22	35 66	66 34	26 35	91 23
49 74	37 25	97 26	33 94	42 23	01 28	59 58	92 69	03 66	73 82
20 26	22 43	88 08	19 85	08 12	47 65	65 63	56 07	97 85	56 79
48 87	77 96	43 39	76 93	08 79	22 18	54 55	93 75	97 26	90 77
08 72	87 46	75 73	00 11	27 07	05 20	30 85	22 21	04 67	19 13
95 97	98 62	17 27	31 42	64 71	46 22	32 75	19 32	20 99	94 85
37 99	57 31	70 40	46 55	46 12	24 32	36 74	69 20	72 10	95 93
05 79	58 37	85 33	75 18	88 71	23 44	54 28	00 48	96 23	66 45
55 85	63 42	00 79	91 22	29 01	41 39	51 40	36 65	26 11	78 32
67 28	96 25	68 36	24 72	03 85	49 24	05 69	64 86	08 19	91 21
85 86	94 78	32 59	51 82	86 43	73 84	45 60	89 57	06 87	08 15
40 10	60 09	05 88	78 44	63 13	58 25	37 11	18 47	75 62	52 21
94 55	89 48	90 80	77 80	26 89	87 44	23 74	66 20	20 19	26 52
11 63	77 77	23 20	33 62	62 19	29 03	94 15	56 37	14 09	47 16
64 00	26 04	54 55	38 57	94 62	68 40	26 04	24 25	03 61	01 20
50 94	13 23	78 41	60 58	10 60	88 46	30 21	45 98	70 96	36 89
66 98	37 96	44 13	45 05	34 59	75 85	48 97	27 19	17 85	48 51
66 91	42 83	60 77	90 91	60 90	79 62	57 66	72 28	08 70	96 03
33 58	12 18	02 07	19 40	21 29	39 45	90 42	58 84	85 43	95 67
52 49	40 16	72 40	73 05	50 90	02 04	98 24	05 30	27 25	20 88
74 98	93 99	78 30	79 47	96 92	45 58	40 37	89 76	84 41	74 68
50 26	54 30	01 88	69 57	54 45	69 88	23 21	05 69	93 44	05 32
49 46	61 89	33 79	96 84	28 34	19 35	28 73	39 59	56 34	97 07
19 65	13 44	78 39	73 88	62 03	36 00	25 96	86 76	67 90	21 68
64 17	47 67	87 59	81 40	72 61	14 00	28 28	55 86	23 38	16 15
18 43	97 37	68 97	56 56	57 95	01 88	11 89	48 07	42 60	11 92
65 58	60 87	51 09	96 61	15 53	66 81	66 88	44 75	37 01	28 88
79 90	31 00	91 14	85 65	31 75	43 15	45 93	64 78	34 53	88 02
07 23	00 15	59 05	16 09	94 42	20 40	63 76	65 67	34 11	94 10
90 08	14 24	01 51	95 46	30 32	33 19	00 14	19 28	40 51	92 69
53 82	62 02	21 82	34 13	41 03	12 85	65 30	00 97	56 30	15 48
98 17	26 15	04 50	76 25	20 33	54 84	39 31	23 33	59 64	96 27
08 91	12 44	82 40	30 62	45 50	64 54	65 17	89 25	59 44	99 95
37 21	46 77	84 87	67 39	85 54	97 37	33 41	11 74	90 50	29 62

Each digit is an independent sample from a population in which the digits o to 9 are equally likely, that is each has a probability of $\frac{1}{10}$.

Minimum Values of p which are Significant
at the 5% and 1% level

n	5% level	1% level
4 or less	none	none
5	1.000	none
6	0.886	1.000
7	0.750	0.893
8	0.714	0.857
9	0.683	0.833
10	0.648	0.794

For values of n greater than 10 use the minimum values
of r which are significant at the 5% and 1% levels

Significance of the correlation coefficient

Minimum value of r if correlation is significant at:

n	5% level	1% level
10	0.63	0.76
11	0.60	0.73
12	0.58	0.71
13	0.55	0.68
14	0.53	0.66
15	0.51	0.64
16	0.50	0.62
17	0.48	0.60
18	0.47	0.59
19	0.46	0.58
20	0.44	0.56

Index